Business Mathematics

Business Mathematics

Jerry Funk
Brazosport College

Charles E. Merrill Publishing Company
A Bell & Howell Company
Columbus • Toronto • London • Sydney

Published by Charles E. Merrill Publishing Co.
A Bell & Howell Company
Columbus, Ohio   43216

This book was set in Univers.
Production Editor: Francie Margolin.
Cover Design Coordination: Cathy Watterson.
Cover Photo: *Blue Fugue* by Larry Hamill.
Production Coordination: Rex Davidson.

Major portions of this text were originally published under the title *Business Mathematics,* by Allyn and Bacon, Inc.

Library of Congress Catalog Card Number: 84–62257
International Standard Book Number: 0–675–20307–4
Printed in the United States of America
   2   3   4   5   6   7   8   9   10—88   87

Preface

This text provides the fundamental mathematics necessary to prepare students in all areas of business. The explanatory material is extensive, so that the students have a thorough understanding of when and where the concepts are applied, and not just how they are applied. The workbook problems are presented in two degrees of difficulty. Drill problems provide the repetition necessary for some students. Application problems are word problems that require students to assimilate the information from written form.

Each unit begins with situational statements or problems to arouse the reader's interest. These statements illustrate some of the practical applications of the material in that unit, and they are followed by detailed learning objectives that pinpoint exactly what the student is expected to be able to do and to know. Each chapter begins with a Preview of Terms, where the special terms the student will encounter in that chapter are defined. Within the chapters of each unit the concepts are explained and then illustrated with clear examples. Practice problems follow the examples, and detailed step-by-step solutions for the practice problems are found at the back of each chapter, preceding the chapter exercises. This provides immediate reinforcement to students. The practice problems can be utilized in the classroom by the teacher or by the students at home. At the end of each unit is a "Test Yourself" quiz, with answers in the back of the unit. The questions are a representative sampling of the problems in that unit. Each of these quizzes is a practice test for that unit, which means that the student knows what to expect and also how well prepared he or she is for that unit. For the problems designed as homework assignments, the traditional odd-numbered problems with answers in the back of the book are included in this text as "Self-Check Exercises" and numbered as "A" problems. For these, the answers appear on the back of each assignment page. The "B" problems have no answers provided. The text also contains a Practice Set to allow the student to apply what he or she has learned to a practical business situation. The Practice Set features Ed Carson, whose small business grows to the stage of incorporation.

The author would like to acknowledge the following for their assistance and helpful comments with this manuscript: Mary M. Blyth, James W. Cox, Joyce T. Davis, John Drury, Sally Preissig, Margaret E. Sprenz, and Sue B. Weaver.

Contents

Chapter 5:   Payroll

Unit 3   Interest

Chapter 6:   Simple Interest Loans and Discounted Loans

Chapter 7:   Compound Interest, Present Value, and Annuities

Chapter 8:   Installment Interest and Rebates

Chapter 9:   Charge Account Interest and Repayment Schedules

Unit 7   Insurance, Taxes, Stocks, and Bonds

Mini Practice Set

Unit

1

Review of Arithmetic, Fractions, and Solving for the Unknown

A Preview of What's Next

The invoice has several lines that have been added to arrive at the amount you owe. Can you add them correctly yourself to verify the accuracy of the total?

If the manager in charge of payroll asked you to be sure to round all overtime pay to hundredths, could you?

At work, you are given the job of completing the report on absenteeism. Could you determine the total number of workers out of 3,432 who were absent in a three-month period if 3/12 were absent the first month, 1/6 the second month, and 5/24 the third month?

When dividing 813,467 by 1,000 do you know the short-cut method of moving the decimal?

The Sales Department has 40 employees, which is 4 more than 3 times as many as the Marketing Department. How many employees are in the Marketing Department?

Objectives

The ability to perform the fundamental operations of arithmetic with reasonable skill is essential to success in the study of business mathematics. It requires knowledge of the methods for each situation and practice in using them. When you have successfully completed this unit, you will be able to:

1. Express in writing a number that is in numerical form.
2. Express numerically a number that is in written form.
3. Determine the correct place value for any digit in a number.
4. Perform the fundamental operations of arithmetic including the placement of decimals—add, subtract, multiply, divide.
5. Recognize decimal fractions and their equivalent decimal numbers.
6. Perform fundamental operations of arithmetic for all computations involving fractions—add, subtract, multiply, divide.
7. Reduce fractions to their lowest terms.
8. Convert mixed numbers to improper fractions and improper fractions to mixed numbers.
9. Find the Least Common Denominator.
10. Solve for the unknown in an equation.
11. Set up and solve equations when given the problem in word form.

Chapter

Arithmetic

Preview of Terms

Addend
In addition, the numbers being added are called *addends.* In the problem 8 + 3 = 11, the 8 and 3 are addends.

Difference
The results of subtraction are called the *difference.* In the problem 10 − 3 = 7, the 7 is the difference.

Dividend
The number being divided is called the *dividend.* In the problem 20/5 = 4, the 20 is the dividend.

Hundredths
The value of the second place to the right of the decimal point. **Example:** .14 would be 14 hundredths or $\frac{14}{100}$.

Minuend
The number being subtracted from is called the *minuend.* In the problem 7 − 5 = 2, the 7 is the minuend.

Multiplicand
The number being multiplied is called the *multiplicand.* In the problem 6 × 3 = 18, the 6 is the multiplicand.

Product
The results of multiplication are called the *product.* In the problem 4 × 3 = 12, the 12 is the product.

Quotient
The result of division is called the *quotient.* In the problem 10/2 = 5, the 5 is the quotient.

Subtrahend
The number subtracted from the minuend is called the *subtrahend.* In the problem 9 − 6 = 3, the 6 is the subtrahend.

Tenths
The value of the first place to the right of the decimal point. **Example:** .8 would be 8 tenths or $\frac{8}{10}$.

Thousandths
The value of the third place to the right of the decimal point. **Example:** .125 would be 125 thousandths or $\frac{125}{1000}$.

Although you have learned to add, subtract, multiply, and divide, unless you constantly use those skills they will need to be sharpened again. The following chapter, then, is a review of these basic arithmetic processes. This chapter will also allow you to begin to think mathematically again and regain your confidence that you have not forgotten everything you used to know.

PLACE VALUES AND THEIR NAMES

Our base ten number system is illustrated in Table 1.1. Notice that each successive place from the left of the decimal is 10 times greater than the one preceding it. The same is true moving to the right from the decimal.

TABLE 1.1

Millions	Hundred thousands	Ten thousands	Thousands	Hundreds	Tens	Ones	Tenths	Hundredths	Thousandths	Ten-thousandths
3 ,	4	5	6 ,	7	8	1 .	1	2	3	4

Example 1: The number "425,894" is read "four hundred twenty-five thousand eight hundred ninety-four."

Example 2: The number "817.413" is read "eight hundred seventeen and four hundred thirteen thousandths."

Practice Problems

1. Using the number 325,041:
   A. The 2 is in the _____ position.
   B. The 4 is in the _____ position.
   C. The digit in the hundreds position is _____ .

2. Write out the number "six million eighty-one thousand one hundred twenty-two" using digits. _____

3. Write out the number 38,416.27.

   _____

4. Using the number 3.1057:
   A. The digit in the tenths position is _____ .
   B. The digit in the thousandths position is _____ .
   C. The zero is in the _____ position.

ADDITION

You may add from top to bottom or from bottom to top. A good practice would be to add in one direction and check your answer by re-adding in the opposite direction.

The quickest way, of course, is to find two or three digits that are easily recognized as a sum of 10 — such as 2 + 8 or 7 + 2 + 1 or 5 + 5 — and then add the rest of the digits to that sum. These numbers are called *addends.*

Example 3:

Add:
```
   4 8 2
   2 2 5
   3 3 4
 + 7 4 1
 ─────────
 1 7 8 2
```

When adding or subtracting numbers that have decimals, be careful to line up the decimals to ensure that you are adding equivalent place values.

Example 4:

Add 24.015, .36, 107.205, and 3.4.

```
  24.015 ⎫
    .36  ⎪   Notice that the decimals
 107.205 ⎬   are lined up.
 +  3.4  ⎭
 ────────
 134.98
```

Practice Problems

Add the following:

5.	1062	6.	15.107	7.	4,384.24
	385		368.091		6,615.75
	+ 623		+500.112		+3,009.27

SUBTRACTION

When subtracting, remember when you "borrow" one from the place value to the left, it decreases the digit borrowed "from" by one but increases the digit borrowed "for" by ten.

Example 5:

```
 2 14
 3 4̸      The digit borrowed "from" (3) is decreased by 1
-1 9      but the digit borrowed "for" (4) is increased by 10.
────
 1 5
```

Note:

The number doing the subtracting—19 in Example 5—is called the subtrahend. *The number being subtracted from—34 in Example 5—is called the* minuend.

Generally, the easiest way to check subtraction is to add the answer (the "difference") to the subtrahend.

Example 6:

```
  2 8    Minuend
 -1 7    Subtrahend
 ────
  1 1    Difference
```

```
 +1 7    Subtrahend
 ────              Check
  2 8    Minuend
```

Subtract the following:

8.   3,142
         − 856

9.   506.02
         − 37.88

10.   623.93
          −477.94

MULTIPLICATION

Multiplication is nothing more than repeated addition. In the problem 4×3, the answer could be found either by adding 4 three times ($4 + 4 + 4 = 12$) or by adding 3 four times ($3 + 3 + 3 + 3 = 12$). Of course, it is quicker to actually multiply.

The number doing the multiplying is called the *multiplier,* the number being multiplied is the *multiplicand,* and the answer is the *product.* The intermediate results are called *partial products.* Remember, as you multiply, each successive partial product begins one place value to the left of the previous one.

Example 7: $34 \times 24 =$

3 4	Multiplicand
× 2 4	Multiplier
1 3 6	Partial product
6 8	Partial product
8 1 6	Product

Note: *A common means of checking multiplication is to reverse the multiplier and multiplicand.*

Check:

2 4	(Although the partial products are not the same, the product
× 3 4	will be the same.)
9 6	Partial product
7 2	Partial product
8 1 6	Product

Note: *You can also check multiplication by dividing the product by the multiplier to see if you get the multiplicand, or vice-versa. (816 divided by 24 = 34.)*

When multiplying by 10 or by multiples of 10 (20, 30, 10, etc.), you can start multiplying with the first nonzero digit and then add to the right end of the product the number of zeros found on the right of the multiplier and multiplicand.

Example 8:   2340 → 234 + (1 zero to the right)
                 × 200 → × 2 + (2 zeros to the right)
                         468 + (3 zeros to the right)
                         468,000

When **multiplying** by 10, 100, 1000, etc., you can simply move the decimal point to the **right** the same number of places as there are zeros in those multipliers—$3.57 \times 10 = 35.7$ or $3.57 \times 100 = 357$.

When a multiplier or a multiplicand has a zero in it, but not at the end of the number, be sure to include zeros as partial products so the next partial product will be moved to the proper place value.

Example 9:

```
      4 8 2
    × 3 0 1
      4 8 2
    0 0 0   ←Used as place value holders.
  1 4 4 6
  1 4 5,0 8 2
```

Practice Problems

Multiply the following:

11.   35
         ×27

12.   416
         × 50

13.   802
         × 43

DIVISION

Division is finding how many times one number will go into another number. The number doing the dividing is called the *divisor.* The number being divided is called the *dividend.* The answer is called the *quotient.* If the divisor does not divide evenly into the dividend, what is left is called the *remainder.*

Example 10:

```
                    2 ← Quotient
Divisor → 20 )4 2 ← Dividend
            4 0
              2 ← Remainder
```

Division is the opposite of multiplication, so division problems can be checked by reversing the process.

Check:

quotient × divisor + remainder = dividend.
  2   ×   20   +   2   =   42.

When **dividing** a number by 10, 100, 1000, etc., you can simply move the decimal point to the **left** the same number of places as there are zeros in the divisor. (55.123 divided by 100 = .55123 or 7 divided by 10 = .7) When the division is by .01, .001, .0001, etc., the decimal will have to be moved to the **right** the same number of places to the right of the decimal in the divisor (843.7 divided by .01 = 84,370 or .0841 divided by .001 = 84.1).

Practice Problems

Divide the following:

14. 20)848

15. 41)533

16. 1560 ÷ 65

THE ROUNDING RULE

In rounding off decimal numbers, you need to be instructed how many places after the decimal to round off to. The place value immediately following the desired place value should then be analyzed. If the number following the desired place value is 5 or more, round up (increase the desired place value by 1); otherwise, leave the desired place value as it is and drop the remaining numbers.

Example 11:   Round .014 to two places after the decimal.

.014 = .01

Example 12:   Round 2.1578 to three places after the decimal.

2.1578 = 2.158

Example 13:   Round 525.452 to one place after the decimal.

525.452 = 525.5

Example 14:   Round .0649 to two places after the decimal.

.0649 = .06

Practice Problems

Round to:

	Tenths	Hundredths	Thousandths
17. 1.4782			
18. .0566			
19. 240.5494			

DECIMAL PLACEMENT

In multiplication and division problems involving decimals, you must take particular care to place the decimal accurately in the answer.

In multiplication of decimal numbers, the digits in the multiplier and multiplicand are multiplied as if there were no decimals. Count the total number of digits to the **right** of the decimal in the multiplier and the multiplicand. Then, in the product, count from the **right** that same total number of places and place the decimal there.

Example 15:

$$
\begin{array}{r}
23.5 \\
\times\ 4.21 \\
\hline
2\ 3\ 5 \\
4\ 7\ 0\ \\
9\ 4\ 0\ \\
\hline
9\ 8.9\ 3\ 5
\end{array}
$$

23.5 → 1 place to right of decimal
× 4.21 → + 2 places to right of decimal
3 total places to right of decimal

98.935 ← In this product, there should be 3 places to the right of the decimal.

In division you first move the decimal in the divisor to the extreme right of the number. For example, 2.14 would become 214. In this example, the decimal in the divisor was moved two places to the right. The decimal in the dividend must also be moved the same number of places to the right. (64.2 would become 6420.) Then divide as you would normally.

Example 16:

$$2.14\overline{)64.2} \quad = \quad 214.\overline{)6420.}$$

$$\begin{array}{r} 30. \\ 214.\overline{)6420.} \\ \underline{642} \\ 00 \end{array}$$

Note: *When a number shows no decimal, it is understood to be at the extreme right. That is, 480 is understood to be 480.*

Practice Problem Solutions

1. A. Ten thousandths   **B.** Tens   **C.** 0

2. 6,081,122   **3.** Thirty-eight thousand four hundred sixteen and twenty-seven hundredths

4. A. 1   **B.** 5   **C.** Hundredths

5.
```
  1062
   385
+  623
 2,070
```

6.
```
  15.107
 368.091
+500.112
 883.310
```

7.
```
  4,384.24
  6,615.75
+ 3,009.27
 14,009.26
```

8.
```
  3,142
 −  856
  2,286
```

9.
```
 506.02
 − 37.88
 468.14
```

10.
```
 623.93
 −477.94
 145.99
```

11.
```
  35
 ×27
 245
 70
 945
```

12.
```
  416
 ×50 → 1 zero to the right
 2080 + 1 zero to the right
 20,800
```

13.
```
   802
 ×  43
 2 406
 32 08
 34,486
```

14.
```
       42.4
 20)848.0
     80
     48
     40
      8 0
      8 0
        0
```

15.
```
     13
 41)533
    41
    123
    123
      0
```

16.
```
     24
 65)1560
    130
    260
    260
      0
```

	Tenths	Hundredths	Thousandths
17. 1.4782	1.5	1.48	1.478
18. .0566	.1	.06	.057
19. 240.5494	240.5	240.55	240.549

20.
```
  31.4 →   1 place to right of decimal
 ×2.33 → +2 places to right of decimal
  942      3 places to right of decimal
  9 42
  62 8
 73.162 →  3 places to right of decimal
```

21.
```
   .16 →   2 places to right of decimal
 ×.04 → +2 places to right of decimal
 .0064 ←   4 places to right of decimal
```

22.
$$.015 \rightarrow \quad \text{3 places to right of decimal}$$
$$\times 1.203 \rightarrow \underline{+3} \text{ places to right of decimal}$$
$$45 \qquad \text{6 places to right of decimal}$$
$$0\,00$$
$$3\,0$$
$$\underline{15}$$
$$.018\,045 \leftarrow \quad \text{6 Places to right of decimal}$$

23.
```
           2 12
4.43)939.16
      886
       53 1
       44 3
        8 86
        8 86
           0
```

24.
```
        2 0.
3.1)62.0
    62
     0
```

25.
```
          2.46
.001).002 46
        2
        0 4
          4
          06
           6
           0
```

"Self-Check Exercises"

For all of the "Self-Check Exercises" labeled "A", the answers are on the back of that page. You can conveniently match your answers with those on the back by folding the back of the sheet over as shown here.

Exercise 21:
Course/Section No.:
Instructions:

(1) _132_	(1) 132
(2) _18.4_	(2) 18.4
(3) _107_	(3) 107
(4) _61_	(4) 61
(5) _1.44_	(5) 1.44
(6) _4018_	(6) 4018
(7) _62_	(7) 62
(8) _.807_	(8) .807
(9) _311_	(9) 311
(10) _1723_	(10) 1723

Answers

Exercise 1A: Place Values and their Names

Name_____

Course/Sect. No._____   Score_____

Instructions: Answer each as instructed.

Self-Check Exercise (Answers on back)

(1) _____

(2) _____

(3) _____

(4) _____

(5) _____

(6) _____

(7) _____

_____

(8) _____

_____

_____

Using the number 35,276:
(1) The 5 is in the _____ position.

(2) The 2 is in the _____ position.

Using the number 4,087,631:
(3) What digit is in the ten-thousands position? _____

(4) What digit is in the hundred-thousands position? _____

Express the following in numerical form.
(5) Nine hundred seven and fourteen hundredths. _____

(6) Seventeen thousandths. _____

Express the following numbers in written form.
(7) 8.05 _____

(8) 36.119 _____

(1) _____Thousands_____

(2) _____Hundreds_____

(3) _____8_____

(4) _____0_____

(5) _____907.14_____

(6) _____.017_____

(7) ____Eight and____

____five hundredths____

(8) __Thirty-six and one__

__hundred nineteen__

____thousandths____

Exercise 1B: Place Values and their Names

Name _____

Course/Sect. No. _____   Score _____

Instructions: Answer each as instructed.

Using the number 35,276:

(1) The 7 is in the _____ position.

(2) The 6 is in the _____ position.

Using the number 4,087,631:

(3) What digit is in the hundreds position? _____

(4) What digit is in the millions position? _____

Express the following in numerical form.

(5) Four thousand two hundred eighty-one _____

(6) Twenty-six million fourty-four thousand seventy _____

Express the following numbers in written form.

(7) 4,321.2 _____

(8) 974,551 _____

Exercise 2A: Addition

Name_____

Course/Sect. No._____   Score_____

Instructions: Answer each as instructed.

Self-Check Exercise (Answers on back)

(1) _____

(2) _____

(3) _____

(4) _____

(5) _____

(6) _____

(7) _____

(8) _____

(9) _____

(10) _____

(1)
```
  2528
  4887
  1109
+ 4149
```

(2)
```
  2227
  2654
  1876
+ 2361
```

(3)
```
   552
    61
  4767
+  102
```

(4)
```
  4499
  2490
  1833
+ 2556
```

(5)
```
  3164
   646
  2908
+ 4253
```

(6)
```
  4154
  3801
   427
+ 3777
```

(7)
```
  1232
  4730
  3600
+ 2586
```

(8)
```
  4213
  3411
   316
+   56
```

(9) Harold averaged grades for his students and Chuck had the following grades: 87, 93, 79, 60, 81, 97. What would be the total of those six grades?

_____

(10) The computer was programmed to an accuracy of five places after the decimal. The latest printout on departmental margins were: Dept. 1— 32.0463, Dept. 2—1,666.66667, Dept. 3—194.00302, Dept. 4—6.1 Add the numbers. _____

(1) _____12,673_____

(2) _____9,118_____

(3) _____5,482_____

(4) _____11,378_____

(5) _____10,971_____

(6) _____12,159_____

(7) _____12,148_____

(8) _____7,996_____

(9) _____497_____

(10) _____1,898.81599_____

Exercise 2B: Addition

Name_____

Course/Sect. No._____ Score_____

Instructions: Answer each as instructed.

(1) _____

(2) _____

(3) _____

(4) _____

(5) _____

(6) _____

(7) _____

(8) _____

(9) _____

(10) _____

(1)
```
  1151
  4162
  1775
 +2207
```

(2)
```
  4017
  4403
   257
 +2976
```

(3)
```
  4860
  2404
  1300
 + 644
```

(4)
```
  3674
     2
  3990
 +4773
```

(5)
```
  1114
  4640
  4400
 + 117
```

(6)
```
  4404
   338
  2279
 +3331
```

(7)
```
   774
  3363
  3988
 +4662
```

(8)
```
  4065
  2406
  1801
 +1533
```

(9) The company found scraps of the following lengths: 130.2 inches, 18.46 inches, .008 inches, 3.11 inches, and 101.015 inches. How many total inches did they have? _____

(10) Stephanie was computing the payroll, which included overtime. What would be the sum of these overtime rates? $8.425, $8.50, $6.4125, $9.35, and $11.005. _____

Exercise 3A: Subtraction

Name _____

Course/Sect. No. _____   Score _____

Instructions: Answer each as instructed.

---

(1) _____

(2) _____

(3) _____

(4) _____

(5) _____

(6) _____

(7) _____

(8) _____

(9) _____

(10) _____

Self-Check Exercises (Answers on back)

(1) 1642
    − 983

(2) 28.427
    −23.548

(3) 16.0407
    −12.1349

(4) 8,347
    −4,252

(5) 115,433
    −109,525

(6) 11,007
    − 9,396

(7) 2,483,000
    − 684,917

(8) $ 13.15
      −12.66

(9) The credit slip showed that the customer had purchased 8 items at 40¢ each, 4 items at $1.25 each, and 2 items at 62¢ each. A tax of 47¢ had been added. If the 4 items at $1.25 each were being returned, what would the new total be? (Disregard any credit for tax on the returned items.) _____

(10) Three dozen pens were purchased. After 6 months there were 17 pens left. How many had been used? _____

(1) _____659_____

(2) _____4,879_____

(3) _____3.9058_____

(4) _____4,095_____

(5) _____5,908_____

(6) _____1,611_____

(7) ____1,798,083____

(8) _____$.49_____

(9) _____$4.91_____

(10) _____19_____

Exercise 3B: Subtraction

Name _____

Course/Sect. No. _____   Score _____

Instructions: Answer each as instructed.

(1) _____

(2) _____

(3) _____

(4) _____

(5) _____

(6) _____

(7) _____

(8) _____

(9) _____

(10) _____

---

(1)   482
        −295

(2)   68
        −59

(3)   14,927
        −13,928

(4)   23.4
        −17.37

(5)   16,994
        −13,846

(6)   2,472.1
        − 848.4

(7)   14.40
        − 3.88

(8) $ 9.58
        −9.49

(9) The warehouse inventory count showed 832,195 boxes of Brand X last year and only 595,388 this year. How many were sold? _____

(10) In September the enrollment was 3,438; in October the enrollment was 2,966; in November the enrollment was 2,784. What was the decrease in enrollment from September to November? _____

Exercise 4A: Multiplication

Name _____

Course/Sect. No. _____   Score_____

Instructions: Multiply the following numbers.

(1) _____

(2) _____

(3) _____

(4) _____

(5) _____

(6) _____

(7) _____

(8) _____

(9) _____

(10) _____

Self Check Exercises (Answers on back)

(1) 63
     $\times$ 9

(2) 42
     $\times$30

(3) 214
     $\times$ 10

(4) 485
     $\times$ 5

(5) 300
     $\times$ 50

(6) 1,724
      $\times$ 25

(7) 366
     $\times$ 77

(8) 643
     $\times$643

(9) Company A had 23 times the employees that Company B had and Company B had 11 times the employees that Company C had. How many employees did Company A have if Company C had 8 employees? _____

(10) What is the product of 82 times 307 times 410? _____

(1)   567

(2)   1,260

(3)   2,140

(4)   2,425

(5)   15,000

(6)   43,100

(7)   28,182

(8)   413,449

(9)   2,024

(10)   10,321,340

Exercise 4B: Multiplication

Name _____

Course/Sect. No. _____   Score_____

Instructions: Multiply the following numbers.

(1) _____	**(1)** 78 × 7
(2) _____	
(3) _____	**(3)** 16 × 8
(4) _____	
(5) _____	**(5)** 203 × 31
(6) _____	
(7) _____	**(7)** 8041 × 67
(8) _____	
(9) _____	
(10) _____	

(1)   78
    × 7

(2)   80
    ×20

(3)   16
    × 8

(4)   91
    ×12

(5)   203
    × 31

(6)   606
    ×202

(7)   8041
    × 67

(8)   499
    ×347

(9) The invoice showed the number of items and the price of each item. What would 313 items at $8 each cost? _____

(10) It took 20 boxes of nails with 430 nails in each box to make 1 carton. How many nails would be in 10 cartons? _____

Exercise 5A: Division

Name _____

Course/Sect. No. _____   Score _____

Instructions: Divide the following numbers.

Self-Check Exercise (Answers on back)

(1) _____   **(1)** 6)‾942‾   **(2)** 17)‾527‾

(2) _____

(3) _____   **(3)** 26)‾1170‾   **(4)** 55)‾1210‾

(4) _____

(5) _____   **(5)** 25)‾10,075‾   **(6)** 87)‾7569‾

(6) _____

(7) _____   **(7)** 89)‾8010‾   **(8)** 60)‾120,240‾

(8) _____

(9) _____   **(9)** The 5 test grades totaled 465. What would the average test grade be?

(10) _____   **(10)** How many times can 161,403 be divided by 201?

(1) _____157_____

(2) _____31_____

(3) _____45_____

(4) _____22_____

(5) _____403_____

(6) _____87_____

(7) _____90_____

(8) _____2,004_____

(9) _____93_____

(10) _____803_____

Exercise 5B: Division

Instructions: Divide the following numbers.

(1) _____   **(1)** 12$\overline{)72}$ **(2)** 23$\overline{)2300}$

(2) _____

(3) _____   **(3)** 14$\overline{)4,228}$ **(4)** 34$\overline{)306}$

(4) _____

(5) _____   **(5)** 10$\overline{)1,070}$ **(6)** 171$\overline{)4959}$

(6) _____

(7) _____   **(7)** 108$\overline{)7344}$ **(8)** 123$\overline{)5658}$

(8) _____

(9) _____   **(9)** The 228 samples were to be divided into 6 equal groups. How many should be in each group?

(10) _____   **(10)** How many times can 4,991 be divided by 217?

Exercise 6A: Rounding Decimal Numbers

Name_____

Course/Sect. No._____   Score_____

Instructions: Round each problem as instructed.

Self-Check Exercises (Answers on back)

(1) _____

(1) Round .084 to tenths.   **(2)** Round 275.0483 to a whole number.

(2) _____

(3) _____

(3) Round 34.477 to hundredths.   **4.** Round .0005 to thousandths.

(4) _____

(5) _____

(5) Round .9945 to hundredths.   **(6)** Round $324.6666 to cents (hundredths).

(6) _____

(7) _____

(7) Round 564.4562 to tenths.   **(8)** Round .3347 to thousandths.

(8) _____

(1) _____.1_____

(2) _____275_____

(3) _____34.48_____

(4) _____.001_____

(5) _____.99_____

(6) _____$324.67_____

(7) _____564.5_____

(8) _____.335_____

Exercise 6B: Rounding Decimal Numbers

Name_____

Course/Sect. No. _____   Score_____

Instructions: Round each problem as instructed.

(1) _____

(2) _____

(3) _____

(4) _____

(5) _____

(6) _____

(7) _____

(8) _____

(1) Round 3.146 to tenths

(2) Round 87.4962 to a whole whole number.

(3) Round .1234 to hundredths.

(4) Round 5.0307 to thousandths.

(5) Round 1.274 to hundredths.

(6) Round .98743 to hundredths.

(7) Round $7.13333 to cents (hundredths).

(8) Round 12.4499 to tenths.

Chapter

Fractions

Preview of Terms

Cancel
To remove a common factor from the numerator and denominator of a

fraction. **Example:** $\dfrac{\overset{3}{\cancel{12}}}{\underset{5}{\cancel{20}}}$ The common factor 4 was divided into both the

numerator, 12, and the denominator, 20.

Complex Fraction
A fraction that has a fraction in it. **Example:** $\dfrac{\frac{5}{8}}{\frac{6}{7}}$ and $\dfrac{3}{5} \div \dfrac{4}{9}$ are both

complex fractions.

Decimal Fraction
Has a denominator that is 10, or some power of 10 ($\frac{1}{10}$, $\frac{23}{100}$, $\frac{17}{1000}$, etc.)

Denominator
The number below the line in a fraction; the quantity that you divide into the

numerator. **Example:** $\dfrac{3}{4}$ You would divide 4 into 3; the 4 is the

denominator.

Equivalent Decimal Number
A decimal number less than 1 (.043, .96, .5758, etc.)

Factor
One or more of the quantities that make up some designated product.
Example: $2 \times 3 = 6$ 2 and 3 are factors of 6.

Improper Fraction
A fraction in which the numerator is larger than the denominator. **Example:** $\dfrac{8}{3}$

Inverted	When the positions of the numerator and denominator of a fraction are reversed. **Example:** $\frac{4}{5}$ would be $\frac{5}{4}$ when inverted.
LCD	Stands for *least (or lowest) common denominator.*
Mixed Decimal Number	Has both a whole number and a decimal fraction (2.1, 25.75, etc.)
Mixed Number	A number containing a whole number and a fraction. **Example:** $4\frac{1}{3}$.
Numerator	The number above the line in a fraction; the quantity that is being divided. **Example:** $\frac{2}{5}$ You would divide the numerator, 2, by the denominator, 5.
Prime Number	Any number that can be divided evenly only by itself or the number 1.
Proportion	An equation stating the equality of two ratios.
Quotient	The amount obtained (the answer) when you divide one number into another number. **Example:** $20 \div 5 = 4$ The 4 is the quotient.

The numerator of a fraction indicates the number of parts contained in the fraction. The denominator of a fraction expresses the number of "total parts" that could be contained in that fraction.

Example 1: The fraction "3/4" indicates that there are 3 parts out of 4 total parts (in this situation).

Decimal fractions have a denominator that is ten, or some power of ten (100, 1,000, etc.). These decimal fractions have equivalent decimal numbers. In an **equivalent decimal number** the numerator is found to the right of the decimal point. The denominator is not written in decimal form, but its value is determined by the number of places to the right of the decimal point. The fractions below are decimal fractions with their equivalent decimal numbers shown to the right.

Decimal Fraction		Equivalent Decimal Number
$\frac{5}{10}$	=	.5
$\frac{7}{100}$	=	.07
$\frac{19}{1000}$	=	.019

A **mixed decimal number** has both a whole number and a decimal fraction.

Number	=	5	4 .	3	6	2
Place values	=	(10's)	(1's).	(1/10ths)	(1/100ths)	(1/1000ths)

Note: *(1/10ths) position means 1/10 of the whole number 1; (1/100ths) position means 1/100 of the whole number 1; (1/1000ths) position means 1/1000 of the whole number 1.*

The decimal fraction for the equivalent decimal number just illustrated would be 54 362/1000. Notice the numerator of the fraction is the number found to the right of the decimal while the denominator is determined by the number of **place values** to the right of the decimal (3 places to the right is thousandths).

Practice Problems

Express the following equivalent decimal numbers as decimal fractions.

1. 3.4 =   **2.** 15.02 =   **3.** .105 =

Express the following decimal fractions as equivalent decimal numbers.

4. 18 12/100 =   **5.** 7/1000 =   **6.** 7 1/10 =

REDUCING FRACTIONS

BY FACTORING

A consistent and time-tested procedure for reducing fractions is factoring with prime numbers. A *prime number* is a number that is divisible evenly only by itself and the number 1. Therefore, the only even prime number is the number 2. Here are prime numbers from 0 to 100.

2, 3, 5, 7, 11, 13, 17, 19, 23, 29, 31, 37, 41,
43, 47, 53, 59, 61, 67, 71, 73, 79, 83, 89, 97

To reduce a fraction, the numerator and denominator are factored into their prime factors. *Like* factors in the numerator and denominator are crossed out (canceled). The remaining *unlike* factors are then multiplied. The resulting fraction will be the reduced form.

Example 2:

$$\frac{12}{54} = \frac{2 \times 2 \times 3}{2 \times 3 \times 3 \times 3} = \frac{2}{9}$$

$$
\begin{array}{r|r}
2 & 12 \\
2 & 6 \\
3 & 3 \\
& 1
\end{array}
\qquad
\begin{array}{r|r}
2 & 54 \\
3 & 27 \\
3 & 9 \\
3 & 3 \\
& 1
\end{array}
$$

When dividing to try to find the prime factors of a number, you need to know when to stop trying new divisors and accept the number as prime.

Example 3:   Factor the number 127.

Solution: Start with the lowest prime number and work your way up the prime numbers in the order from low to high. You will, of course, reject those that do not divide evenly. This eliminates the trial-and-error method of skipping around.

127 ÷ "2" = 63.5	127 ÷ "7" = 18.142857
127 ÷ "3" = 42.333333	127 ÷ "11" = 11.545454
127 ÷ "5" = 25.4	127 ÷ "13" = 9.7692307

After testing all prime numbers in order from low to high, you can accept the number as prime as soon as the square of the last prime number exceeds the number you are testing.

Remember, a number "squared" is that number multiplied by itself ($5 \times 5 = 25$, $17 \times 17 = 289$, etc.). So in Example 3, since all the prime numbers tested were unsuccessful, you can begin to square those prime divisors used.

$$7 \times 7 = 49 \qquad 11 \times 11 = 121 \qquad 13 \times 13 = 169$$

Since 11 squared is 121, which is not larger than the test number (127), it was necessary to try the next largest prime number, 13. When 13 is squared, the product (169) is larger than the test number (127). By the rule, then, you can stop trying other numbers and accept the number 127 as a prime number.

Practice Problems

Reduce the following fractions by factoring.

7. $\dfrac{14}{36}$ 8. $\dfrac{27}{63}$ 9. $\dfrac{266}{2135}$

BY CANCELING

Canceling is nothing more than a shortcut to factoring by prime numbers. Each part of the fraction should be divided by the same number (not necessarily a prime number). Each time you cancel, you reduce the fraction by a common factor. Continue this canceling until all common factors are exhausted. The remaining fraction is then in its reduced form. (The size of the factor or the number of divisions is not important. Continue to divide [cancel] by a common factor until the numerator and denominator no longer have a common factor.)

Example 4:

$$\frac{12}{54} = \frac{\overset{\overset{2}{\cancel{6}}}{\cancel{12}}}{\underset{\underset{9}{\cancel{27}}}{\cancel{54}}} = \frac{2}{9}$$

CONVERTING IMPROPER FRACTIONS

An improper fraction is a fraction whose numerator is the same size as or larger than its denominator. To convert an improper fraction into a mixed number (a whole number and a fraction), simply divide the denominator into the numerator. This will give you the whole number. The remainder, if any, will be the numerator of the fraction for the mixed number. The denominator will be the same as the original denominator.

Example 5:

$$\frac{29}{7} = 4\frac{1}{7} \qquad \begin{array}{r} 4 \\ 7\overline{)29} \\ \underline{28} \\ 1 \end{array}$$

CONVERTING MIXED NUMBERS

It is much easier to work problems with fractions in them by converting the mixed numbers to improper fractions. This procedure is the reverse of that used to convert improper fractions to mixed numbers. Since the opposite of division is multiplication, you must multiply the denominator by the whole number, then add the numerator. This will give you the new numerator. The denominator for the fraction will be the same as the denominator of the mixed-number fraction.

Example 6:

$$4\frac{3}{5} = \frac{(4 \times 5) + 3}{5} = \frac{20 + 3}{5} = \frac{23}{5}$$

Note:

In all fraction problems involving mixed numbers, you should first convert all of the mixed numbers into improper fractions, and then work the problems as the signs indicate.

Practice Problems

Convert the following improper fractions to mixed numbers.

10. $\frac{16}{3}$ **11.** $\frac{38}{12}$ **12.** $\frac{117}{20}$

Convert the following mixed numbers to improper fractions.

13. $5\frac{2}{3}$ **14.** $16\frac{1}{2}$ **15.** $8\frac{4}{5}$

MULTIPLYING FRACTIONS

When multiplying fractions, multiply all numerators together, multiply all denominators together, and reduce. A shortcut method is to cancel first and then multiply. You may cancel *any* numerator number with *any* denominator number. Continue canceling until all like factors are eliminated. After multiplying the numerators and denominators, the resulting fraction is in its reduced form.

Example 7:

$$\frac{10}{12} \times \frac{3}{8} \times \frac{6}{5} = \frac{\overset{1}{\overset{5}{\cancel{10}}}}{\underset{2}{\cancel{12}}} \times \frac{3}{\underset{4}{8}} \times \frac{\overset{1}{\cancel{6}}}{\underset{1}{\cancel{5}}} = \frac{3}{8}$$

Practice Problems

Multiply the following fractions and reduce.

16. $\frac{6}{9} \times \frac{3}{8} \times \frac{10}{15}$ **17.** $\frac{2}{7} \times \frac{15}{20} \times 1\frac{3}{4}$

DIVISION OF FRACTIONS

All division of fractions should be solved in two steps. The first step is to change the sign of operation from division to multiplication. The second step is to invert the divisor. (In a complex fraction, the dividend is the fraction in the numerator, and the divisor is the fraction in the denominator. In Example 8, 3/8 is the dividend and 5/6 is the divisor.) You then have a multiplication problem, which is solved as before.

Note: *Be careful not to invert the dividend number, as this will give you an incorrect answer.*

Example 8:

$$\frac{\frac{3}{8}}{\frac{5}{6}} = \frac{3}{8} \div \frac{5}{6} = \frac{3}{8} \times \frac{6}{5} = \frac{3}{8} \times \frac{\overset{3}{6}}{\underset{4}{5}} = \frac{9}{20}$$

step 1 / step 2

Practice Problems

Divide the following fractions and reduce.

18. $\dfrac{3}{8} \div \dfrac{4}{9}$ **19.** $1\dfrac{1}{5} \div \dfrac{7}{10}$

LEAST COMMON DENOMINATORS

To add or subtract fractions, all of the fractions must have the same denominator. You must, then, convert all of the denominators to common denominators. It is traditional for the conversion to be to least common denominators (LCDs), sometimes called least common multiples (LCMs).

You can find the LCD by dividing the denominators of each fraction by *prime factors.* If a division cannot be made evenly, bring the denominator down as it is. Continue to divide by prime numbers until all the denominators are equal to 1. Then the prime factors, when multiplied, will give you the LCD.

Example 9: Find the LCD for each fraction in the following problem.

$$\frac{7}{8} + \frac{3}{12} + \frac{8}{15} =$$

Denominators

	8	12	15
2	4	6	15
2	2	3	15
2	1	3	15
3	1	1	5
5	1	1	1

LCD = 2 × 2 × 2 × 3 × 5 = 120

(To check your LCD, divide each of the original denominators into the LCD. If they all divide evenly, you have found the LCD.) Once you find the LCD, the new denominators of all the fractions will be the same as the LCD.

When you increase part of a fraction—such as the denominator—you must increase the other part of the fraction—in this case, the numerator—in the same proportion. Therefore, if the denominator of a fraction increases from 8 to 120 (15 times larger), then the numerator must also increase by 15 times so that the fraction remains in the same proportion.

Example 10:

$$\frac{7}{8} = \frac{105}{120}$$

Both the numerator, 7, and the denominator, 8, increased 15 times.

Practice Problems

Find the LCD for each set of numbers.

20. 3, 16, 20   **21.** 8, 12, 14   **22.** 42, 114, 399

To determine the new numerators, multiply the old numerators by the number of times the old denominators can be divided into the new denominator—the LCD.

Example 11:

$$\frac{7}{8} + \frac{3}{12} + \frac{8}{15} = \frac{105 + 30 + 64}{LCD} = \frac{199}{120} = 1\frac{79}{120}$$

Note:

To help eliminate the possibility of a careless mistake, it is safer to show all new numerators over the single LCD. The number 105 was found by dividing 8 into the LCD (120), and multiplying the quotient of 15 by the old numerator (7). The second numerator, 30, was found by dividing the second denominator, 12, into the LCD (120), and multiplying the quotient of 10 by the old numerator (3). The same procedure was followed for the last fraction (120 was divided by 15, and the quotient multiplied by 8). After the new numerator conversions, the numerators are added or subtracted as the signs indicate.

Practice Problems

Work the following addition and subtraction problems and reduce.

23. $\frac{1}{3} + \frac{2}{7} + \frac{5}{9}$ **24.** $\frac{2}{8} + \frac{5}{10} - \frac{7}{12}$ **25.** $\frac{15}{24} + \frac{2}{9}$

ADDITION OF MIXED FRACTIONS

When a problem contains both whole numbers and fractions and you are not using a calculator, it is probably easier to set the problem up the traditional way.

Example 12: $3\frac{1}{2} + \frac{5}{8} + 6\frac{2}{3} =$

$$3\frac{1}{2} = 3\frac{12}{24}$$

$$\frac{5}{8} = \frac{15}{24}$$

$$+6\frac{2}{3} = +6\frac{16}{24}$$

$$9\frac{43}{24}$$

or

$$10\frac{19}{24}$$

$$\begin{array}{r|rrr} 2 & 2 & 8 & 3 \\ 2 & 1 & 4 & 3 \\ 2 & 1 & 2 & 3 \\ 3 & 1 & 1 & 3 \\ \hline & 1 & 1 & 1 \end{array}$$

$$\text{LCD} = 2 \times 2 \times 2 \times 3 = 24$$

Note: *It is still necessary to find the LCD when adding, but the new fractions with common denominators will not be improper fractions because the whole numbers are added separately. The final value of 9 43/24 is reduced to 10 19/24.*

Practice Problems

Using the traditional way, add in the following.

26. $10\frac{1}{2}$

      $8\frac{2}{5}$

   $+12\frac{6}{9}$

27. $5\frac{7}{8}$

      $1\frac{9}{10}$

    $+\ \frac{4}{6}$

SUBTRACTION OF MIXED FRACTIONS

Aside from the fact that you are subtracting, basically the only difference from adding the traditional way is the occasional need to "borrow" from the whole number so the numerator value in the answer will not be a negative number.

Example 13:

          Step 1   Step 2   Step 3

$$(\text{LCD} = 10) \quad 8\frac{1}{5} = \overset{7}{8}\overset{12}{\frac{2}{10}} = 7\frac{12}{10}$$

$$-3\frac{1}{2} = -3\frac{5}{10} = -3\frac{5}{10}$$

$$4\frac{7}{10}$$

Note: *In Step 2 the numerator 5 could not be subtracted from the numerator 2, so "one" had to be borrowed from the whole number 8 and added to the fraction*

2/10. (The fractional value of 1 can be expressed as 10/10; 10/10 added to 2/10 equals 12/10).

Practice Problems

Using the traditional way, subtract the following.

28. $16\dfrac{3}{4}$

    $-\ 9\dfrac{2}{3}$

29. $4\dfrac{1}{6}$

    $-3\dfrac{2}{3}$

Practice Problem Solutions

1. $3.4 = 3\dfrac{4}{10}$

2. $15.02 = 15\dfrac{2}{100}$

3. $.105 = \dfrac{105}{1000}$

4. $18\dfrac{12}{100} = 18.12$

5. $\dfrac{7}{1000} = .007$

6. 7.1

7.
$$\dfrac{14}{36} = \dfrac{\cancel{2} \times 7}{\cancel{2} \times 2 \times 3 \times 3} = \dfrac{7}{18}$$

$$
\begin{array}{c|c}
2 & 14 \\\hline
7 & 7 \\\hline
 & 1
\end{array}
\qquad
\begin{array}{c|c}
2 & 36 \\\hline
2 & 18 \\\hline
3 & 9 \\\hline
3 & 3 \\\hline
 & 1
\end{array}
$$

8.
$$\dfrac{27}{63} = \dfrac{\cancel{3} \times \cancel{3} \times 3}{\cancel{3} \times \cancel{3} \times 7} = \dfrac{3}{7}$$

$$
\begin{array}{c|c}
3 & 27 \\\hline
3 & 9 \\\hline
3 & 3 \\\hline
 & 1
\end{array}
\qquad
\begin{array}{c|c}
2 & 63 \\\hline
3 & 21 \\\hline
7 & 7 \\\hline
 & 1
\end{array}
$$

9.
$$\dfrac{266}{2135} = \dfrac{2 \times 7 \times 19}{5 \times 7 \times 61} = \dfrac{38}{305}$$

$$
\begin{array}{c|c}
2 & 266 \\\hline
7 & 133 \\\hline
19 & 19 \\\hline
 & 1
\end{array}
\qquad
\begin{array}{c|c}
5 & 2135 \\\hline
7 & 427 \\\hline
61 & 61 \\\hline
 & 1
\end{array}
$$

10. $\dfrac{16}{3} = 3\overline{)16} = 5\dfrac{1}{3}$

$$
\begin{array}{r}
5 \\
3\overline{)16} \\
\underline{15} \\
1
\end{array}
$$

11. $\dfrac{38}{12} = 12\overline{)38} = 3\dfrac{2}{12} = 3\dfrac{1}{6}$

$$
\begin{array}{r}
3 \\
12\overline{)38} \\
\underline{36} \\
2
\end{array}
$$

12. $\dfrac{117}{20} = 20\overline{)117} = 5\dfrac{17}{20}$

$$
\begin{array}{r}
5 \\
20\overline{)117} \\
\underline{100} \\
17
\end{array}
$$

13. $5\dfrac{2}{3} = \dfrac{(5 \times 3) + 2}{3} = \dfrac{15 + 2}{3} = \dfrac{17}{3}$

14. $16\dfrac{1}{2} = \dfrac{(16 \times 2) + 1}{2} = \dfrac{32 + 1}{2} = \dfrac{33}{2}$

15. $8\dfrac{4}{5} = \dfrac{(8 \times 5) + 4}{5} = \dfrac{40 + 4}{5} = \dfrac{44}{5}$

16. $\dfrac{6}{9} \times \dfrac{3}{8} \times \dfrac{10}{15} = \dfrac{\cancel{6}}{\cancel{9}} \times \dfrac{\cancel{3}}{\cancel{8}} \times \dfrac{\cancel{10}}{\cancel{15}} = \dfrac{1}{6}$

17. $\dfrac{2}{7} \times \dfrac{15}{20} \times 1\dfrac{3}{4} = \dfrac{2}{7}^{1} \times \dfrac{\overset{3}{15}}{20}_{4} \times \dfrac{\overset{1}{7}}{4}_{2} = \dfrac{3}{8}$

18. $\dfrac{3}{8} \div \dfrac{4}{9} = \dfrac{3}{8} \times \dfrac{9}{4} = \dfrac{27}{32}$

19. $1\dfrac{1}{5} \div \dfrac{7}{10} = \dfrac{6}{5} \times \dfrac{\overset{2}{10}}{7} = \dfrac{12}{7} = 1\dfrac{5}{7}$

20. 3, 16, 20

2	3	16	20
2	3	8	10
2	3	4	5
2	3	2	5
3	3	1	5
5	1	1	5
	1	1	1

LCD $= 2 \times 2 \times 2 \times 2 \times 3 \times 5 = 240$

21. 8, 12, 14

2	8	12	14
2	4	6	7
2	2	3	7
3	1	3	7
7	1	1	7
	1	1	1

LCD $= 2 \times 2 \times 2 \times 3 \times 7 = 168$

22. 42, 114, 399

2	42	114	399
3	21	57	399
7	7	19	133
19	1	19	19
	1	1	1

LCD $= 2 \times 3 \times 7 \times 19 = 798$

23. $\dfrac{1}{3} + \dfrac{2}{7} + \dfrac{5}{9} = \dfrac{21 + 18 + 35}{LCD} = \dfrac{74}{63} = 1\dfrac{11}{63}$

3	3	7	9
3	1	7	3
7	1	7	1
	1	1	1

LCD $= 3 \times 3 \times 7 = 63$

24. $\dfrac{2}{8} + \dfrac{5}{10} - \dfrac{7}{12} = \dfrac{30 + 60 - 70}{LCD} = \dfrac{20}{120} = \dfrac{1}{6}$

2	8	10	12
2	4	5	6
2	2	5	3
3	1	5	3
5	1	5	1
	1	1	1

LCD $= 2 \times 2 \times 2 \times 3 \times 5 = 120$

25. $\dfrac{15}{24} + \dfrac{2}{9} = \dfrac{45 + 16}{\text{LCD}} = \dfrac{61}{72}$

3	24	9
3	8	3
2	8	1
2	4	1
2	2	1
	1	1

$\text{LCD} = 3 \times 3 \times 2 \times 2 \times 2 = 72$

26.
$$10\tfrac{1}{2} = 10\tfrac{45}{90}$$
$$8\tfrac{2}{5} = 8\tfrac{36}{90}$$
$$+12\tfrac{6}{9} = +12\tfrac{60}{90}$$
$$30\tfrac{141}{90} = 31\tfrac{51}{90} = 31\tfrac{17}{30}$$

2	2	5	9
3	1	5	9
3	1	5	3
5	1	5	1
	1	1	1

$\text{LCD} = 2 \times 3 \times 3 \times 5 = 90$

27.
$$5\tfrac{7}{8} = 5\tfrac{105}{120}$$
$$1\tfrac{9}{10} = 1\tfrac{108}{120}$$
$$+\tfrac{4}{6} = +\tfrac{80}{120}$$
$$6\tfrac{293}{120} = 8\tfrac{53}{120}$$

2	8	10	6
2	4	5	3
2	2	5	3
3	1	5	3
5	1	5	1
	1	1	1

$\text{LCD} = 2 \times 2 \times 2 \times 3 \times 5 = 120$

28.
$$16\tfrac{3}{4} = 16\tfrac{9}{12}$$
$$-9\tfrac{2}{3} = -9\tfrac{8}{12}$$
$$7\tfrac{1}{12}$$

2	4	3
2	2	3
3	1	3
	1	1

$\text{LCD} = 2 \times 2 \times 3 = 12$

29.
$$4\tfrac{1}{6} = 4\overset{3}{\tfrac{\overset{7}{1}}{6}} = 3\tfrac{7}{6}$$
$$-3\tfrac{2}{3} = -3\tfrac{4}{6} = -3\tfrac{4}{6}$$
$$\tfrac{3}{6} = \tfrac{1}{2}$$

2	6	3
3	3	3
	1	1

$\text{LCD} = 2 \times 3 = 6$

Exercise 7A: Decimal Numbers and Fractions

Name_____

Course/Sect. No._____   Score_____

Instructions: Express the following as instructed.

---

(1) _____

(2) _____

(3) _____

(4) _____

(5) _____

(6) _____

(7) _____

(8) _____

Self-Check Exercise (Answers on back)

For problems 1 through 4, express the equivalent decimal numbers as decimal fractions.

(1) 20.2

(2) 3.46

(3) .025

(4) .01

For problems 5 through 8, express the decimal fractions as equivalent decimal numbers.

(5) $1\frac{3}{10}$

(6) $25\frac{3}{100}$

(7) $530\frac{941}{1000}$

(8) $12\frac{7}{10}$

(1) _____ $20\frac{2}{10}$ _____

(2) _____ $3\frac{46}{100}$ _____

(3) _____ $\frac{25}{1000}$ _____

(4) _____ $\frac{1}{100}$ _____

(5) _____ 1.3 _____

(6) _____ 25.03 _____

(7) _____ 530.941 _____

(8) _____ 12.7 _____

Exercise 7B: Decimal Numbers and Fractions

Name _____

Course/Sect. No. _____   Score_____

Instructions: Express the following as instructed.

---

(1) _____

(2) _____

(3) _____

(4) _____

(5) _____

(6) _____

(7) _____

(8) _____

For problems 1 through 4, express the equivalent decimal numbers as decimal fractions.

(1) 5.6

(2) .04

(3) 181.541

(4) 55.62

For problems 5 through 8, express the decimal fractions as equivalent decimal numbers.

(5) $\dfrac{9}{10}$

(6) $\dfrac{15}{100}$

(7) $\dfrac{27}{1000}$

(8) $106\dfrac{5}{100}$

Exercise 8A: Reducing Fractions

Name_____

Course/Sect. No._____   Score_____

Instructions: Reduce the following by factoring.

Self-Check Exercise (Answers on back)

(1) _____

(2) _____

(3) _____

(4) _____

(5) _____

(6) _____

(7) _____

(8) _____

(9) _____

(10) _____

(1) $\dfrac{25}{110}$

(2) $\dfrac{84}{216}$

(3) $\dfrac{271}{522}$

(4) $\dfrac{110}{121}$

(5) $\dfrac{48}{160}$

(6) $\dfrac{561}{935}$

(7) $\dfrac{66}{297}$

(8) $\dfrac{864}{1080}$

(9) Dennis drove 60 miles from Pearsall on the way to Houston, a 240-mile trip. In lowest terms, what fractional part of the trip had been completed?

(10) Of 210 people surveyed, 84 said that they liked to play tennis. Of every 5 people surveyed, how many said they liked to play tennis?

(1) $\dfrac{5}{22}$ _____

(2) _____ $\dfrac{7}{18}$

(3) _____ $\dfrac{271}{522}$

(4) _____ $\dfrac{10}{11}$

(5) _____ $\dfrac{3}{10}$

(6) _____ $\dfrac{3}{5}$

(7) _____ $\dfrac{2}{9}$

(8) _____ $\dfrac{4}{5}$

(9) _____ $\dfrac{1}{4}$

(10) _____ 2

Exercise 8B: Reducing Fractions

Name_____

Course/Sect. No._____   Score_____

Instructions: Reduce the following by factoring.

(1) _____

(2) _____

(3) _____

(4) _____

(5) _____

(6) _____

(7) _____

(8) _____

(9) _____

(10) _____

(1) $\dfrac{16}{22}$

(2) $\dfrac{3}{24}$

(3) $\dfrac{56}{420}$

(4) $\dfrac{130}{144}$

(5) $\dfrac{24}{1008}$

(6) $\dfrac{64}{114}$

(7) $\dfrac{63}{210}$

(8) $\dfrac{1534}{4238}$

(9) Eighteen of the 64 customers of a store had complaints about service. In lowest terms, what fraction of the customers complained?

(10) Ken found that 33 of his 44 students could identify an unknown chemical. In lowest terms, what fraction of his students could identify it?

Exercise 9A: Converting Improper Fractions

Name _____

Course/Sect. No. _____   Score _____

Instructions: Convert the following improper fractions to mixed numbers. (Do not reduce in the answer.)

(1) _____	
(2) _____	
(3) _____	
(4) _____	
(5) _____	
(6) _____	
(7) _____	
(8) _____	
(9) _____	
(10) _____	

Self-Check Exercise (Answers on back)

(1) $\dfrac{55}{8}$ **(2)** $\dfrac{44}{7}$

(3) $\dfrac{17}{3}$ **(4)** $\dfrac{19}{9}$

(5) $\dfrac{29}{12}$ **(6)** $\dfrac{16}{5}$

(7) $\dfrac{687}{21}$ **(8)** $\dfrac{67}{17}$

(9) A baseball pitcher retired 23 batters in a row. How many innings did the streak cover since there are 3 outs for each inning? (Use a mixed number.)

(10) How many minutes of commercial time did an advertiser buy if he bought 26 different 10-second spots?

(1) _____ $6\frac{7}{8}$ _____

(2) _____ $6\frac{2}{7}$ _____

(3) _____ $5\frac{2}{3}$ _____

(4) _____ $2\frac{1}{9}$ _____

(5) _____ $2\frac{5}{12}$ _____

(6) _____ $3\frac{1}{5}$ _____

(7) _____ $32\frac{15}{21}$ _____

(8) _____ $3\frac{16}{17}$ _____

(9) _____ $7\frac{2}{3}$ _____

(10) _____ $4\frac{20}{60}$ _____

Exercise 9B: Converting Improper Fractions

Name _____

Course/Sect. No. _____   Score _____

Instructions: Convert the following improper fractions to mixed numbers. (Do not reduce in the answer.)

(1) _____

(2) _____

(3) _____

(4) _____

(5) _____

(6) _____

(7) _____

(8) _____

(9) _____

(10) _____

(1) $\dfrac{12}{7}$

(2) $\dfrac{110}{25}$

(3) $\dfrac{139}{11}$

(4) $\dfrac{241}{15}$

(5) $\dfrac{172}{8}$

(6) $\dfrac{481}{24}$

(7) $\dfrac{333}{33}$

(8) $\dfrac{39}{14}$

(9) There are 24 bottles of soda pop to a case. Expressed as a mixed number, how many cases will 97 bottles make?

(10) Expressed as a mixed number, how many dozen doughnuts must be purchased if 27 persons eat 2 doughnuts each?

Exercise 10A: Converting Mixed Numbers

Name_____

Course/Sect. No._____   Score_____

Instructions: Convert the following mixed numbers to improper fractions.

Self-Check Exercise (Answers on back)

(1) _____   **(1)** $4\frac{2}{9}$ **(2)** $8\frac{5}{8}$

(2) _____

(3) _____   **(3)** $10\frac{1}{9}$ **(4)** $16\frac{13}{15}$

(4) _____

(5) _____   **(5)** $31\frac{9}{13}$ **(6)** $260\frac{3}{4}$

(6) _____

(7) _____   **(7)** $25\frac{4}{5}$ **(8)** $62\frac{11}{12}$

(8) _____

(9) _____   **(9)** A salesperson receives \$1 for each $\frac{1}{5}$ of a case sold. If $9\frac{4}{5}$ cases are sold, how many dollars will she earn?

(10) _____   **(10)** Every quarter hour an employee works, he is paid \$1. If an employee works $37\frac{1}{4}$ hours, how much money will he receive?

(1) _____ $\dfrac{38}{9}$ _____

(2) _____ $\dfrac{69}{8}$ _____

(3) _____ $\dfrac{91}{9}$ _____

(4) _____ $\dfrac{253}{15}$ _____

(5) _____ $\dfrac{412}{13}$ _____

(6) _____ $\dfrac{1043}{4}$ _____

(7) _____ $\dfrac{129}{5}$ _____

(8) _____ $\dfrac{755}{12}$ _____

(9) _____ $49 _____

(10) _____ $149 _____

Exercise 10B: Converting Mixed Numbers

Name _____

Course/Sect. No. _____   Score_____

Instructions: Convert the following mixed numbers to improper fractions.

---

(1) _____

(2) _____

(3) _____

(4) _____

(5) _____

(6) _____

(7) _____

(8) _____

(9) _____

(10) _____

(1) $3\frac{1}{7}$

(2) $10\frac{2}{5}$

(3) $7\frac{21}{24}$

(4) $22\frac{5}{7}$

(5) $104\frac{3}{9}$

(6) $14\frac{1}{2}$

(7) $5\frac{31}{39}$

(8) $1\frac{117}{120}$

(9) A retailer has $6\frac{3}{5}$ cases of an item. Expressed as an improper fraction, how many cases does she have?

(10) Larry picked $5\frac{2}{3}$ boxes of apples and his friend picked $4\frac{1}{3}$ boxes. How many $\frac{1}{3}$'s did they pick?

Exercise 11A: Multiplying Fractions

Name_____

Course/Sect. No. _____   Score_____

Instructions: Multiply the following fractions and reduce.

---

(1) _____

(2) _____

(3) _____

(4) _____

(5) _____

(6) _____

(7) _____

(8) _____

(9) _____

(10) _____

Self-Check Exercise (Answers on back)

(1) $\dfrac{7}{36} \times \dfrac{2}{13}$

(2) $\dfrac{5}{7} \times \dfrac{18}{36}$

(3) $\dfrac{8}{12} \times \dfrac{5}{11} \times \dfrac{22}{25}$

(4) $\dfrac{14}{15} \times \dfrac{21}{24}$

(5) $5\dfrac{3}{4} \times 6\dfrac{1}{6}$

(6) $3 \times 4\dfrac{2}{3} \times 2$

(7) $\dfrac{7}{8} \times 4\dfrac{1}{2} \times \dfrac{10}{15}$

(8) $\dfrac{5}{9} \times \dfrac{1}{3} \times \dfrac{12}{15} \times \dfrac{3}{4}$

(9) Betty sold $\frac{4}{5}$ of the total number of tickets in a raffle. Terry sold $\frac{1}{3}$ of the number of tickets Betty sold. What fraction of the total tickets did Terry sell?

(10) The ABC Co. used $\frac{4}{7}$ as much computer time as the XYZ Co., which used $10\frac{1}{2}$ hours. How much time did the ABC Co. use?

(1) $\dfrac{7}{234}$

(2) $\dfrac{5}{14}$

(3) $\dfrac{4}{15}$

(4) $\dfrac{49}{60}$

(5) $35\dfrac{11}{24}$

(6) 28

(7) $2\dfrac{5}{8}$

(8) $\dfrac{1}{9}$

(9) $\dfrac{4}{15}$

(10) 6

Exercise 11B: Multiplying Fractions

Name _____

Course/Sect. No. _____   Score _____

Instructions: Multiply the following fractions and reduce.

(1) _____   (1) $\frac{1}{12} \times \frac{14}{21}$ (2) $\frac{6}{10} \times \frac{3}{20}$

(2) _____

(3) _____   (3) $\frac{1}{9} \times \frac{3}{15} \times \frac{5}{6}$ (4) $\frac{17}{70} \times \frac{10}{51} \times \frac{19}{57}$

(4) _____

(5) _____   (5) $1\frac{1}{2} \times \frac{3}{5} \times \frac{2}{5}$ (6) $17\frac{1}{2} \times 2\frac{7}{10}$

(6) _____

(7) _____   (7) $5\frac{6}{7} \times 4$ (8) $17\frac{1}{6} \times 2\frac{1}{4}$

(8) _____

(9) _____   (9) Jill worked $\frac{2}{3}$ as much as Bill, who worked $\frac{3}{4}$ of an hour. How many hours did Jill work?

(10) _____   (10) John's wage rate is $8 per hour. Mark's rate is $\frac{5}{16}$ of John's. How much does Mark make per hour?

Exercise 12A: Dividing Fractions

Name _____

Course/Sect. No. _____   Score _____

Instructions: Divide the following fractions and reduce.

Self-Check Exercise (Answers on back)

(1) _____   **(1)** $\dfrac{9}{10} \div \dfrac{12}{15}$ **(2)** $\dfrac{13}{29} \div \dfrac{15}{35}$

(2) _____

(3) _____   **(3)** $\dfrac{5}{12} \div \dfrac{25}{42}$ **(4)** $\dfrac{9}{15} \div \dfrac{1}{9}$

(4) _____

(5) _____   **(5)** $1\dfrac{2}{3} \div \dfrac{4}{5}$ **(6)** $\dfrac{7}{8} \div 1\dfrac{10}{12}$

(6) _____

(7) _____   **(7)** $6 \div \dfrac{1}{4}$ **(8)** $19\dfrac{1}{2} \div \dfrac{1}{3}$

(8) _____

(9) _____   **(9)** In a circle, the diameter is equal to the circumference divided by pi. Assuming pi is approximately $3\frac{1}{7}$, what is the diameter of a circle with a circumference of $2\frac{10}{28}$?

(10) _____   **(10)** How many $\frac{2}{3}$'s are in $\frac{1}{9}$?

(1) _____ $1\frac{1}{8}$ _____

(2) _____ $1\frac{4}{87}$ _____

(3) _____ $\frac{7}{10}$ _____

(4) _____ $5\frac{2}{5}$ _____

(5) _____ $2\frac{1}{12}$ _____

(6) _____ $\frac{21}{44}$ _____

(7) _____ 24 _____

(8) _____ $58\frac{1}{2}$ _____

(9) _____ $\frac{3}{4}$ _____

(10) _____ $\frac{1}{6}$ _____

Exercise 12B: Dividing Fractions

Name _____

Course/Sect. No. _____   Score_____

Instructions: Divide the following fractions and reduce.

(1) _____

(2) _____

(3) _____

(4) _____

(5) _____

(6) _____

(7) _____

(8) _____

(9) _____

(10) _____

(1) $\dfrac{9}{21} \div \dfrac{10}{20}$

(2) $\dfrac{1}{11} \div \dfrac{1}{13}$

(3) $\dfrac{15}{28} \div \dfrac{1}{2}$

(4) $\dfrac{35}{36} \div \dfrac{35}{41}$

(5) $\dfrac{3}{14} \div \dfrac{5}{7}$

(6) $5\dfrac{2}{6} \div 1\dfrac{5}{8}$

(7) $\dfrac{15}{28} \div 1\dfrac{5}{10}$

(8) $18 \div 2\dfrac{1}{2}$

(9) A farmer plans to pack his peaches in $\frac{2}{3}$-bushel baskets. How many of these baskets will he pack from 24 full bushels of peaches?

(10) Alice worked $44\frac{3}{4}$ hours, which was $\frac{7}{8}$ as much as Bob worked. How many hours did Bob work?

Exercise 13A: Finding the Least Common Denominator

Name_____

Course/Sect. No. _____   Score_____

Instructions: Find the LCD in each set of numbers.

---

(1) _____	Self-Check Exercise (Answers on back)	

(1) _____   **(1)** 6, 10, 5   **(2)** 4, 6, 8

(2) _____

(3) _____   **(3)** 24, 26, 28   **(4)** 22, 26, 33

(4) _____

(5) _____   **(5)** 30, 70, 77   **(6)** 39, 51, 85

(6) _____

(7) _____   **(7)** 442, 494, 342   **(8)** 2346, 2737

(8) _____

(1) _____30_____

(2) _____24_____

(3) _____2,184_____

(4) _____858_____

(5) _____2,310_____

(6) _____3,315_____

(7) _____75,582_____

(8) _____16,422_____

Exercise 13B: Finding the Least Common Denominator

Name _____

Course/Sect. No. _____   Score _____

Instructions: Find the LCD in each set of numbers.

(1) _____   **(1)** 8, 12, 9   **(2)** 5, 15, 12

(2) _____

(3) _____   **(3)** 21, 9, 15   **(4)** 16, 6, 14

(4) _____

(5) _____   **(5)** 19, 12, 38   **(6)** 78, 70, 42

(6) _____

(7) _____   **(7)** 245, 420, 140   **(8)** 1225, 2205

(8) _____

Exercise 14A: Adding Fractions

Name_____

Course/Sect. No. _____   Score_____

Instructions: Add the following fractions and reduce.

Self-Check Exercise (Answers on back)

(1) _____

(2) _____

(3) _____

(4) _____

(5) _____

(6) _____

(7) _____

(8) _____

(9) _____

(10) _____

(1) $\dfrac{3}{10} + \dfrac{2}{15}$

(2) $\dfrac{5}{22} + \dfrac{1}{2}$

(3) $\dfrac{1}{24} + \dfrac{2}{13}$

(4) $\begin{array}{r} 4\frac{1}{5} \\ +3\frac{3}{10} \\ \hline \end{array}$

(5) $\begin{array}{r} 10\frac{3}{5} \\ 6\frac{2}{3} \\ +\ 4\frac{5}{9} \\ \hline \end{array}$

(6) $\begin{array}{r} 7\frac{1}{3} \\ \frac{5}{9} \\ +1\frac{1}{6} \\ \hline \end{array}$

(7) $\begin{array}{r} \frac{6}{15} \\ \frac{5}{16} \\ +\ \frac{3}{8} \\ \hline \end{array}$

(8) $\begin{array}{r} 2\frac{15}{16} \\ 12\frac{7}{8} \\ +\ 5\frac{20}{24} \\ \hline \end{array}$

(9) The secretary noticed that $\frac{5}{8}$ of the time was spent typing and $\frac{2}{10}$ of it was spent filing. What fraction of the time was spent doing these activities?

(10) A carpenter needs to glue boards together to make a total thickness of 6 inches. If he glues boards of $2\frac{5}{8}$ inches, $2\frac{3}{4}$ inches, and $\frac{5}{6}$ inches, what would be the total thickness?

(1) ____ $\dfrac{13}{30}$ ____

(2) ____ $\dfrac{8}{11}$ ____

(3) ____ $\dfrac{61}{312}$ ____

(4) ____ $7\dfrac{1}{2}$ ____

(5) ____ $21\dfrac{37}{45}$ ____

(6) ____ $9\dfrac{1}{18}$ ____

(7) ____ $1\dfrac{7}{80}$ ____

(8) ____ $21\dfrac{31}{48}$ ____

(9) ____ $\dfrac{33}{40}$ ____

(10) ____ $6\dfrac{5}{24}$ ____

Exercise 14B: Adding Fractions

Name _____

Course/Sect. No. _____   Score _____

Instructions: Add the following fractions and reduce.

(1) _____

(2) _____

(3) _____

(4) _____

(5) _____

(6) _____

(7) _____

(8) _____

(9) _____

(10) _____

(1) $\dfrac{1}{30} + \dfrac{6}{12}$

(2) $\dfrac{5}{11} + \dfrac{2}{3}$

(3) $\dfrac{1}{4} + \dfrac{6}{15}$

(4) $\begin{array}{r} 3\dfrac{1}{2} \\ +1\dfrac{5}{8} \\ \hline \end{array}$

(5) $\begin{array}{r} \dfrac{5}{6} \\ 1\dfrac{2}{3} \\ +3\dfrac{7}{12} \\ \hline \end{array}$

(6) $\begin{array}{r} 6\dfrac{7}{8} \\ +4\dfrac{5}{6} \\ \hline \end{array}$

(7) $\begin{array}{r} 2\dfrac{2}{9} \\ 10\dfrac{5}{12} \\ +\ 4 \\ \hline \end{array}$

(8) $\begin{array}{r} 1\dfrac{8}{9} \\ 1\dfrac{2}{3} \\ +1\dfrac{5}{6} \\ \hline \end{array}$

(9) A painter finished $\frac{3}{5}$ of a house one day and $\frac{2}{10}$ of it the next. What fraction of the house did he finish in the two days?

(10) The interest rate on a loan was a basic $9\frac{3}{4}\%$ plus $2\frac{1}{2}\%$ for special considerations. What was the total rate?

Exercise 15A: Subtracting Fractions

Name _____

Course/Sect. No. _____   Score _____

Instructions: Subtract the following fractions and reduce.

Self-Check Exercise (Answers on back)

(1) _____

(1) $\dfrac{8}{10} - \dfrac{1}{18}$

(2) $\dfrac{3}{4} - \dfrac{11}{25}$

(2) _____

(3) _____

(3) $\dfrac{11}{12} - \dfrac{8}{21}$

(4) $\begin{array}{r} 10\frac{5}{12} \\ -\ 3\frac{3}{24} \\ \hline \end{array}$

(4) _____

(5) _____

(5) $\begin{array}{r} 8\frac{4}{5} \\ -3\frac{3}{20} \\ \hline \end{array}$

(6) $\begin{array}{r} 7\frac{4}{9} \\ -3\frac{2}{7} \\ \hline \end{array}$

(6) _____

(7) _____

(7) $\begin{array}{r} 3\frac{7}{9} \\ -2\frac{1}{3} \\ \hline \end{array}$

(8) $\begin{array}{r} 12\frac{8}{9} \\ -\ 7\frac{4}{27} \\ \hline \end{array}$

(8) _____

(9) _____

(9) Of the bank statements last year, $\frac{3}{8}$ were prepared by Agnes. Agnes and Betty together prepared $\frac{2}{3}$ of the statements. What fraction of the statements did Betty do?

(10) _____

(10) The meat cutter had $5\frac{1}{4}$ pounds of ground beef. He used $2\frac{1}{4}$ pounds in one package and $1\frac{3}{5}$ pounds in another package. How much meat did he have left for the last package?

(1) _____ $\dfrac{67}{90}$ _____

(2) _____ $\dfrac{31}{100}$ _____

(3) _____ $\dfrac{15}{28}$ _____

(4) _____ $7\dfrac{7}{24}$ _____

(5) _____ $5\dfrac{13}{20}$ _____

(6) _____ $4\dfrac{10}{63}$ _____

(7) _____ $1\dfrac{4}{9}$ _____

(8) _____ $5\dfrac{20}{27}$ _____

(9) _____ $\dfrac{7}{24}$ _____

(10) _____ $1\dfrac{2}{5}$ _____

Exercise 15B: Subtracting Fractions

Name _____

Course/Sect. No. _____   Score _____

Instructions: Subtract the following fractions and reduce.

(1) _____

(2) _____

(3) _____

(4) _____

(5) _____

(6) _____

(7) _____

(8) _____

(9) _____

(10) _____

(1) $\dfrac{5}{6} - \dfrac{1}{7}$

(2) $\dfrac{8}{9} - \dfrac{5}{25}$

(3) $\dfrac{9}{25} - \dfrac{6}{30}$

(4) $\begin{array}{r} 5\frac{6}{7} \\ -2\frac{2}{8} \\ \hline \end{array}$

(5) $\begin{array}{r} 8\frac{1}{3} \\ -2\frac{1}{2} \\ \hline \end{array}$

(6) $\begin{array}{r} 6\frac{2}{3} \\ -1\frac{1}{8} \\ \hline \end{array}$

(7) $\begin{array}{r} 18\frac{1}{5} \\ -10\frac{3}{10} \\ \hline \end{array}$

(8) $\begin{array}{r} \frac{8}{10} \\ -\frac{3}{4} \\ \hline \end{array}$

(9) Mary lost $1\frac{1}{2}$ inch in her waist measurement, which was $26\frac{5}{6}$ inches. What was her new waist size?

(10) Sam needed a board $7\frac{7}{8}$ inches wide. The board he had was $10\frac{1}{6}$ inches wide. How much did he have to trim off the wide board?

Chapter

Solving for
the Unknown

Preview of Terms

Equation   A mathematical statement separated into left and right sides that are conditionally equal to each other. **Example:** $x + 2 = 8$. The condition that would make this equation equal on both sides would be $x = 6$.

Product   The answer obtained when numbers are multiplied together. **Example:** $3 \times 8 = 24$. The 24 is the product.

An equation is a mathematical statement in which the terms on the left side of the equal sign have the same value as the terms on the right side of the equal sign. It is particularly useful in business math to be able to take an equation and solve for the part that is not known. In many instances, you need know only one equation instead of several. There are few rules that you should learn.

SOLVING FOR THE UNKNOWN

The objective in solving for an unknown is to get the unknown on one side of the equation by itself and everything else on the other side. It is customary that the unknown end up on the left side, but it is not imperative.

Rule 1:   Perform the inverse (opposite) operation on the unknown; that is, undo what has already been done to the unknown. Addition is the inverse operation of subtraction, and multiplication is the inverse operation of division.

Example 1: $A + 2 = 8$ The unknown in this problem is A. Since 2 is being added to A, the inverse operation would be to subtract 2 from A.

Rule 2:

> To keep an equation balanced, whatever you do to one side you must also do to the other side.

Example 2:

$$\begin{array}{rcr} A + 2 &=& 8 \\ -\ 2 &=& -\ 2 \\ \hline A &=& 6 \end{array}$$

Using Rule 1, you subtract 2 from the left side of the equation. Using Rule 2, you subtract 2 from the right side of the equation. The 2s on the left side cancel each other out, and 8 minus 2 on the right side gives you the solution to the unknown, A.

Example 3:

$$\begin{array}{rcr} B - 3 &=& 9 \\ +\ 3 &=& +\ 3 \\ \hline B &=& 12 \end{array}$$

Applying Rules 1 and 2 to this equation, you added 3 (the inverse of -3) to *both* sides of the equation.

There are several acceptable ways of indicating multiplication. Some of the alternatives available follow.

5×6 This is the standard method where the times sign, the $\times$, separates the numbers being multiplied. This sign is usually not used in equations containing letters since it may become confused with the letter X, often used to represent an unknown.

$2 \cdot 4$ Another way of showing multiplication is by using a decimal-like dot between two numbers. You must be careful to locate the dot high enough that it will not be confused with a decimal point.

$3(7)$ or $(3)(7)$ When parentheses are used as shown, the numbers should be multiplied together.

$2X$ If there is no sign between a number and a letter, multiplication is understood to be the sign of operation.

Example 4:

$$\frac{C}{4} = 20$$

$$4 \cdot \frac{C}{4} = 20 \cdot 4$$

$$C = 80$$

When you apply Rules 1 and 2 to this equation, you multiply *both* sides of the equation by 4 (the inverse of division by 4).

Example 5:

$$5X = 30$$

$$\frac{5X}{5} = \frac{30}{5}$$

$$X = 6$$

When you apply Rules 1 and 2 to this equation, you divide both sides of the equation by 5 (the inverse of multiplication by 5).

Rule 3:

> When more than one operation is indicated on the unknown, you should undo them one at a time. Always undo addition and subtraction first, then multiplication and division.

Example 6:

$$2X + 4 = 20$$
$$\underline{ - 4 = -4}$$
$$2X = 16$$
$$\frac{2X}{2} = \frac{16}{2}$$
$$X = 8$$

Since the unknown, X, is being multiplied by 2, and 4 is being added to the product, you should undo addition of 4 first.

Next, you should undo multiplication by 2.

Rule 4: When the unknowns are already on the same side of the equation, just add or subtract as the signs indicate and proceed.

Example 7:

$$4C + 2C = 24$$
$$6C = 24$$
$$\frac{6C}{6} = \frac{24}{6}$$
$$C = 4$$

Rule 5: When the unknown is on both sides of an equation, you must group all parts of the unknown on the same side. Always eliminate the smallest number of unknowns by performing the inverse operation.

Example 8:

$$X = 30 + .5X$$
$$\underline{-.5X = - .5X}$$
$$.5X = 30$$
$$\frac{.5X}{.5} = \frac{30}{.5}$$
$$X = 60$$

The left side has $1X$; the right side, $.5X$ ($1/2\ X$). Because .5 is smaller than 1, to eliminate the Xs on the right side, perform the inverse operation of subtracting $.5X$ from both sides.

Next, undo multiplication of .5 by dividing both sides by .5.

Rule 6: To remove parentheses from an equation, multiply the number outside the parentheses by everything inside the parentheses, being careful to include the sign of operation. After removing the parentheses, solve the equation as instructed previously.

Example 9:

$$2(X + 4) = 20$$

Multiply 2 times X **and** 2 times 4.

$$2(X + 4) = 20$$
$$2X + 8 = 20$$

2 times $X = 2X$, and 2 times $4 = 8$.

CROSS-MULTIPLICATION

A shortcut technique for solving for the unknown can be used when you have a fraction on both sides of an equation. To simplify the equation and eliminate the fractions, cross-multiply through the equals sign. (Multiply the denominator of the first fraction by the numerator of the second fraction; then multiply the second denominator by the numerator of the first fraction.) You must remember to separate the products by the equal sign since they were separated by the equal sign when they were fractions.

Example 10:

$$\frac{2}{X} = \frac{3}{8}$$

$$\frac{2}{X} \diagdown \frac{3}{8}$$

$$(X)(3) = (2)(8)$$

$$3X = 16$$

$$X = 5\frac{1}{3}$$

Example 11:

$$\frac{X}{5} = \frac{7}{12}$$

$$\frac{X}{5} \diagdown \frac{7}{12}$$

$$(12)(X) = (5)(7)$$

$$12X = 35$$

$$X = 2\frac{11}{12}$$

Example 12:

$$\frac{4}{X} = 16$$

$$\frac{4}{X} \diagdown \frac{16}{1}$$

Note: *Placing the number 1 under a whole number will not change the value of the number.*

$$(X)(16) = (4)(1)$$

$$16X = 4$$

$$X = \frac{1}{4}$$

Practice Problems

Solve for the unknown.

1. $X + 4 = 24$ **2.** $A - 3 = 7$ **3.** $\dfrac{B}{5} = 6$

4. $2C = 12$ **5.** $13X + 7 = 9X + 43$ **6.** $2(X - 3) = 18$

7. $\dfrac{2X}{3} = 18$ **8.** $SP = 24 + .4SP$ **9.** $5X + 8X = 20 + X + 4$

SETTING UP EQUATIONS

Most word problems consist of statements that can each be converted to a mathematical statement (an equation). To translate the information from words to mathematical symbols is sometimes quite difficult, but some general guidelines will usually help.

GUIDELINES FOR SETTING UP EQUATIONS

1. Read the problem very carefully. Try to understand clearly what is given and what is wanted. You may have to re-read it several times.

The purpose here is to learn what the problem is about. Ask yourself questions like:

- What is the general idea of the problem?
- What is known and what is unknown?
- What facts or clues will help solve the problem?
- What is wanted?

As you read the problem, do not worry about details. When a statement says, "Joe earned $6,000," think of it as saying "Joe earned some money." Or if it says, "Bill earned 3/4 of what Sara earned," you should think of it as saying "Bill earned less than Sara." You can fill in the details later. Here you are trying to grasp the overall meaning of the problem.

2. Let some symbol represent what is wanted (an unknown) and then express the other unknown quantities in terms of that same symbol. ("*X*" is the traditional unknown, but it may help to select a symbol that more closely relates to the problem.)

Example 13:

Joe earns $6000, which is $200 less than 2 times as much as Henry. What does Henry earn?

Analysis: Given—Joe's earnings = $6000
  Joe's earnings = 2 times Henry's less $200
  Wanted—Henry's earnings

Solution: Let H = Henry's earnings, then
  Joe's earnings = 2 times H, less $200

3. Certain words usually give clues to the solution of word problems.

- "Less than" and "decrease" indicate subtraction.
- "More than" and "increase" indicate addition.
- "Goes into" indicates division.
- "Of" usually indicates multiplication.
- "Is" indicates equals.

Use these clues to develop the equation you think the problem represents. Do not attempt to solve it at this stage. Re-read the problem and compare it to your equation. Does your equation express in mathematical terms what the statement says? If not, revise it and check it again.

Example 13 (cont.):

"Less" means subtraction, therefore
Joe's earnings = $2H - 200 and
Joe's earnings = $6000
Solution: $6000 = 2H - 200

4. Solve the equation to find the value of the unknown.

Example 13 (cont.):

$$\begin{aligned} \$6000 &= 2H - \$200 \\ + 200 &= + 200 \\ \hline \$6200 &= 2H \\ \$6200/2 &= 2H/2 \\ \$3100 &= H \end{aligned}$$

5. Check your solution by substituting the value you solved for into your equation. If the equation balances, you assume your solution is correct. As a last step on checking your answer, ask yourself if your solution actually answers the questions you had. If your answer is not a reasonable solution, retrace the steps to determine what needs to be changed so that it will be reasonable.

Example 13 (cont.): $6000 = 2H - \$200$ ← Original equation
                         $H = \$3100$ ← Solution

Check: Substituting $3100 for H in the original equation,
$6000 = 2(\$3100) - \$200$
$6000 = \$6200 - \$200$
$6000 = \$6000$

Practice Problems

Restate each word problem in the form of an equation and solve for the unknown.

10. Eight times a number less three times the same number is twenty. What is the number?

11. Doris is four times as old as April. The sum of Doris' and April's ages is forty-five. Find each of their ages.

Practice Problem Solutions

1.
$$X + 4 = 24$$
$$\underline{ -4 \quad -4}$$
$$X \quad = 20$$

2.
$$A - 3 = 7$$
$$\underline{ +3 \quad +3}$$
$$A \quad = 10$$

3.
$$\frac{B}{5} = 6$$
$$5 \cdot \frac{B}{5} = 6 \cdot 5$$
$$B = 30$$

4.
$$2C = 12$$
$$\frac{2C}{2} = \frac{12}{2}$$
$$C = 6$$

5.
$$13X + 7 = 9X + 43$$
$$\underline{ -7 \qquad -7}$$
$$13X \quad = 9X + 36$$
$$\underline{-9X \quad = -9X}$$
$$4X \quad = \quad 36$$
$$\frac{4X}{4} = \frac{36}{4}$$
$$X = 9$$

6.
$$2(X - 3) = 18$$
$$2X - 6 = 18$$
$$\underline{ +6 \quad +6}$$
$$2X \quad = 24$$
$$\frac{2X}{2} = \frac{24}{2}$$
$$X = 12$$

7.
$$\frac{2X}{3} = 18$$
$$3 \cdot \frac{2X}{3} = 18 \cdot 3$$
$$2X = 54$$
$$\frac{2X}{2} = \frac{54}{2}$$
$$X = 27$$

8.
$$SP = 24 + .4SP$$
$$\underline{-.4SP \qquad -.4SP}$$
$$.6SP = 24$$
$$\frac{.6SP}{.6} = \frac{24}{.6}$$
$$SP = 40$$

9.
$$5X + 8X = 20 + X + 4$$
$$13X = 24 + X$$
$$\underline{-X \qquad -X}$$
$$12X = 24$$
$$\frac{12X}{12} = \frac{24}{12}$$
$$X = 2$$

10.
$$8N - 3N = 20$$
$$5N = 20$$
$$\frac{5N}{5} = \frac{20}{5}$$
$$N = 4$$

11. a. Doris = 4 April $\rightarrow D = 4A$
b. Doris + April = 45 $\rightarrow D + A = 45$

Substitute in equation (**b**) what D is equal to from equation (**a**).

a. $D = 4A$
b.
$$D + A = 45$$
$$4A + A = 45$$
$$5A = 45$$
$$\frac{5A}{5} = \frac{45}{5}$$
$$A = 9$$

Then substitute in equation (**b**) what A is now equal to.

$$A = 9$$
$$D + A = 45$$
$$D + 9 = 45$$
$$\underline{ -9 = -9}$$
$$D \quad = 36$$

Exercise 16A: Solving by Addition and Subtraction

Name_____

Course/Sect. No._____   Score_____

Instructions: Solve for the unknown.

(1) _____	

Self-Check Exercises (Answers on back)

(1) $B - 3 = 15$ **(2)** $3 + D = 12$

(2) _____

(3) _____   **(3)** $11 = 5 + F$ **(4)** $H + 6 = 16$

(4) _____

(5) _____   **(5)** $J - 8 = 1$ **(6)** $1 = L - 1$

(6) _____

(7) _____   **(7)** $N = 16 - 7$ **(8)** $R - 3 = 3$

(8) _____

(1) _____18_____

(2) _____9_____

(3) _____6_____

(4) _____10_____

(5) _____9_____

(6) _____2_____

(7) _____9_____

(8) _____6_____

Exercise 16B: Solve by Addition and Subtraction

Name _____

Course/Sect. No. _____   Score _____

Instructions: Solve for the unknown.

(1) _____	**(1)** $A + 4 = 24$ **(2)** $C - 2 = 9$
(2) _____	
(3) _____	**(3)** $20 = E - 4$ **(4)** $G = 20 + 2$
(4) _____	
(5) _____	**(5)** $I - 12 = 0$ **(6)** $4 + K = 10$
(6) _____	
(7) _____	**(7)** $5 = 5 + M$ **(8)** $P + 2 = 3$
(8) _____	

Exercise 17A: Multiplication and Division

Name_____

Course/Sect. No._____   Score_____

Instructions: Solve for the unknown.

(1) _____	
(2) _____	
(3) _____	
(4) _____	
(5) _____	
(6) _____	
(7) _____	
(8) _____	

Self-Check Exercises (Answers on back)

(1) $6X = 36$ **(2)** $\dfrac{X}{2} = 1$

(3) $7X = 42$ **(4)** $\dfrac{X}{10} = 4$

(5) $\dfrac{X}{5} = 5$ **(6)** $9.5X = 209$

(7) $17X = 59.5$ **(8)** $\dfrac{X}{15} = 2.2$

(1) _____6_____

(2) _____2_____

(3) _____6_____

(4) _____40_____

(5) _____25_____

(6) _____22_____

(7) _____3.5_____

(8) _____33_____

Exercise 17B: Multiplication and Division

Name_____

Course/Sect. No._____   Score_____

Instructions: Solve for the unknown.

(1) _____   **(1)** $2X = 4$ **(2)** $\dfrac{X}{4} = 2$

(2) _____

(3) _____   **(3)** $3X = 18$ **(4)** $\dfrac{X}{3} = 5$

(4) _____

(5) _____   **(5)** $6X = 6$ **(6)** $12X = 216$

(6) _____

(7) _____   **(7)** $\dfrac{X}{1} = 42$ **(8)** $7.5X = 56.25$

(8) _____

Exercise 18A: Combining Unknowns

Name _____

Course/Sect. No. _____   Score _____

Instructions: Solve for the unknown.

---

(1) _____

(2) _____

(3) _____

(4) _____

(5) _____

(6) _____

(7) _____

(8) _____

Self-Check Exercise (Answers on back)

(1) $6M - 2M = 28$

(2) $10C - 9C = 15$

(3) $10B = 20 + 5B$

(4) $5E - 3 = 7 + 4E$

(5) $60 = 7N - 2N$

(6) $10C = 32 + 2C + 6C$

(7) $A = \$29.70 + .7A$

(8) $24 + 3V = 19V$

(1) _____7_____

(2) _____15_____

(3) _____4_____

(4) _____10_____

(5) _____12_____

(6) _____16_____

(7) _____$99_____

(8) _____$1\frac{1}{2}$_____

Exercise 18B: Combining Unknowns

Name _____

Course/Sect. No. _____   Score _____

Instructions: Solve for the unknown.

(1) _____	**(1)** $2Y + 3Y = 40$ **(2)** $8A + A = 108$
(2) _____	
(3) _____	**(3)** $3X = 7 + X$ **(4)** $6T + 2 = 26 - 2T$
(4) _____	
(5) _____	**(5)** $12 = 3A + 3A$ **(6)** $26 = 3X + 4X - 5X$
(6) _____	
(7) _____	**(7)** $X = 2 + .8X$ **(8)** $5N + N = 48 - 2N$
(8) _____	

Exercise 19A: Removing Parentheses and Cross-Multiplication

Name _____

Course/Sect. No. _____   Score _____

Instructions: Solve for the unknown.

---

(1) _____

(2) _____

(3) _____

(4) _____

(5) _____

(6) _____

(7) _____

(8) _____

Self-Check Exercises (Answers on back)

(1) $3(X - 3) = 18$

(2) $7(X - 2) = 42$

(3) $\dfrac{2}{X} = 16$

(4) $\dfrac{5}{X} = 40$

(5) $\dfrac{X}{13} = 1$

(6) $\dfrac{39}{X} = 6.5$

(7) $\dfrac{2X}{8} = \dfrac{1}{2}$

(8) $\dfrac{3X}{20} = \dfrac{6}{5}$

(1) _____9_____

(2) _____8_____

(3) _____$\frac{1}{8}$_____

(4) _____15_____

(5) _____13_____

(6) _____6_____

(7) _____2_____

(8) _____8_____

Exercise 19B: Removing Parentheses and Cross-Multiplication

Name _____

Course/Sect. No. _____   Score_____

Instructions: Solve for the unknown.

(1) _____	**(1)** $2(X + 2) = 34$ **(2)** $2(X) + 2 = 34$
(2) _____	
(3) _____	**(3)** $\dfrac{X}{2} = 16$ **(4)** $\dfrac{X}{5} = 8.2$
(4) _____	
(5) _____	**(5)** $4(3 + X) = 24$ **(6)** $(X - 9) = 13$
(6) _____	
(7) _____	**(7)** $10(2X + 4) = 120$ **(8)** $6(3X - 5) = 150$
(8) _____	

Exercise 20A: Miscellaneous Problems

Name _____

Course/Sect. No. _____   Score _____

Instructions: Solve for the unknown.

Self-Check Exercises (Answers on back)

(1) _____   **(1)** $C + C = 16$ **(2)** $5N - 3N = 16$

(2) _____

(3) _____   **(3)** $5ABC = 20$ (Solve for C)   **(4)** $C = \$100 + .2C$

(4) _____

(5) _____   **(5)** $3(G + 1) = 18$ **(6)** $\dfrac{D}{4} = 52$

(6) _____

(7) _____   **(7)** $5(2X + 6) = 60$ **(8)** $5X - 6 = +54$

(8) _____

(9) _____   **(9)** Does $2B - 10 = 2 - 2B$ when B is equal to 3?

(10) _____   **(10)** Does $4 - \frac{1}{2}X - 7 = 0$ when X is equal to 2?

(1) _____8_____

(2) _____8_____

(3) ___$\frac{4}{AB}$___

(4) _____$125_____

(5) _____5_____

(6) _____208_____

(7) _____3_____

(8) _____12_____

(9) _____yes_____

(10) _____yes_____

Exercise 20B: Miscellaneous Problems

Name _____

Course/Sect. No. _____   Score _____

Instructions: Solve for the unknown.

(1) _____	

Self-Check Exercise (Answers on back)

(1) $\dfrac{X}{5} = \dfrac{4}{5}$ **(2)** $-5P = 9 + 2P$

(2) _____

(3) _____   **(3)** $\dfrac{X}{4} - 2 = 10$ **(4)** $XY = \dfrac{M - N}{P}$ (Solve for M)

(4) _____

(5) _____   **(5)** $\dfrac{2N}{9} = 10$ **(6)** $\dfrac{ABC}{2} = 20$ (Solve for B)

(6) _____

(7) _____   **(7)** $\dfrac{M + N}{ABC} = 2X$ (Solve for B)   **(8)** $5A + A = 3 + 2A + 3A$

(8) _____

(9) _____   **(9)** Does $6X - 8 = 2$ when X is equal to 4?

(10) _____   **(10)** Does $6A + 3 = 3A + 2A + 6$ when A is equal to 3?

Exercise 21A: Setting up and Solving Equations

Name _____

Course/Sect. No. _____   Score_____

Instructions: Set up an equation for each statement and solve it.

Self-Check Exercises (Answers on back)

(1) _____

(1) The sum of two numbers is 20. One of the numbers is 4 more than the other number. What are the two numbers?

(2) _____

(2) Carol has 19 dresses, which is 1 fewer than twice as many as her neighbor, May. How many dresses does May have?

(3) _____

(3) Bill is 3 times as old as Davis. If the sum of their ages is 52, find their ages.

(4) _____

(4) Barbara spent $\frac{2}{3}$ as much money as her sister Sue. If together they spent $900, how much did each spend?

(1) _8, 12_

(2) _10_

(3) _13, 39_

(4) _$360, $540_

Exercise 21B: Setting up and Solving Equations

Name _____

Course/Sect. No. _____   Score _____

Instructions: Set up an equation for each statement and solve it.

(1) _____   **(1)** Sharon made 84 house calls distributing pamphlets, which was 4 more than 2 times as many as Ted. How many house calls did Ted make?

(2) _____   **(2)** Danny worked 108 hours, which was 3 times as long as Leonard. How many hours did Leonard work?

(3) _____   **(3)** One number is 4 times another number. If their sum is 60, what are the numbers?

(4) _____   **(4)** A number is multiplied by 5. When the product is added to 13, the sum is 123. What is the number?

Test Yourself—Unit 1

(1) In the number 972,183, what digit is in the ten-thousand position?

(2) Express 40.123 in written form.

(3) Find the sum of 18.5, 304.46, .031, 91.45, and 3.607.

(4) Find the difference between 830.004 and 564.495.

(5) Find the product of 806.05 and 40.06 rounded to hundredths.

(6) Find the quotient of 1,521.3 divided by 23.05.

(7) Express 1.8 as a decimal fraction. (Do not reduce.)

(8) Express 5/100 as an equivalent decimal number.

(9) Reduce 756/1071 by factoring.

(10) Convert 18/5 to a mixed number.

(11) Convert 17 3/5 to an improper fraction.

(12) Multiply 3 1/2 × 14/15 × 5 1/4 and reduce.

(13) Divide 6 1/3 by 7/18 and reduce.

(14) Add, subtract, and reduce: 1 2/3 − 4/9 + 5 2/15.

(15) Solve for the unknowns:

   (A) $2X + 3 = 29$

   (B) $5A - 2 = 2A + 19$

   (C) $3(N + 4) = 24$

   (D) $3/X = 4/9$

   (E) $I = PRT$ (solve for R)

   (F) $X = \$500 + .8X$

(16) Two less than 3 times a certain number is 22. Set up an equation and solve for the unknown.

Note: *The answers to the "test yourself" problems can be found following each test.*

TEST YOURSELF ANSWERS

(1) 7   **(2)** Forty and one hundred twenty-three thousandths   **(3)** 418.048
(4) 265.509   **(5)** 32,290.36   **(6)** 66   **(7)** 1 8/10   **(8)** .05   **(9)** 12/17
(10) 3 3/5   **(11)** 88/5   **(12)** 17 3/20   **(13)** 16 2/7   **(14)** 6 16/45
(15) (A) $X = 13$ **(16)** $3N - 2 = 22$
      (B) $A = 7$ $N = 8$
      (C) $N = 4$
      (D) $X = 6\ 3/4$
      (E) $R = I/PT$
      (F) $X = \$2,500$

Unit 2

Percentage and Payroll

A Preview of What's Next

You were promised a 15% raise this year. Last year, you made $13,570, and your contract this year is for $15,200. Did you get what you were promised?

The department manager says the company had allocated $282 to her department. She wants you to check to see if the overhead of $846 was allocated correctly if the ratio is 5:1:3, and your portion is the last one.

The new payroll clerk is constantly off a little on the total payroll and cannot figure out why. He thinks it has something to do with Charles Jones's gross pay this week of $251.12. Charles's pay was based on 43.5 hours at $5.55 an hour. Why is the payroll off?

The State Unemployment Tax and Federal Unemployment Tax are due at the end of this month. Can you determine the proper tax due from the Employee's Earnings Records?

Objectives

A thorough understanding of the use of percentages is necessary in many types of business problems, including the preparation and maintenance of payroll records. After successfully completing this unit, you will be able to:

1. Change any whole number, mixed number, decimal, fraction, or percentage to any of the other forms.
2. Use the basic equation $B \times R = P$ to find the base, rate, or percentage when two of the three values are known.
3. Compute the rate of change from two different time periods.
4. Distribute overhead using a predetermined ratio.
5. Compute the regular and overtime earnings for employees.
6. Calculate earnings based on a piece-rate payroll system.
7. Compute social security tax deductions.
8. Calculate earnings based on a commission payroll system.
9. Determine the proper income tax using a tax table.
10. Calculate the employer's payroll taxes.
11. Complete an Employee's Earnings Record.
12. Complete a Payroll Register.

Chapter

Percentages, Proportions, and Ratios

Preview of Terms

Allocate
To divide a single amount into several amounts (usually in some predetermined ratio) that are distributed in several places.

Base
The number used as a basis for comparison in a percentage problem. **Example:** In $B \times R = P$, the B is the base; in $\$500 \times 20\% = \$100$, the $\$500$ is the base.

Decimal Equivalent
The decimal value of a fractional amount. **Example:** $4/5 = .8$ The .8 is the decimal equivalent of 4/5.

Percent
A fraction with a denominator of 100 or a decimal number expressed in hundredths. A percent sign (%) means division by 100. **Example:** $38\% = \frac{38}{100} = .38$.

Percentage
The generally used noun form of *percent,* used without numbers. Sometimes *percent* and *percentage* are used interchangeably.

Proportion
The statement of the equality of two ratios. **Example:** $3 \div 6 = \frac{3}{6}$, which, when reduced, is $\frac{1}{2}$. Thus $\frac{3}{6} = \frac{1}{2}$ means the fraction $\frac{3}{6}$ is in the same proportion as the fraction $\frac{1}{2}$.

Rate
A ratio expressed as a percentage. **Example:** 3/5 or 3:5 $= .6 = 60\%$.

Rate of Change
The percentage of change from one quantity to another.

Ratio
A comparison of one number to another by division; a fraction.

Overhead
Those operating costs of a business that are not specifically associated with the production process. Examples include rent and taxes.

Percentages are used in such business applications as payroll, discounts, markup, and statement analysis. A percentage is simply an extension of a fraction expressed in hundredths. Conversion of numbers to percentages is particularly useful because it allows comparisons to be made since all percentages have a common base — 100.

Percentage involves a comparison of one number to another by division. This division is nothing more than a fraction with the denominator equal to 100.

CONVERTING FRACTIONS TO DECIMALS

All fractions can be converted into percentages by dividing the denominator into the numerator. This division will yield the *decimal equivalent* of that fraction. This decimal equivalent then can be converted into a percentage.

Example 1: $\quad \dfrac{13}{25} = .52 \quad$ The quotient, .52, is the decimal equivalent of the fraction.

Practice Problems

Convert the following fractions into their decimal equivalents.

1. $\dfrac{3}{4}$

2. $\dfrac{15}{60}$

CONVERTING DECIMALS TO PERCENTAGES

To convert a decimal number to a percentage, move the decimal two places to the *right* and add a percent sign.

Example 2: $\quad .52 = .52 = 52\%.$

Practice Problems

Convert the following decimals to percentages.

3. .75

4. .25

5. 4.016

6. .004

CONVERTING PERCENTAGES TO DECIMALS

To convert a percentage to a decimal, move the decimal two places to the *left* and remove the percent sign (the opposite of converting from a decimal to a percentage).

Example 3: $\quad 15\% = 15.\% = .15$

Note: *In the special case of a fractional percentage, such as 3/5%, it is better to convert the fraction to its decimal equivalent and then convert the percentage to a pure decimal number. Thus, 3/5% would first become .6%, and then .006 as you move the decimal two places to the left and remove the percent sign.*

Practice Problems

Convert the following percentages to decimals.

7. 85% **8.** 1.72% **9.** $\frac{3}{10}$%

CONVERTING PERCENTAGES TO FRACTIONS

To convert a percentage to a fraction, remove the percent sign, divide the number by 100, and reduce to lowest terms.

Example 4: $125\% = \frac{125}{100} = 1\frac{25}{100} = 1\frac{1}{4}$

Example 5: $5\frac{1}{4}\% = \frac{5\frac{1}{4}}{100} = \frac{5\frac{1}{4}}{\frac{100}{1}} = \frac{\frac{21}{4}}{\frac{100}{1}} = \frac{21}{4} \times \frac{1}{100} = \frac{21}{400}$

Example 6: $.5\% = \frac{1}{2}\% = \frac{\frac{1}{2}}{100} = \frac{\frac{1}{2}}{\frac{100}{1}} = \frac{1}{2} \times \frac{1}{100} = \frac{1}{200}$

Practice Problems

Convert the following percentages to fractions.

10. 72% **11.** $6\frac{1}{2}$% **12.** 14%

TABLE OF REFERENCE FOR PERCENTAGES

Fractional Equivalent	Percent	Decimal Equivalent
$\frac{1}{1,000}$	$\frac{1}{10}$%	.001
$\frac{1}{100}$	1%	.01
$\frac{1}{10}$	10%	.1
$\frac{100}{100}$	100%	1.0
$\frac{1,000}{100}$	1,000%	10.0

THE PERCENTAGE EQUATION

In a percentage problem, there are three basic components: the *base,* the *rate,* and the *product.* The formula for these elements is given as:

$$B \times R = P$$

where B = base
R = rate
P = product

As long as any two of the three components are known, you can find the third by substituting into the equation what is known and solving for the unknown. The rate is always the percent or the number with the % sign. The only difficulty is determining the base from the product. The base is always the number with which the comparison is being made. It may be larger, smaller, or the same as the product. The size of the number does *not* determine the base. Almost without exception, the base is identified because it is preceded by the words *of* or *than.*

Example 7: What percentage of 250 is 75? In this example, the word *percentage* is the rate (what you are trying to find in this example), the word *of* precedes the number 250, identifying it as the base, and by the process of elimination, the number 75 is the product.

   R   B   P
What percentage of 250 is 75?

$$B \times R = P$$
$$250 \times R = 75$$
$$\frac{250 \times R}{250} = \frac{75}{250}$$
$$R = .3, \quad \text{or} \quad 30\%$$

Note: *When solving for R, the last step should always be to convert the decimal equivalent to its percentage by moving the decimal two places to the right and adding a percent sign.*

Example 8: What number is 4% of 80? In this example, the rate is identified by the % sign, the base is 80 (because it is preceded by the word *of*), and the product is what you are trying to find.

   P   R   B
What number is 4% of 80?

$$B \times R = P$$
$$80 \times .04 = P$$
$$3.2 = P$$

Note: *Before substituting the rate into the equation, convert the percent to its decimal equivalent.*

Example 9: 42 is 6% of what number? Here, the rate is 6% (.06), the unknown is the base (it is preceded by the word *of*), and the product is 42.

$$P \quad R \qquad B$$

42 is 6% of what number?

$$B \times R = P$$
$$B \times .06 = 42$$
$$\frac{B \times .06}{.06} = \frac{42}{.06}$$
$$B = 700$$

Practice Problems

Substitute the known values in the basic equation ($B \times R = P$) and solve for the unknown.

13. What percentage of $875 is $25? (Round to a tenth of 1%.)

14. What number is 23% of 1.700?

15. 19.25 is 3.5% of what number?

RATE OF CHANGE

Many times, it is necessary to find the rate of change, that is, the percentage of increase or decrease. To do so, the following equation should be used:

$$\frac{\text{Amount of change (increase or decrease)}}{\text{Earliest quantity (base) given}} = \frac{\text{Rate of change (percent of increase or decrease)}}{}$$

Example 10: This year, we had 140 students taking Business Math. Last year, we had 100. What was the percentage of increase?

$$\frac{\text{Amount of change}}{\text{Earliest quantity}} = \text{Rate of change}$$

$$\frac{40}{100} = .4, \text{ or } 40\%$$

Example 11: In 1975, the sales for the company were $125,000; in 1976, they fell to $105,000. What was the percentage of decrease?

$$\frac{-\$20,000}{\$125,000} = -.16 = -16\%$$

Note: *As long as you indicate its meaning, an asterisk (*) or parentheses ( ) can be used to represent a decrease instead of a minus sign.*

Practice Problems

Find the rate of change (percentage of increase or decrease).

16. You got a 86 on test one and a 96 on test two. What was your percentage of increase? (Round to a whole percent.)

17. Last week, Velma received $182.50. This week, she received $166. What percentage of decrease does this represent? (Round to a tenth of 1%.)

SPECIAL CASES

Sometimes a special case requires that the information given be rearranged so that the previous methods of solution can be used.

MULTIPLE ITEMS

When two items are being compared, determine the unknown and always set it equal to 100%. The other item then can be given a percentage based on its relationship in the problem to the unknown. Reword the information, and solve by using the basic equation.

Example 12: A coat sells for $29.95, which is 40% *more* than the cost of a matching purse. What is the cost of the purse?

$$100\% = \text{Purse} = \$ \underline{\hspace{2cm}}$$
$$140\% = 100\% + 40\% = \text{Coat} = \$29.95$$

In this example, you are looking for the price of the purse, so it is the unknown. Set it equal to 100%. If it equals 100% and the coat sells for 40% *more*, the coat is equal to 140%. Now, the problem can be reworded: 140% of what is $29.95?

$$B \times R = P$$
$$B \times 1.4 = \$29.95$$
$$\frac{B \times 1.4}{1.4} = \frac{\$29.95}{1.4}$$
$$B = \$21.39$$

Example 13: A used VW sells for $2,125, which is 15% *less* than the cost of a used Volvo. For how much does the Volvo sell?

$$100\% = \text{Volvo} = \$ \underline{\hspace{2cm}}$$
$$85\% = 100\% - 15\% = \text{VW} = \$2,125$$

The information can be reworded: 85% of what is $2,125?

$$B \times R = P$$
$$B \times .85 = \$2,125$$
$$\frac{B \times .85}{.85} = \frac{\$2,125}{.85}$$
$$B = \$2,500$$

Note: *This time, the key word in the problem is* less. *Therefore, you had to subtract from 100%.*

Practice Problems

Find the unknown.

18. Washing machine *X* sells for $420, which is 30% more than washing machine *Y* costs. For how much does washing machine *Y* sell?

19. A three-bedroom home sells for $43,200, which is 10% less than the cost of a four-bedroom home. What is the cost of the four-bedroom home?

MULTIPLE PERCENTAGES

When two percentages are given in a problem, you should multiply them together to find the single equivalent percentage. Then proceed with the solution as before.

Example 14:   15% of 80% of what amount is $720?

$$15\% = .15 \quad \text{and} \quad 80\% = .8$$
$$.15 \times .8 = .12$$
$$B \times R = P$$
$$B \times .12 = \$720$$
$$\frac{B \times \cancel{.12}}{\cancel{.12}} = \frac{\$720}{.12}$$
$$B = \$6,000$$

Note: *Remember, the percent sign labels the rate (R), the word* of *precedes the base (B), and therefore the remaining number is your product (P).*

Practice Problems

Given multiple percentages, find the unknown.
20. 30% of 50% of what amount is 600?
21. What is 3.5% of 17% of $200?

RATIOS AND PROPORTIONS

A ratio is the comparison of two or more numbers. When two ratios are equal, they are proportional (in the same proportion to each other). A large company could have $200,000 in profits compared to $800,000 in sales (a ratio of $200,000 to $800,000). A smaller company could have $10,000 in profits compared to $40,000 in sales (a ratio of $10,000 to $40,000). Since 200,000/800,000 = .25 and 10,000/40,000 = .25, these ratios are in the same proportion to each other.

Often it is desirable to express a ratio as some comparison "to 1." You can accomplish this by dividing the numerator by the denominator (as shown above) and expressing the quotient "to 1." In the illustration above, the ratio expressed "to 1" would be .25 to 1.

Example 15:   2:5 = 2/5 = .4 to 1. This means that 2 is in the same ratio to 5 as .4 is to 1. This type of conversion is more meaningful to the casual reader or the busy executive because larger numerical relationships are reduced to smaller relationships that can be comprehended at a glance. (For instance, the relationship of $27,381 to $433,899 is much easier to grasp if expressed as $.06 to $1.00.) For simplification of illustration, it is usually desirable to round all of these ratios to tenths.

Example 16:   The company's profit was $77,500, and its sales were $338,420. What is the ratio of profit to sales?

$$\frac{\text{Profit}}{\text{Sales}} = \frac{\$77,500}{\$338,420} = .2290053 = .2 \text{ to } 1$$

Note: *If you wanted to state it in terms of dollars and cents, you could say that for every sales dollar, approximately $.23 is returned as profit.*

In a ratio statement, the *first* item mentioned will always be the numerator in the comparison, and the *second* item the denominator.

Example 17:
(a) What is the ratio of 8 to 15? The *first* item mentioned is 8, the *second* is 15. Therefore, the ratio is 8/15, which is .5 to 1.

(b) What is the ratio of 4,800 boys to 2,000 girls? The *first* item is boys (4,800), the *second* is girls (2,000). Therefore, the ratio is 4,800/2,000, which is 2.4 to 1.

Practice Problems

Find the ratio "to 1" in each of the following (round to tenths).

22. Find the ratio of 18,200 to 25,700.

23. Find the ratio of current assets of $125,400 to current liabilities of $58,950.

DISTRIBUTION OF OVERHEAD

If a business wants to divide certain general expenses, called *overhead,* between departments, the most convenient way is to use ratios. If the overhead is distributed in a ratio of three departments, it could be stated like this:

$$2:3:5$$

The numbers would be added (2 + 3 + 5 = 10). This means that the overhead would be divided into 10 parts; and 2 of the total 10 parts would be distributed, or *allocated,* to one department, 3 of the total 10 parts to the second department, and 5 of the total 10 parts to the last department.

Example 18:   The store had overhead costs of $3,000 to distribute (allocate) to departments A, B, and C. If the overhead costs are distributed in the ratio of 2:3:5, respectively, how much should each department receive?

$$\text{Department A:} \quad 2 = \frac{2}{10} \times \$3,000 = \quad \$ \ 600$$

$$\text{Department B:} \quad 3 = \frac{3}{10} \times \$3,000 = \quad 900$$

$$\text{Department C:} \quad \frac{+ \ 5}{10} = \frac{5}{10} \times \$3,000 = \frac{+ \ 1,500}{\$3,000}$$

Practice Problems

Distribute the overhead according to the ratios given.

24. The branch had overhead costs of $1,350 last month. What portion of the overhead should departments M, N, and O receive if they share in the ratio of 3:4:8, respectively?

25. The ratio for sharing overhead costs was based on floor space occupied. Department 1 had 2,000 ft^2 of floor space, department 2 had 3,400 ft^2, and department 3 had 5,600 ft^2. With overhead costs of $3,300 and the ratio 2,000:3,400:5,600, allocate the overhead costs to the departments.

Practice Problem Solutions

1. $\dfrac{3}{4} = .75$

2. $\dfrac{15}{60} = .25$

3. $.75 = .75 = 75\%$

4. $.25 = .25 = 25\%$

5. $4.016 = 4.016 = 401.6\%$

6. $.004 = .004 = .4\%$

7. $85\% = 85\% = .85$

8. $1.72\% = 1.72\% = .0172$

9. $\dfrac{3}{10}\% = .3\% = .003$

10. $72\% = \dfrac{72}{100} = \dfrac{18}{25}$

11. $6\dfrac{1}{2}\% = \dfrac{6\dfrac{1}{2}}{100} = 6\dfrac{1}{2} \div \dfrac{100}{1} = \dfrac{13}{2} \div \dfrac{100}{1} = \dfrac{13}{2} \times \dfrac{1}{100} = \dfrac{13}{200}$

12. $14\% = \dfrac{14}{100} = \dfrac{7}{50}$

13. $B \times R = P$ $B = \$875$
    $R = $ (unknown)
    $P = \$25$

    $\$875 \times R = \$25$
    $R = \dfrac{\$25}{\$875}$
    $R = .0285 = 2.9\%$

14. $B \times R = P$ $B = 1{,}700$
    $R = 23\%$
    $P = $ (unknown)
    $1{,}700 \times .23 = P$
    $391 = P$

15. $B \times R = P$ $B = $ (unknown)
    $R = 3.5\%$
    $P = 19.25$
    $B \times .035 = 19.25$
    $B = \dfrac{19.25}{.035}$
    $B = 550$

16. $\dfrac{\text{Amount of change}}{\text{Earliest quantity}} = \dfrac{10}{86}$
    $= .116$
    $= 12\%$

17. $\dfrac{\text{Amount of change}}{\text{Earliest quantity}} = \dfrac{-\$16.50}{\$182.50}$
    $= -.0904$
    $= 9.0\%$
    decrease

18. $100\% = $ Machine $Y = \$\_\_\_\_$
    $130\% = $ Machine $X = \$420$
    $B \times R = P$
    $B \times 1.3 = \$420$
    $B = \dfrac{\$420}{1.3} = \$323.08$

19. $100\% = $ 4-bedroom $= \$\_\_\_\_$
    $(100\% - 10\%) = 90\% = $ 3-bedroom $= \$43{,}200$

"90% of what is \$43,200?"

$B \times R = P$

$B \times .9 = \$43,200$

$$B = \frac{\$43,200}{.9} = \$48,000$$

20. 30% = .3   and   50% = .5

.3 × .5 = .15

$B \times R = P$

$B \times .15 = 600$

$$B = \frac{600}{.15} = 4,000$$

21. 3.5% = .035   and   17% = .17

.035 × .17 = .00595

$B \times R = P$

\$200 × .00595 = P

\$1.19 = P

22. $\frac{18,200}{25,700} = .7081712 = .7$ to 1

23. $\frac{\text{Current Assets}}{\text{Current Liabilities}} = \frac{\$125,400}{\$58,950}$

= 2.1272264

= 2.1 to 1

24. Department M: $3 = \frac{3}{15} \times \$1,350 = \$\ 270$

Department N: $4 = \frac{4}{15} \times \$1,350 = 360$

Department O: $\underline{+\ 8} = \frac{8}{15} \times \$1,350 = \underline{+\quad 720}$

15   \$1,350

25. Department 1: $2,000 = \frac{2,000}{11,000} \times \$3,300 = \$\ 600$

Department 2: $3,400 = \frac{3,400}{11,000} \times \$3,300 = 1,020$

Department 3: $\underline{+\ 5,600} = \frac{5,600}{11,000} \times \$3,300 = \underline{+\ 1,680}$

11,000   \$3,300

Exercise 22A: Converting Fractions to Decimals

Name _____

Course/Sect. No. _____   Score _____

Instructions: Convert the following fractions to their decimal equivalents.

Self-Check Exercise (Answers on back)

(1) _____   (1) $\dfrac{3}{5}$ (2) $\dfrac{24}{80}$

(2) _____

(3) _____   (3) $\dfrac{1}{5}$ (4) $\dfrac{7}{10}$

(4) _____

(5) _____   (5) $\dfrac{9}{20}$ (6) $\dfrac{15}{40}$

(6) _____

(7) _____   (7) $\dfrac{39}{50}$ (8) $\dfrac{105}{420}$

(8) _____

(9) _____   **(9)** 5 of every 8 customers bought one of Joe's Super Gadgets. Expressed as a decimal, what part of these customers bought a Super Gadget?

(10) _____   **(10)** Tanya bought a Dynamic Digital watch on sale for $11.99, marked down from the regular $15.00. Expressed as a decimal, rounded off to a hundredth, what part of the original price did Tanya pay?

(1) _____.6_____

(2) _____.3_____

(3) _____.2_____

(4) _____.7_____

(5) _____.45_____

(6) _____.375_____

(7) _____.78_____

(8) _____.25_____

(9) _____.625_____

(10) _____.80_____

Exercise 22B: Converting Fractions to Decimals

Name _____

Course/Sect. No. _____   Score _____

Instructions: Convert the following fractions to their decimal equivalents.

(1) _____	**(1)** $\dfrac{1}{25}$ **(2)** $\dfrac{3}{8}$
(2) _____	
(3) _____	**(3)** $\dfrac{17}{25}$ **(4)** $\dfrac{18}{90}$
(4) _____	
(5) _____	**(5)** $\dfrac{1}{4}$ **(6)** $\dfrac{11}{25}$
(6) _____	
(7) _____	**(7)** $\dfrac{4}{5}$ **(8)** $\dfrac{9}{30}$
(8) _____	
(9) _____	**(9)** Sam got 6 hits in 17 times at bat. What was his hitting percentage? Round off to thousandths.
(10) _____	**(10)** Donna bought an Acme battery that was guaranteed for 48 months. It lasted 19 months. Expressed as a decimal rounded off to hundredths, what part of the guarantee life did Donna get from the battery?

Exercise 23A: Converting Decimals to Percentages

Name_____

Course/Sect. No._____   Score_____

Instructions: Convert the following decimals to percentages.

	Self-Check Exercise (Answers on back)

(1) _____

(2) _____

(3) _____

(4) _____

(5) _____

(6) _____

(7) _____

(8) _____

(9) _____

(10) _____

(1) .42   **(2)** .031

(3) .006   **(4)** .68

(5) .084   **(6)** .049

(7) .475   **(8)** .10

(9) Sammy sold 2.5 times as many Zippies as his cosalesperson, June. What percentage of June's total did Sammy sell?

(10) The test average turned out to be .87643127. Rounded to a tenth of 1 percent, what would the percent be?

(1)   42%

(2)   3.1%

(3)   .6%

(4)   68%

(5)   8.4%

(6)   4.9%

(7)   47.5%

(8)   10%

(9)   250%

(10)   87.6%

Exercise 23B: Converting Decimals to Percentages

Name_____

Course/Sect. No._____   Score_____

Instructions: Convert the following decimals to percentages.

---

(1) _____   **(1)** 2.17   **(2)** 1.874

(2) _____

(3) _____   **(3)** .002   **(4)** 3.05

(4) _____

(5) _____   **(5)** 5.50   **(6)** .1005

(6) _____

(7) _____   **(7)** .880   **(8)** .1652

(8) _____

(9) _____   **(9)** Nancy takes the temperature of .875 of her patients. What percentage is this?

(10) _____   **(10)** The volume of shares traded on Wall Street today was .40 of yesterday's volume. What percentage of yesterday's volume is that?

Exercise 24A: Converting Percentages to Decimals

Name_____

Course/Sect. No._____   Score_____

Instructions: Convert the following percentages to decimals.

Self-Check Exercise (Answers on back)

(1) _____   **(1)** 9%   **(2)** 17%

(2) _____

(3) _____   **(3)** $\frac{1}{5}$%   **(4)** .4%

(4) _____

(5) _____   **(5)** 20%   **(6)** .008%

(6) _____

(7) _____   **(7)** .06%   **(8)** 40.02%

(8) _____

(9) _____   **(9)** Kelly won 65% of his bets. Expressed as a decimal, how many was that?

(10) _____   **(10)** Scott scored in one-fourth of his soccer games. This is 25% of them. Express it as a decimal.

(1) _____.09_____

(2) _____.17_____

(3) _____.002_____

(4) _____.004_____

(5) _____.2_____

(6) _____.00008_____

(7) _____.0006_____

(8) _____.4002_____

(9) _____.65_____

(10) _____.25_____

Exercise 24B: Converting Percentages to Decimals

Name _____

Course/Sect. No. _____   Score _____

Instructions: Convert the following percentages to decimals.

(1) _____	**(1)** 5%
(2) _____	
(3) _____	**(3)** $30\frac{1}{2}$%
(4) _____	
(5) _____	**(5)** 15.5%
(6) _____	
(7) _____	**(7)** 27.2%
(8) _____	
(9) _____	**(9)** Ted made 70.5% of his first tennis serves good. Expressed as a decimal, what part of them were good?
(10) _____	**(10)** There were 900 football games each Friday night. The underdog won 35% of them. Express the percentage as a decimal.

(2) $2\frac{1}{2}$%

(4) 110%

(6) .8%

(8) 1.05%

Exercise 25A: Converting Percentages to Fractions

Name_____

Course/Sect. No._____   Score_____

Instructions: Convert the following percentages to fractions and reduce.

Self-Check Exercise (Answers on back)

(1) _____

(2) _____

(3) _____

(4) _____

(5) _____

(6) _____

(7) _____

(8) _____

(9) _____

(10) _____

(1) 3%   **(2)** 4.5%

(3) 110%   **(4)** 2.05%

(5) 40%   **(6)** $16\frac{2}{5}$%

(7) .02%   **(8)** $12\frac{1}{4}$%

(9) If 93% of the U.S. population live in households with color televisions, what fraction of these people have a color television?

(10) Of all television programs, $7\frac{2}{5}$% last longer than five years. What fractional number of programs last longer than five years?

(1) $\dfrac{3}{100}$

(2) $\dfrac{9}{200}$

(3) $1\dfrac{1}{10}$

(4) $\dfrac{41}{2000}$

(5) $\dfrac{2}{5}$

(6) $\dfrac{41}{250}$

(7) $\dfrac{1}{5000}$

(8) $\dfrac{49}{400}$

(9) $\dfrac{93}{100}$

(10) $\dfrac{37}{500}$

Exercise 25B: Converting Percentages to Fractions

Name_____

Course/Sect. No._____   Score_____

Instructions: Convert the following percentages to fractions and reduce.

---

(1) _____

(1) 22%

(2) 71%

(2) _____

(3) _____

(3) $5\dfrac{1}{8}\%$

(4) $1\dfrac{1}{10}\%$

(4) _____

(5) _____

(5) $13\dfrac{1}{2}\%$

(6) 200%

(6) _____

(7) _____

(7) 8.4%

(8) 50.1%

(8) _____

(9) _____

(9) Of all television programs, 27.25% were judged to have unacceptable sex or violence and were canceled. What fraction of these programs were rejected?

(10) _____

(10) Of the cattle on Roy's ranch, 14.4% are rated "choice." Expressed as a fraction in lowest terms, what part of these cattle are "choice"?

Exercise 26A: The Percentage Equation

Name _____

Course/Sect. No. _____   Score _____

Instructions: Substitute the known values and solve for the unknown. Use $B \times R = P$.

---

(1) _____

(2) _____

(3) _____

(4) _____

(5) _____

(6) _____

(7) _____

(8) _____

(9) _____

(10) _____

Self-Check Exercise (Answers on back)

(1) What percent of 622 is 72.152?   **(2)** What number is .035% of $10,000?

(3) $73.50 is 110% of what number? **(4)** What number is .3% of $2500?

(5) What percent of 3,025 is 1,663.75?

(6) 216 is 24% of what number?   **(7)** .51 is what percent of 8.5?

(8) 225% of 7,700 is what?

(9) Mary sold $494 worth of bridal apparel, and Deron sold $556 worth of groom's apparel. What percent of Deron's sales did Mary have? (Round to a tenth of 1%.)

(10) A survey revealed that 32.3% of all Burp Burgers were rated "terrible." If 370 hamburgers were tried, how many of them were "terrible"? (Round to a whole number.)

(1) _____11.6%_____

(2) _____$3.50_____

(3) _____$66.82_____

(4) _____$7.50_____

(5) _____55%_____

(6) _____900_____

(7) _____6%_____

(8) _____17,325_____

(9) _____88.8%_____

(10) _____120_____

Exercise 26B: The Percentage Equation

Name _____

Course/Sect. No. _____   Score _____

Instructions: Substitute the known values and solve for the unknown. Use $B \times R = P$.

---

(1) _____

(2) _____

(3) _____

(4) _____

(5) _____

(6) _____

(7) _____

(8) _____

(9) _____

(10) _____

(1) What percent of 2,015 is 100.75?   **(2)** What number is $37\frac{1}{2}\%$ of 100?

(3) 46.2 is 42% of what number?   **(4)** What percent of 512 is 588.8?

(5) 288 is 16% of what number?   **(6)** What number is 35.2% of 6100?

(7) 1.3% of $84 is what?   **(8)** $8.70 is 29.5% of what?

(9) The number of new tires produced by the ABC Tire Co. was 32% of the old total of 12 million. How many tires did the ABC Co. make that were new?

(10) Of the students in West School, 76.3% went to the county fair. If 550 students went to the fair, how many students are there in West School? (Round to a whole number.)

Exercise 27A: Rate of Change

Name _____

Course/Sect. No. _____   Score _____

Instructions: Find the rate of change, accurate to one-tenth of 1%.

Self-Check Exercise (Answers on back)

(1) _____

(1) July sales = $650
August sales = $900

(2) April sales = $2000
May sales = $1500

(2) _____

(3) _____

(3) Nov. sales = $989.50
Dec. sales = $775.75

(4) June sales = $28,000
July sales = $42,000

(4) _____

(5) _____

(5) Aug. sales = $18,000
Sept. sales = $20,000

(6) Jan. sales = $315.82
Feb. sales = $350.11

(6) _____

(7) _____

(7) Last year = $16,151.89
This year = $16,700.33

(8) Last month = $26,452
This month = $28,377

(8) _____

(9) _____

(9) During a gas shortage, the price per gallon jumped from 95.9¢ to 109.9¢. Find the percentage of increase.

(10) _____

(10) Since the computer was installed, the number of employees in the Payroll Department has decreased from 28 to 21. Find the percentage of decrease.

(1) _____38.5%_____

(2) _____(25%)_____

(3) _____(21.6%)_____

(4) _____50.0%_____

(5) _____11.1%_____

(6) _____10.9%_____

(7) _____3.4%_____

(8) _____7.3%_____

(9) _____14.6%_____

(10) _____(25%)_____

Exercise 27B: Rate of Change

Name _____

Course/Sect. No. _____   Score_____

Instructions: Find the rate of change, accurate to one-tenth of 1%.

---

(1) _____

(2) _____

(3) _____

(4) _____

(5) _____

(6) _____

(7) _____

(8) _____

(9) _____

(10) _____

(1) Jan. sales = $500
Mar. sales = $700

(2) Oct. sales = $1200
Nov. sales = $1300

(3) June sales = $3,350
July sales = $5,000

(4) Feb. sales = $8,000
Mar. sales = $17,000

(5) Aug. sales = $150,000
Sept. sales = $120,000

(6) April sales = $95,000
May sales = $100,000

(7) Last month = $240,000
This month = $230,000

(8) Oct. sales = $51,000
Nov. sales = $71,000

(9) The office supplies cost $11.50 this month and only $7.50 last month. What is the percentage of increase?

(10) The XYZ Co. sold their major product for $200 last year and reduced it to $189.95 this year. Find the percentage of decrease.

Exercise 28A: Miscellaneous Percentage Problems

Name_____

Course/Sect. No._____   Score_____

Instructions: Find the new amount that includes the change needed.

Self-Check Exercises (Answers on back)

		Present Amount Is	Change Needed Is
(1) _____		**(1)** $151.30	25% more
(2) _____		**(2)** 48 gallons	10% more
(3) _____		**(3)** 30 liters	2% less
(4) _____		**(4)** $1,200	30% more
(5) _____		**(5)** $3.50	75% more
(6) _____		**(6)** 15 gallons	20% less
(7) _____		**(7)** 300 meters	125% more
(8) _____		**(8)** $.82	17.5% less

(9) _____   **(9)** A set of steel belted tires cost $500, which is 25% more than a set of radial tires. How much does the set of radial tires cost?

(10) _____   **(10)** Your English average is 95, which is 4.5% less than your history average. What is your history average? (Round to a whole number.)

(1) _____$189.13_____

(2) _____52.8 gal._____

(3) _____29.4 liters_____

(4) _____$1,560_____

(5) _____$6.13_____

(6) _____12 gal._____

(7) _____675 meters_____

(8) _____$.68_____

(9) _____$400_____

(10) _____99_____

Exercise 28B: Miscellaneous Percentage Problems

Name_____

Course/Sect. No._____   Score_____

Instructions: Find the new amount that includes the change needed.

	Present Amount Is	Change Needed Is
(1) _____	**(1)** $18.50	50% more
(2) _____	**(2)** 522 pounds	90% less
(3) _____	**(3)** 8.4 meters	85% more
(4) _____	**(4)** 75.5 yards	200% more
(5) _____	**(5)** 51.5 feet	6% less
(6) _____	**(6)** $35.50	75% more
(7) _____	**(7)** $250,125.30	5% more
(8) _____	**(8)** 3.5 feet	40% more

(9) _____   **(9)** The refrigerator sold for $780, which was 45% more than the dryer. What was the price of the dryer?

(10) _____   **(10)** The new microwave oven cost $385, which is 12% less than the old microwave oven. What did the old microwave cost?

Exercise 29A: Multiple Percentages

Name_____

Course/Sect. No._____   Score_____

Instructions: Find the unknown.

Self-Check Exercises (Answers on back)

(1) _____   **(1)** 120% of 70% of what is 16? (Round to a whole number.)

(2) _____   **(2)** 22% of 22% of $500 is what?

(3) _____   **(3)** 13.5% of 100% of 450 is what?

(4) _____   **(4)** 50% of 150% of what is 80.5? (Round to a tenth.)

(5) _____   **(5)** 110% of 120% of $10 is what?

(6) _____   **(6)** 3.3% of 20% of $1,452.80 is what?

(7) _____   **(7)** 90% of 90% of what is 707? (Round to a tenth.)

(8) _____   **(8)** 68% of 105% of what is $555?

(9) _____   **(9)** The Watsons spend 38% of their income on food. Of that, 22% is spent on meat. If the Watsons' income was $1,450 in November, what amount did they spend on meat?

(10) _____   **(10)** The Conglomerate, Inc., produced 3,500 pounds in May, which was 6% of the total pounds of production at Plant A during May. Plant A produced 75% of the total pounds for the corporation. What was the total production for the corporation for May? (Round to a whole number.)

(1) _____19_____

(2) _____$24.20_____

(3) _____60.75_____

(4) _____107.3_____

(5) _____$13.20_____

(6) _____$9.59_____

(7) _____872.8_____

(8) _____$777.31_____

(9) _____$121.22_____

(10) _____77,778_____

Exercise 29B: Multiple Percentages

Name _____

Course/Sect. No. _____   Score _____

Instructions: Find the unknown.

---

(1) _____   **(1)** 10% of 40% of what is 200?

(2) _____   **(2)** 6% of 30% of $120 is what?

(3) _____   **(3)** 15.5% of 3% of what is $25?

(4) _____   **(4)** 80% of 60% of what is 3,200? (Round to a whole number.)

(5) _____   **(5)** 4% of 42% of 190 is what? (Round to a tenth.)

(6) _____   **(6)** 27.5% of 70% of what is 1500? (Round to a whole number.)

(7) _____   **(7)** 33% of 50% of $1.50 is what?

(8) _____   **(8)** 210% of 40% of $5,000 is what?

(9) _____   **(9)** Statusburg has a 5% sales tax. Of this tax revenue, 7.4% is to be spent on sewer improvement. If the sales for October were $685,000, what amount was to be spent on the sewers that month?

(10) _____   **(10)** In September, Bigsville spent $6,000 on their streets. What were the sales for September if the street improvement costs were 10% of sales?

Exercise 30A: Ratios

Name_____

Course/Sect. No._____ Score_____

Instructions: Find the ratio of the amount in column 1 to the amount in column 2. Express the ratio "to 1" rounded to one-tenth.

	Column 1	Column 2	Column 1	Column 2
(1) _____	**(1)** 20	18	**(2)** $14	$4
(2) _____				
(3) _____	**(3)** 35	39	**(4)** 30	40
(4) _____				
(5) _____	**(5)** 80	72	**(6)** 1,500	900
(6) _____				
(7) _____	**(7)** $27.50	$12.50	**(8)** 16.5	27.05
(8) _____				

(9) _____

(9) Find the ratio of Kenneth's savings of $5,150 to Terrell's savings of $2,600.

(10) _____

(10) The cost of goods sold is $40,000; gross profit is $110,000. What is the ratio of gross profit to cost of goods sold?

(1) _1.1 to 1_

(2) _3.5 to 1_

(3) _.9 to 1_

(4) _.8 to 1_

(5) _1.1 to 1_

(6) _1.7 to 1_

(7) _2.2 to 1_

(8) _.6 to 1_

(9) _2.0 to 1_

(10) _2.8 to 1_

Exercise 30B: Ratios

Name _____

Course/Sect. No. _____   Score _____

Instructions: Find the ratio of the amount in column 1 to the amount in column 2. Express the ratio "to 1" rounded to one-tenth.

	Column 1	Column 2	Column 1	Column 2
(1) _____	**(1)** 12	6	**(2)** 24	16
(2) _____				
(3) _____	**(3)** 1.2	.8	**(4)** .6	2.4
(4) _____				
(5) _____	**(5)** $2.12	$3.50	**(6)** 6.7	12
(6) _____				
(7) _____	**(7)** 801	910	**(8)** $1,000	$1,200
(8) _____				

(9) _____

(9) Find the ratio of Joe's earnings of $33,500 to Sally's earnings of $37,800.

(10) _____

(10) What is the ratio of profit to expenses if profits are $105,000 this month and expenses are $175,000?

Chapter

Payroll

Preview of Terms

Cumulative Earnings Total earnings from the beginning of the year.

Employee's Earnings Record A form used to record the individual payroll information of each employee.

FICA Taxes Social security taxes (*FICA* stands for the Federal Insurance Contribution Act, which established social security taxes).

FUT Shortened term for Federal Unemployment Tax.

Gross Pay Total earnings before any deductions.

Mandatory Deductions Payroll deductions the employer is required by law to deduct each pay period; that is, income taxes and social security taxes.

Optional Deductions Those payroll deductions each employee must approve in advance.

Overtime Any hours worked over 40 in one week are classified as "overtime hours."

Payroll Register A form used to summarize the total payroll for a single pay period, usually for a week.

Piece-rate Pay A method of paying employees that is based on the amount of work completed.

Salary Fixed compensation, usually on annual basis.

SUT Short for State Unemployment Tax.

Wages Variable compensation, usually on an hourly basis.

Withholding Taxes Another term for income taxes deducted.

Payroll computation and the resulting tax records are an integral part of any business operation. The proper methods and forms will be discussed in this section.

Tax laws are constantly changing. The rates and maximum taxable amounts used in this section were current at the time of printing, but may not be current as you now read it. However, the concepts are the same regardless of the specific rates and amounts.

PIECEWORK PAY

Some types of jobs lend themselves to a pay system whereby the worker is paid by the amount of work done rather than for the amount of time spent on the job. When work can be measured by the number of pieces produced or completed, a "piece-rate pay system" can be implemented. Under this system, a standard is set per piece of product produced or completed. The worker is then paid for each of these pieces. The more pieces completed, the more the pay.

Example 1:

Joe completed the following number of pieces this week. He receives 41¢ for each piece completed.

Mon.	Tues.	Wed.	Thurs.	Fri.	Sat.	Total
61	72	59	67	57	32	348

Therefore, he earned 348 pieces × 41¢ each = $142.68.

Practice Problems

Find the piecework pay in the following situations.

1. Edith completed the following number of pieces of work this week at $1.16 per piece: 16, 22, 24, 18, 20. What was her pay?

2. Herb works on an assembly line installing crodges on pirtles. He is paid 85¢ for each crodge installed. During the week, he installed 312 crodges. How much did he earn?

EARNINGS COMPUTATIONS

The most common method for calculating payroll is based on an hourly wage rate, whereby the worker is paid based on the amount of time spent on the job. The wage and hour law (the Fair Labor Standards Act) requires that employees be paid at a minimum of 1.5 times (time and one-half) the regular hourly rate for each hour worked in excess of 40 hours for each week. The hours that exceed 40 are called *overtime hours,* and the adjusted rate is called the *overtime rate.* The computations follow:

$$\text{Regular hours} \times \text{Regular rate} = \text{Regular pay}$$
$$\text{Overtime hours} \times \text{Overtime rate} = +\underline{\text{Overtime pay}}$$
$$\text{Gross pay}$$
$$-\underline{\text{Deductions}}$$
$$\text{Net pay}$$

REGULAR PAY

Employees who work 40 hours or less do not receive overtime pay. Their pay is calculated as follows.

Example 2:   Mr. A receives $4.50 an hour and works 37 hours this week. Find his gross pay.

$$\$4.50 \times 37 \text{ hours} = \$166.50 \text{ gross pay}$$

OVERTIME PAY

There are two approaches to calculating overtime pay, the overtime rate approach and the overtime premium approach. The *overtime rate* approach emphasizes the total number of hours and dollars paid for overtime. The *overtime premium* approach emphasizes the *difference* in overtime costs and straight-time costs. The product of the additional hours and the additional rate (the overtime bonus) reflects the premium paid above the amount additional workers would receive working straight time.

OVERTIME RATE APPROACH

An employee who works more than 40 hours is entitled to overtime pay in addition to regular pay. The overtime rate is generally based on 1.5 times the regular rate and is calculated as follows.

Example 3:   Ms. B receives $8.20 an hour. What would be her overtime rate?

$$\$8.20 \times 1.5 = \$12.30 \text{ an hour}$$

Example 4:   Mr. C receives $7.85 an hour. What would be his overtime rate?

$$\$7.85 \times 1.5 = \$11.775 \text{ an hour}$$

Note: *The overtime rate should not be rounded. Always wait until you multiply the overtime rate by the overtime hours before you round off.*

To find the gross pay, calculate the regular pay and the overtime pay, and add.

Example 5:   Mr. D receives $3.89 an hour and worked 43.5 hours this week. What was his gross pay?

$$\text{Regular pay} = 40 \text{ hours} \times \$3.89 = \$155.60$$
$$\text{Overtime pay} = 3.5 \text{ hours} \times \$5.835 = \underline{+\ 20.42^*}$$
$$\text{Gross pay} = \$176.02$$

(*Remember that the overtime *rate* is not rounded (1.5 × $3.89), but the overtime *pay* of $20.4225 is rounded.)

OVERTIME PREMIUM APPROACH

In the overtime premium approach, it is assumed that the total hours could have all been worked at the straight-time rate if additional employees had been hired. This means that no employee would have to work over 40 hours in any week. Any additional cost over the total hours worked times the straight-time rate represents a premium for paying overtime.

The straight-time cost is calculated for the total hours and added to the overtime bonus to yield the gross pay. The overtime bonus is therefore the premium for paying overtime.

To illustrate the computation, assume a worker earns $6 an hour and works 44 hours this week.

Step 1: Calculate the straight-time cost for the total hours.

44 total hours × $6 straight-time rate = $264

Step 2: Calculate the overtime bonus. With a straight-time rate of $6 an hour, at time and one-half, the overtime bonus would be the one-half portion, or $3 in this example, times the number of overtime hours.

$3 × 4 overtime hours = $12

Step 3: Add the overtime bonus and the straight-time cost.

$264	Straight-tome cost
+ 12	Overtime bonus
$276	Gross pay

The gross pay for Mr. D in Example 5 (above) would be calculated as follows using the overtime premium approach:

$169.22	Straight-time cost (43.5 × $3.89 an hour)
+ 6.81	Overtime bonus [($3.89 × .5) × 3.5 hours]
$176.03*	Gross pay

(*The 1¢ difference is due to rounding.)

EMPLOYEE DEDUCTIONS

The gross pay is reduced by the amount of the mandatory and optional deductions from each employee's earnings.

Mandatory Deductions
$\begin{cases} \text{Income taxes (withholding taxes)} \\ \text{FICA tax (social security tax)} \end{cases}$

Possible Optional Deductions
$\begin{cases} \text{Hospitalization} \\ \text{Union dues} \\ \text{Credit union payments} \\ \text{United Fund contributions} \\ \text{Payment for bonds} \end{cases}$

INCOME TAXES

The amount of income tax deducted from each paycheck depends on the amount of the gross pay and the number of dependents the person claims. There is no maximum amount or ceiling on income taxes withheld.

Generally, the employer determines the income tax to be withheld by referring to Circular E, "Employer's Tax Guide," distributed by the Internal Revenue Service. This publication contains withholding tax tables for weekly, biweekly, semimonthly, monthly, and daily or miscellaneous payroll periods. The four withholding tax tables that follow are from Circular E for single and married taxpayers being paid weekly. (All problems will assume a weekly payroll period.)

LOCATING THE INCOME TAX

Assume a single person claiming one dependent and earning $327.50 per week. (Refer to the withholding tax tables on the following pages.)

Step 1: Locate the correct table.

Step 2: Locate the wages between the two amounts in the columns headed "At least" and "But less than." In this case, $327.50 falls between "At least" $320 and "Less than" $330.

Step 3: Go down the withholding allowance for one dependent. The amount of $57.80 represents the income tax that should be withheld this pay period.

Practice Problems

Using the tax tables, locate the correct withholding tax.

7. Single person with wages of $405 claiming 1 dependent.

8. Married person with wages of $215 claiming 3 dependents.

9. Married person with wages of $900 claiming 2 dependents.

SINGLE Persons — WEEKLY Payroll Period

And the wages are—		And the number of withholding allowances claimed is—										
At least	But less than	0	1	2	3	4	5	6	7	8	9	10
		The amount of income tax to be withheld shall be—										
$0	$28	$0	$0	$0	$0	$0	$0	$0	$0	$0	$0	$0
28	29	.20	0	0	0	0	0	0	0	0	0	0
29	30	.30	0	0	0	0	0	0	0	0	0	0
30	31	.40	0	0	0	0	0	0	0	0	0	0
31	32	.60	0	0	0	0	0	0	0	0	0	0
32	33	.70	0	0	0	0	0	0	0	0	0	0
33	34	.90	0	0	0	0	0	0	0	0	0	0
34	35	1.00	0	0	0	0	0	0	0	0	0	0
35	36	1.10	0	0	0	0	0	0	0	0	0	0
36	37	1.30	0	0	0	0	0	0	0	0	0	0
37	38	1.40	0	0	0	0	0	0	0	0	0	0
38	39	1.60	0	0	0	0	0	0	0	0	0	0
39	40	1.70	0	0	0	0	0	0	0	0	0	0
40	41	1.80	0	0	0	0	0	0	0	0	0	0
41	42	2.00	0	0	0	0	0	0	0	0	0	0
42	43	2.10	0	0	0	0	0	0	0	0	0	0
43	44	2.30	0	0	0	0	0	0	0	0	0	0
44	45	2.40	0	0	0	0	0	0	0	0	0	0
45	46	2.50	0	0	0	0	0	0	0	0	0	0
46	47	2.70	0	0	0	0	0	0	0	0	0	0
47	48	2.80	.10	0	0	0	0	0	0	0	0	0
48	49	3.00	.30	0	0	0	0	0	0	0	0	0
49	50	3.10	.40	0	0	0	0	0	0	0	0	0
50	51	3.20	.60	0	0	0	0	0	0	0	0	0
51	52	3.40	.70	0	0	0	0	0	0	0	0	0
52	53	3.50	.80	0	0	0	0	0	0	0	0	0
53	54	3.70	1.00	0	0	0	0	0	0	0	0	0
54	55	3.80	1.10	0	0	0	0	0	0	0	0	0
55	56	3.90	1.30	0	0	0	0	0	0	0	0	0
56	57	4.10	1.40	0	0	0	0	0	0	0	0	0
57	58	4.20	1.50	0	0	0	0	0	0	0	0	0
58	59	4.40	1.70	0	0	0	0	0	0	0	0	0
59	60	4.50	1.80	0	0	0	0	0	0	0	0	0
60	62	4.70	2.00	0	0	0	0	0	0	0	0	0
62	64	5.00	2.30	0	0	0	0	0	0	0	0	0
64	66	5.30	2.60	0	0	0	0	0	0	0	0	0
66	68	5.60	2.90	.20	0	0	0	0	0	0	0	0
68	70	5.80	3.10	.50	0	0	0	0	0	0	0	0
70	72	6.10	3.40	.70	0	0	0	0	0	0	0	0
72	74	6.40	3.70	1.00	0	0	0	0	0	0	0	0
74	76	6.70	4.00	1.30	0	0	0	0	0	0	0	0
76	78	7.00	4.30	1.60	0	0	0	0	0	0	0	0
78	80	7.30	4.50	1.90	0	0	0	0	0	0	0	0
80	82	7.60	4.80	2.10	0	0	0	0	0	0	0	0
82	84	8.00	5.10	2.40	0	0	0	0	0	0	0	0
84	86	8.30	5.40	2.70	0	0	0	0	0	0	0	0
86	88	8.60	5.70	3.00	.30	0	0	0	0	0	0	0
88	90	8.90	5.90	3.30	.60	0	0	0	0	0	0	0
90	92	9.20	6.20	3.50	.80	0	0	0	0	0	0	0
92	94	9.60	6.50	3.80	1.10	0	0	0	0	0	0	0
94	96	9.90	6.80	4.10	1.40	0	0	0	0	0	0	0
96	98	10.20	7.10	4.40	1.70	0	0	0	0	0	0	0
98	100	10.50	7.40	4.70	2.00	0	0	0	0	0	0	0
100	105	11.10	8.00	5.10	2.50	0	0	0	0	0	0	0
105	110	12.10	8.80	5.80	3.20	.50	0	0	0	0	0	0
110	115	13.00	9.60	6.50	3.90	1.20	0	0	0	0	0	0
115	120	14.00	10.40	7.30	4.60	1.90	0	0	0	0	0	0
120	125	14.90	11.30	8.10	5.30	2.60	0	0	0	0	0	0
125	130	15.90	12.20	8.90	6.00	3.30	.60	0	0	0	0	0
130	135	16.80	13.20	9.70	6.70	4.00	1.30	0	0	0	0	0

(Continued on next page)

And the wages are—		And the number of withholding allowances claimed is—										
At least	But less than	0	1	2	3	4	5	6	7	8	9	10
		The amount of income tax to be withheld shall be—										
$135	$140	$17.80	$14.10	$10.50	$7.40	$4.70	$2.00	$0	$0	$0	$0	$0
140	145	18.70	15.10	11.40	8.20	5.40	2.70	0	0	0	0	0
145	150	19.70	16.00	12.40	9.00	6.10	3.40	.70	0	0	0	0
150	160	21.10	17.50	13.80	10.20	7.20	4.40	1.70	0	0	0	0
160	170	23.00	19.40	15.70	12.10	8.80	5.80	3.10	.40	0	0	0
170	180	24.90	21.30	17.60	14.00	10.40	7.30	4.50	1.80	0	0	0
180	190	27.00	23.20	19.50	15.90	12.20	8.90	5.90	3.20	.50	0	0
190	200	29.40	25.10	21.40	17.80	14.10	10.50	7.40	4.60	1.90	0	0
200	210	31.80	27.20	23.30	19.70	16.00	12.40	9.00	6.00	3.30	.60	0
210	220	34.20	29.60	25.20	21.60	17.90	14.30	10.60	7.50	4.70	2.00	0
220	230	36.60	32.00	27.40	23.50	19.80	16.20	12.50	9.10	6.10	3.40	.80
230	240	39.00	34.40	29.80	25.40	21.70	18.10	14.40	10.70	7.70	4.80	2.20
240	250	41.40	36.80	32.20	27.60	23.60	20.00	16.30	12.60	9.30	6.20	3.60
250	260	43.80	39.20	34.60	30.00	25.50	21.90	18.20	14.50	10.90	7.80	5.00
260	270	46.20	41.60	37.00	32.40	27.80	23.80	20.10	16.40	12.80	9.40	6.40
270	280	48.90	44.00	39.40	34.80	30.20	25.70	22.00	18.30	14.70	11.00	7.90
280	290	51.80	46.40	41.80	37.20	32.60	28.00	23.90	20.20	16.60	12.90	9.50
290	300	54.70	49.10	44.20	39.60	35.00	30.40	25.80	22.10	18.50	14.80	11.20
300	310	57.60	52.00	46.60	42.00	37.40	32.80	28.10	24.00	20.40	16.70	13.10
310	320	60.50	54.90	49.40	44.40	39.80	35.20	30.50	25.90	22.30	18.60	15.00
320	330	63.40	57.80	52.30	46.80	42.20	37.60	32.90	28.30	24.20	20.50	16.90
330	340	66.50	60.70	55.20	49.60	44.60	40.00	35.30	30.70	26.10	22.40	18.80
340	350	69.70	63.60	58.10	52.50	47.00	42.40	37.70	33.10	28.50	24.30	20.70
350	360	72.90	66.70	61.00	55.40	49.80	44.80	40.10	35.50	30.90	26.30	22.60
360	370	76.10	69.90	63.90	58.30	52.70	47.20	42.50	37.90	33.30	28.70	24.50
370	380	79.30	73.10	66.90	61.20	55.60	50.00	44.90	40.30	35.70	31.10	26.50
380	390	82.50	76.30	70.10	64.10	58.50	52.90	47.40	42.70	38.10	33.50	28.90
390	400	85.70	79.50	73.30	67.20	61.40	55.80	50.30	45.10	40.50	35.90	31.30
400	410	88.90	82.70	76.50	70.40	64.30	58.70	53.20	47.60	42.90	38.30	33.70
410	420	92.10	85.90	79.70	73.60	67.40	61.60	56.10	50.50	45.30	40.70	36.10
420	430	95.30	89.10	82.90	76.80	70.60	64.50	59.00	53.40	47.80	43.10	38.50
430	440	98.60	92.30	86.10	80.00	73.80	67.70	61.90	56.30	50.70	45.50	40.90
440	450	102.30	95.50	89.30	83.20	77.00	70.90	64.80	59.20	53.60	48.00	43.30
450	460	106.00	98.90	92.50	86.40	80.20	74.10	67.90	62.10	56.50	50.90	45.70
460	470	109.70	102.60	95.70	89.60	83.40	77.30	71.10	65.00	59.40	53.80	48.30
470	480	113.40	106.30	99.10	92.80	86.60	80.50	74.30	68.20	62.30	56.70	51.20
480	490	117.10	110.00	102.80	96.00	89.80	83.70	77.50	71.40	65.20	59.60	54.10
490	500	120.80	113.70	106.50	99.40	93.00	86.90	80.70	74.60	68.40	62.50	57.00
500	510	124.50	117.40	110.20	103.10	96.20	90.10	83.90	77.80	71.60	65.50	59.90
510	520	128.20	121.10	113.90	106.80	99.70	93.30	87.10	81.00	74.80	68.70	62.80
520	530	131.90	124.80	117.60	110.50	103.40	96.50	90.30	84.20	78.00	71.90	65.70
530	540	135.60	128.50	121.30	114.20	107.10	100.00	93.50	87.40	81.20	75.10	68.90
540	550	139.30	132.20	125.00	117.90	110.80	103.70	96.70	90.60	84.40	78.30	72.10
550	560	143.00	135.90	128.70	121.60	114.50	107.40	100.30	93.80	87.60	81.50	75.30
560	570	146.70	139.60	132.40	125.30	118.20	111.10	104.00	97.00	90.80	84.70	78.50
570	580	150.40	143.30	136.10	129.00	121.90	114.80	107.70	100.60	94.00	87.90	81.70
580	590	154.10	147.00	139.80	132.70	125.60	118.50	111.40	104.30	97.20	91.10	84.90
590	600	157.80	150.70	143.50	136.40	129.30	122.20	115.10	108.00	100.80	94.30	88.10
600	610	161.50	154.40	147.20	140.10	133.00	125.90	118.80	111.70	104.50	97.50	91.30
610	620	165.20	158.10	150.90	143.80	136.70	129.60	122.50	115.40	108.20	101.10	94.50
620	630	168.90	161.80	154.60	147.50	140.40	133.30	126.20	119.10	111.90	104.80	97.70
630	640	172.60	165.50	158.30	151.20	144.10	137.00	129.90	122.80	115.60	108.50	101.40
640	650	176.30	169.20	162.00	154.90	147.80	140.70	133.60	126.50	119.30	112.20	105.10
650	660	180.00	172.90	165.70	158.60	151.50	144.40	137.30	130.20	123.00	115.90	108.80
660	670	183.70	176.60	169.40	162.30	155.20	148.10	141.00	133.90	126.70	119.60	112.50
		37 percent of the excess over $670 plus—										
$670 and over		185.50	178.40	171.30	164.20	157.10	149.90	142.80	135.70	128.60	121.50	114.40

MARRIED Persons — WEEKLY Payroll Period

And the wages are—		And the number of withholding allowances claimed is—										
At least	But less than	0	1	2	3	4	5	6	7	8	9	10
		The amount of income tax to be withheld shall be—										
$0	$47	$0	$0	$0	$0	$0	$0	$0	$0	$0	$0	$0
47	48	.20	0	0	0	0	0	0	0	0	0	0
48	49	.30	0	0	0	0	0	0	0	0	0	0
49	50	.50	0	0	0	0	0	0	0	0	0	0
50	51	.60	0	0	0	0	0	0	0	0	0	0
51	52	.70	0	0	0	0	0	0	0	0	0	0
52	53	.90	0	0	0	0	0	0	0	0	0	0
53	54	1.00	0	0	0	0	0	0	0	0	0	0
54	55	1.20	0	0	0	0	0	0	0	0	0	0
55	56	1.30	0	0	0	0	0	0	0	0	0	0
56	57	1.40	0	0	0	0	0	0	0	0	0	0
57	58	1.60	0	0	0	0	0	0	0	0	0	0
58	59	1.70	0	0	0	0	0	0	0	0	0	0
59	60	1.90	0	0	0	0	0	0	0	0	0	0
60	62	2.10	0	0	0	0	0	0	0	0	0	0
62	64	2.40	0	0	0	0	0	0	0	0	0	0
64	66	2.60	0	0	0	0	0	0	0	0	0	0
66	68	2.90	.20	0	0	0	0	0	0	0	0	0
68	70	3.20	.50	0	0	0	0	0	0	0	0	0
70	72	3.50	.80	0	0	0	0	0	0	0	0	0
72	74	3.80	1.10	0	0	0	0	0	0	0	0	0
74	76	4.00	1.30	0	0	0	0	0	0	0	0	0
76	78	4.30	1.60	0	0	0	0	0	0	0	0	0
78	80	4.60	1.90	0	0	0	0	0	0	0	0	0
80	82	4.90	2.20	0	0	0	0	0	0	0	0	0
82	84	5.20	2.50	0	0	0	0	0	0	0	0	0
84	86	5.40	2.70	.10	0	0	0	0	0	0	0	0
86	88	5.70	3.00	.30	0	0	0	0	0	0	0	0
88	90	6.00	3.30	.60	0	0	0	0	0	0	0	0
90	92	6.30	3.60	.90	0	0	0	0	0	0	0	0
92	94	6.60	3.90	1.20	0	0	0	0	0	0	0	0
94	96	6.80	4.10	1.50	0	0	0	0	0	0	0	0
96	98	7.10	4.40	1.70	0	0	0	0	0	0	0	0
98	100	7.40	4.70	2.00	0	0	0	0	0	0	0	0
100	105	7.90	5.20	2.50	0	0	0	0	0	0	0	0
105	110	8.60	5.90	3.20	.50	0	0	0	0	0	0	0
110	115	9.30	6.60	3.90	1.20	0	0	0	0	0	0	0
115	120	10.00	7.30	4.60	1.90	0	0	0	0	0	0	0
120	125	10.70	8.00	5.30	2.60	0	0	0	0	0	0	0
125	130	11.40	8.70	6.00	3.30	.60	0	0	0	0	0	0
130	135	12.10	9.40	6.70	4.00	1.30	0	0	0	0	0	0
135	140	12.80	10.10	7.40	4.70	2.00	0	0	0	0	0	0
140	145	13.50	10.80	8.10	5.40	2.70	0	0	0	0	0	0
145	150	14.20	11.50	8.80	6.10	3.40	.70	0	0	0	0	0
150	160	15.40	12.50	9.90	7.20	4.50	1.80	0	0	0	0	0
160	170	17.00	13.90	11.30	8.60	5.90	3.20	.50	0	0	0	0
170	180	18.60	15.50	12.70	10.00	7.30	4.60	1.90	0	0	0	0
180	190	20.20	17.10	14.10	11.40	8.70	6.00	3.30	.60	0	0	0
190	200	21.80	18.70	15.60	12.80	10.10	7.40	4.70	2.00	0	0	0
200	210	23.40	20.30	17.20	14.20	11.50	8.80	6.10	3.40	.70	0	0
210	220	25.20	21.90	18.80	15.80	12.90	10.20	7.50	4.80	2.10	0	0
220	230	27.20	23.50	20.40	17.40	14.30	11.60	8.90	6.20	3.50	.80	0
230	240	29.20	25.40	22.00	19.00	15.90	13.00	10.30	7.60	4.90	2.20	0
240	250	31.20	27.40	23.60	20.60	17.50	14.40	11.70	9.00	6.30	3.60	.90
250	260	33.20	29.40	25.50	22.20	19.10	16.00	13.10	10.40	7.70	5.00	2.30
260	270	35.20	31.40	27.50	23.80	20.70	17.60	14.50	11.80	9.10	6.40	3.70
270	280	37.20	33.40	29.50	25.70	22.30	19.20	16.10	13.20	10.50	7.80	5.10
280	290	39.20	35.40	31.50	27.70	23.90	20.80	17.70	14.70	11.90	9.20	6.50
290	300	41.20	37.40	33.50	29.70	25.80	22.40	19.30	16.30	13.30	10.60	7.90
300	310	43.70	39.40	35.50	31.70	27.80	24.00	20.90	17.90	14.80	12.00	9.30

(Continued on next page)

And the wages are—		And the number of withholding allowances claimed is—										
At least	But less than	0	1	2	3	4	5	6	7	8	9	10
		The amount of income tax to be withheld shall be—										
$310	$320	$46.20	$41.40	$37.50	$33.70	$29.80	$26.00	$22.50	$19.50	$16.40	$13.40	$10.70
320	330	48.70	43.80	39.50	35.70	31.80	28.00	24.10	21.10	18.00	14.90	12.10
330	340	51.20	46.30	41.50	37.70	33.80	30.00	26.10	22.70	19.60	16.50	13.50
340	350	53.70	48.80	44.00	39.70	35.80	32.00	28.10	24.30	21.20	18.10	15.00
350	360	56.20	51.30	46.50	41.70	37.80	34.00	30.10	26.30	22.80	19.70	16.60
360	370	58.70	53.80	49.00	44.20	39.80	36.00	32.10	28.30	24.40	21.30	18.20
370	380	61.20	56.30	51.50	46.70	41.90	38.00	34.10	30.30	26.40	22.90	19.80
380	390	63.70	58.80	54.00	49.20	44.40	40.00	36.10	32.30	28.40	24.60	21.40
390	400	66.20	61.30	56.50	51.70	46.90	42.10	38.10	34.30	30.40	26.60	23.00
400	410	68.70	63.80	59.00	54.20	49.40	44.60	40.10	36.30	32.40	28.60	24.80
410	420	71.20	66.30	61.50	56.70	51.90	47.10	42.30	38.30	34.40	30.60	26.80
420	430	73.70	68.80	64.00	59.20	54.40	49.60	44.80	40.30	36.40	32.60	28.80
430	440	76.20	71.30	66.50	61.70	56.90	52.10	47.30	42.50	38.40	34.60	30.80
440	450	78.70	73.80	69.00	64.20	59.40	54.60	49.80	45.00	40.40	36.60	32.80
450	460	81.60	76.30	71.50	66.70	61.90	57.10	52.30	47.50	42.70	38.60	34.80
460	470	84.70	78.80	74.00	69.20	64.40	59.60	54.80	50.00	45.20	40.60	36.80
470	480	87.80	81.90	76.50	71.70	66.90	62.10	57.30	52.50	47.70	42.90	38.80
480	490	90.90	85.00	79.00	74.20	69.40	64.60	59.80	55.00	50.20	45.40	40.80
490	500	94.00	88.10	82.10	76.70	71.90	67.10	62.30	57.50	52.70	47.90	43.10
500	510	97.10	91.20	85.20	79.20	74.40	69.60	64.80	60.00	55.20	50.40	45.60
510	520	100.20	94.30	88.30	82.30	76.90	72.10	67.30	62.50	57.70	52.90	48.10
520	530	103.30	97.40	91.40	85.40	79.50	74.60	69.80	65.00	60.20	55.40	50.60
530	540	106.40	100.50	94.50	88.50	82.60	77.10	72.30	67.50	62.70	57.90	53.10
540	550	109.50	103.60	97.60	91.60	85.70	79.70	74.80	70.00	65.20	60.40	55.60
550	560	112.60	106.70	100.70	94.70	88.80	82.80	77.30	72.50	67.70	62.90	58.10
560	570	116.00	109.80	103.80	97.80	91.90	85.90	80.00	75.00	70.20	65.40	60.60
570	580	119.40	112.90	106.90	100.90	95.00	89.00	83.10	77.50	72.70	67.90	63.10
580	590	122.80	116.30	110.00	104.00	98.10	92.10	86.20	80.20	75.20	70.40	65.60
590	600	126.20	119.70	113.10	107.10	101.20	95.20	89.30	83.30	77.70	72.90	68.10
600	610	129.60	123.10	116.50	110.20	104.30	98.30	92.40	86.40	80.40	75.40	70.60
610	620	133.00	126.50	119.90	113.40	107.40	101.40	95.50	89.50	83.50	77.90	73.10
620	630	136.40	129.90	123.30	116.80	110.50	104.50	98.60	92.60	86.60	80.70	75.60
630	640	139.80	133.30	126.70	120.20	113.70	107.60	101.70	95.70	89.70	83.80	78.10
640	650	143.20	136.70	130.10	123.60	117.10	110.70	104.80	98.80	92.80	86.90	80.90
650	660	146.60	140.10	133.50	127.00	120.50	113.90	107.90	101.90	95.90	90.00	84.00
660	670	150.20	143.50	136.90	130.40	123.90	117.30	111.00	105.00	99.00	93.10	87.10
670	680	153.90	146.90	140.30	133.80	127.30	120.70	114.20	108.10	102.10	96.20	90.20
680	690	157.60	150.50	143.70	137.20	130.70	124.10	117.60	111.20	105.20	99.30	93.30
690	700	161.30	154.20	147.10	140.60	134.10	127.50	121.00	114.40	108.30	102.40	96.40
700	710	165.00	157.90	150.80	144.00	137.50	130.90	124.40	117.80	111.40	105.50	99.50
710	720	168.70	161.60	154.50	147.40	140.90	134.30	127.80	121.20	114.70	108.60	102.60
720	730	172.40	165.30	158.20	151.10	144.30	137.70	131.20	124.60	118.10	111.70	105.70
730	740	176.10	169.00	161.90	154.80	147.70	141.10	134.60	128.00	121.50	115.00	108.80
740	750	179.80	172.70	165.60	158.50	151.40	144.50	138.00	131.40	124.90	118.40	111.90
750	760	183.50	176.40	169.30	162.20	155.10	147.90	141.40	134.80	128.30	121.80	115.20
760	770	187.20	180.10	173.00	165.90	158.80	151.60	144.80	138.20	131.70	125.20	118.60
770	780	190.90	183.80	176.70	169.60	162.50	155.30	148.20	141.60	135.10	128.60	122.00
780	790	194.60	187.50	180.40	173.30	166.20	159.00	151.90	145.00	138.50	132.00	125.40
790	800	198.30	191.20	184.10	177.00	169.90	162.70	155.60	148.50	141.90	135.40	128.80
800	810	202.00	194.90	187.80	180.70	173.60	166.40	159.30	152.20	145.30	138.80	132.20
810	820	205.70	198.60	191.50	184.40	177.30	170.10	163.00	155.90	148.80	142.20	135.60
820	830	209.40	202.30	195.20	188.10	181.00	173.80	166.70	159.60	152.50	145.60	139.00
830	840	213.10	206.00	198.90	191.80	184.70	177.50	170.40	163.30	156.20	149.10	142.40
840	850	216.80	209.70	202.60	195.50	188.40	181.20	174.10	167.00	159.90	152.80	145.80
850	860	220.50	213.40	206.30	199.20	192.10	184.90	177.80	170.70	163.60	156.50	149.40
		37 percent of the excess over $860 plus—										
$860 and over		222.40	215.30	208.10	201.00	193.90	186.80	179.70	172.60	165.50	158.30	151.20

FICA TAXES

Social security taxes are a pay-as-you-go system like withholding taxes. Each time a worker is paid, the employer is required by law to deduct the tax and send it to the Internal Revenue collection center in the area. Unlike withholding taxes, which have no maximum amount that can be deducted, social security taxes have a maximum, or ceiling. Under the present laws, the maximum is $2130.60. This figure is based on the current rate of 6.7% and the current base of $31,800.* The tax rate applies to *first* $31,800 each employee earns each calendar year. Once the cumulative, or total, earnings of an individual have exceeded $31,800, no more deductions for FICA taxes should be made from that employee's wages for that calendar year. [See the end of this section for the schedule from the 1977 Social Security Law.]

Example 6:

Mr. A earned $12,000 last year. What were his FICA taxes?

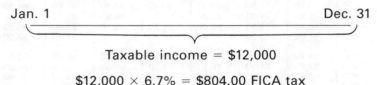

Taxable income = $12,000

$12,000 × 6.7% = $804.00 FICA tax

Mr. A had FICA taxes deducted from his pay the entire year. Since his total earnings did not exceed $31,800, the entire amount earned was taxable.

Example 7:

Ms. B earned $39,000 last year. What were her FICA taxes?

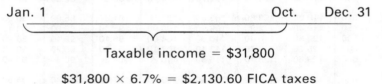

Taxable income = $31,800

$31,800 × 6.7% = $2,130.60 FICA taxes

Assuming that her income was earned evenly throughout the year, Ms. B would have reached the $31,800 ceiling sometime in October. From that time through the rest of the year, no FICA taxes should have been deducted from her pay.

Example 8:

Prior to this week, Mr. C had cumulative earnings of $8,520.16. This week, his gross pay is $250. What should be deducted for FICA taxes?

$250 × 6.7% = $16.75 FICA taxes

Note:

This week's gross pay of $250 will not cause the cumulative total to exceed the $31,800 maximum ($8,520.16 + $250 = $8,770.16). This means that the entire $250 is taxable this week.

Example 9:

Before this week, Ms. D had cumulative earnings of $31,550.50. This week, her gross pay is $375. What should be deducted for FICA taxes?

$249.50 × 6.7% = $16.72 FICA taxes

Note:

Ms. D's earnings for the current week, when added to her previous cumulative earnings, caused her new cumulative earnings to exceed the $31,800 maximum. This means that only a portion of her earnings for the current week are taxable. Remember, the tax is on the first $31,800, and therefore, any earnings over $31,800 are not taxable.

$31,800.00 Maximum taxable earnings
−31,550.50 Previous earnings
$ 249.50 Remaining taxable earnings
× 6.7%
$ 16.72 FICA tax

The rate is multiplied by the gross pay or the remaining taxable earnings, whichever is smaller.

Example 10:

Prior to this week, Mr. E had cumulative earnings of $35,080.72. This week his gross pay is $415. What should be deducted for FICA taxes?

No FICA taxes should be deducted.

Since the cumulative earnings prior to this week were already over the $31,800 maximum, the maximum $2,130.60 for FICA taxes had already been deducted. Therefore, no further FICA taxes should be deducted from Mr. E's pay for this week or the remainder of the year.

Practice Problems

Find the proper weekly FICA tax deductions assuming a 6.7% rate on the first $31,800 earned.

10. Cumulative earnings prior to this week of $14,829.50 and earnings this week of $327.85.

11. Cumulative earnings prior to this week of $31,628.72 and earnings this week of $273.

12. Cumulative earnings prior to this week of $33,772.88 and earnings this week of $197.50.

OTHER DEDUCTIONS

The other, optional deductions for hospitalization, United Fund, and so on are all individually determined and frequently change each pay period. Their amounts will be supplied and not calculated.

EMPLOYEE'S EARNINGS RECORD

To be able to calculate the proper FICA taxes, it is necessary to keep accurate cumulative earnings records for each employee. The following examples of Employee's Earnings Records are completed for the first quarter of the year and the fourth quarter of the year. There are lines for monthly totals and quarterly totals since these totals are used for tax, insurance, and other reports. The information is also needed for the yearly Wage and Tax Statement (W-2) issued to each taxpayer.

Look at Henry Haynes's earnings record in the FICA tax column for October 22. The regular weekly FICA tax was lower ($15.39) that week because only $229.75 was taxable ($31,800 maximum less $31,570.25 previous cumulative earnings). After the October 22 earnings, Henry owed no more FICA taxes for the remainder of the year, since the maximum taxable amount of $31,800 had been reached.

Employee's Earnings Record

Name: Haynes, Henry **Hourly Rate: $7.20** **Dependents: 1 (single)**

Period Ending	Hrs.	Regular (Earnings)	Overtime	Gross	FICA Tax (Deductions)	Inc. Tax	Hosp.	Total	Net Amount	Cumulative Earnings
Oct. 1	38	273.60		273.60	18.33	44.00	5	67.33	206.27	31,019.45
Oct. 8	39.5	284.40		284.40	19.05	46.40	5	70.45	213.95	31,303.85
Oct. 15	37	266.40		266.40	17.85	41.60	5	64.45	201.95	31,570.25
Oct. 22	42	288.00	21.60	309.60	15.39	52.00	5	72.39	237.21	31,879.85
Oct. 29	40	288.00		288.00		46.40		46.40	241.60	32,167.85
Monthly Total		1,400.40	21.60	1,422.00	70.62	230.40	20	321.02	1,100.98	
Nov. 5	44	288.00	43.20	331.20		60.70	5	65.70	265.50	32,499.05
Nov. 12	43.5	288.00	37.80	325.80		57.80	5	62.80	263.00	32,824.85
Nov. 19	40	288.00		288.00		46.40	5	51.40	236.60	33,112.85
Nov. 26	39	280.80		280.80		46.40	5	51.40	229.40	33,393.65
Monthly Total		1,144.80	81.00	1,225.80		211.30	20	231.30	994.50	
Dec. 3	15	108.00		108.00		8.80	5	13.80	94.20	33,501.65
Dec. 10	21	151.20		151.20		17.50	5	22.50	128.70	33,652.85
Dec. 17	40	288.00		288.00		46.40	5	51.40	236.60	33,940.85
Dec. 24	44	288.00	43.20	331.20		60.70	5	65.70	265.50	34,272.05
Dec. 31	48	288.00	86.40	374.40		73.10		73.10	301.30	34,646.45
Monthly Total		1,123.20	129.60	1,252.80		206.50	20	226.50	1,026.30	
Quarterly Total		3,668.40	232.20	3,900.60	70.62	648.20	60	778.82	3,121.78	

Employee's Earnings Record

Name: Montgomery, Mary				Hourly Rate: $3.50					Dependents: 3 (Married)	
		Earnings			**Deductions**					
Period Ending	Hrs.	Regular	Over-time	Gross	FICA Tax	Inc. Tax	Hosp.	Total	Net Amount	Cumulative Earnings
Jan. 8	40	140.00		140.00	9.38	5.40	7.50	22.28	117.72	140.00
Jan. 15	42	140.00	10.50	150.50	10.08	7.20	7.50	24.78	125.72	290.50
Jan. 22	38	133.00		133.00	8.91	4.00	7.50	20.41	112.59	422.50
Jan. 29	40	140.00		140.00	9.38	5.40	7.50	22.28	117.72	563.50
Monthly Total		553.00	10.50	563.50	37.75	22.00	30.00	89.75	473.75	
Feb. 5	40	140.00		140.00	9.38	5.40	7.50	22.28	117.72	703.50
Feb. 12	45	140.00	26.25	166.25	11.14	8.60	7.50	27.24	139.01	869.75
Feb. 19	40	140.00		140.00	9.38	5.40	7.50	22.28	117.72	1,009.75
Feb. 26	41.5	140.00	7.88	147.88	9.91	6.10	7.50	23.51	124.37	1,157.63
Monthly Total		560.00	34.13	594.13	39.81	25.50	30.00	95.31	498.82	
Mar. 5	37	129.50		129.50	8.68	3.30	7.50	19.48	110.02	1,287.13
Mar. 12	40	140.00		140.00	9.38	5.40	7.50	22.28	117.72	1,427.13
Mar. 19	40	140.00		140.00	9.38	5.40	7.50	22.28	117.72	1,567.13
Mar. 26	42	140.00	10.50	150.50	10.08	7.20	7.50	24.78	125.72	1,717.63
Monthly Total		549.50	10.50	560.00	37.52	21.30	30.00	88.82	471.18	
Quarterly Total		1,662.50	55.13	1,717.63	115.08	68.80	90.00	273.88	1,443.75	

Complete the individual Employee's Earnings Record for each of the following. Assume an FICA tax of 6.7% on the first $31,800.

13. Bill Bailey; married claiming 2 dependents; $7.20 per hour; cumulative earnings $0.

> Jan. 7 = 40 hours worked; hospitalization deduction of $15.
> Jan. 14 = 42 hours worked; hospitalization deduction of $15.
> Jan. 21 = 39 hours worked; hospitalization deduction of $15.

Name: **Hourly Rate:**   **Dependents:**

| Period Ending | Hrs. | Earnings | | | Deductions | | | | Net Amount | Cumulative Earnings |
		Regular	Over-time	Gross	FICA Tax	Inc. Tax	Hosp.	Total		

14. Helen Kramer; single claiming 1 dependent; $4.88 per hour; cumulative earnings $31,510.17.

> Dec. 13 = 45 hours worked; hospitalization deduction of $10.
> Dec. 20 = 50 hours worked; hospitalization deduction of $10.
> Dec. 27 = 52 hours worked; hospitalization deduction of $10.

Name: **Hourly Rate:**   **Dependents:**

| Period Ending | Hrs. | Earnings | | | Deductions | | | | Net Amount | Cumulative Earnings |
		Regular	Over-time	Gross	FICA Tax	Inc. Tax	Hosp.	Total		

EMPLOYER'S PAYROLL TAXES

The examples being covered here assume a smaller, single-owner business (sole proprietorship) or partnership rather than a corporation. Only the corporate form of business has to pay income taxes as a business. Therefore, the employer expenses considered here do not include income taxes.

FICA TAXES

Every employer is required by law to make an FICA contribution to each employee's social security account. As of this printing, the employer contribution is 7.0% of the taxable earnings each pay period. This means an employee with weekly FICA taxable earnings of $400 would pay $26.80 in FICA taxes ($400×6.7%), while the employer would pay $28.00 ($400 × 7.0%)

The computations for the employer are the same as for the employees, except it is customary for the employer to total the entire taxable earnings for FICA each pay period and multiply the single amount by the 7.0% rate. This taxable total is determined in a separate column of the Payroll Register and will be shown later in this section.

Example 11:   For one employer, the taxable earnings for FICA purposes totaled $2,520.35 this pay period. If the rate is 7.0%, what would be the employer's FICA tax?

$$\$2,520.35 \times 7.0\% = \$176.42 \text{ FICA tax}$$

Example 12:   The XYZ Co. has two workers, Joe and Bob. Prior to this week, their cumulative earnings were $31,659.75 and $10,500, respectively. If Joe has gross earnings this week of $275 and Bob of $195, what FICA tax would the XYZ Co. have to pay based on their earnings?

Joe:	$31,800.00	Maximum taxable
	− 31,659.75	Cumulative total
	$ 140.25	Taxable earnings
	× 7.0%	
	$ 9.82	FICA tax
Bob:	$31,800.00	Maximum taxable
	− 10,500.00	Cumulative total
	$21,300.00*	Taxable earnings

(*Remember, if the taxable earnings exceed the gross earnings that pay period, the rate is multiplied by the weekly gross earnings.)

$195.00	Weekly gross earnings
× 7.0%	
$ 13.65	FICA tax

Practice Problems

Compute the employer's FICA taxes (at 7.0% on the first $31,800).

15. The taxable earnings for FICA purposes were $872.55 this pay period.

16. Employee A had cumulative earnings prior to this week of $25,572.83 and earnings this week of $325.17.

17. Employee B had cumulative earnings prior to this week of $31,590 and earnings this week of $630.

FEDERAL UNEMPLOYMENT TAX

The federal government requires each state to provide some type of unemployment benefits for those workers that become unemployed and are unsuccessful in finding other jobs. To administer the separate programs, a small contribution is sent to the federal government to cover the expenses. The amount of tax that is sent in by the employer is based on what the employees earn. The employees generally do not contribute to the program, but the employer contributions are based on the employee earnings. The concept is the same as FICA tax contributions, but the rates and maximum taxable amounts are different. For the Federal Unemployment Tax (FUT), the rate is .7%, and it is based on the first $6,000 earned in any one year by each employee. Once the $6,000 cumulative earnings amount is reached by an employee, the employer does not have to pay unemployment taxes based on that employee's earnings until the next year.

Example 13:   Doris had cumulative earnings prior to this pay period of $5,714.55. Her gross pay this week was $202.50. Based on her pay, what FUT would be owed by her employer?

$6,000.00	Maximum amount	$202.50	Weekly gross earnings
− 5,714.55	Cumulative earnings	× .7%	
$ 285.45*	Taxable earnings	$ 1.42	FUT

(*Remember, as with FICA taxes, if the taxable earnings exceed the gross earnings, the rate is multiplied by the weekly gross earnings.)

STATE UNEMPLOYMENT TAX

The actual unemployment compensation benefits are paid out by the states. A majority of the taxes therefore go to state governments to build up the funds necessary to make the payments. The maximum amount that is used as the tax base is the same for the State Unemployment Tax (SUT) as that for FUT—$6,000. The tax rate fluctuates from business to business and from year to year. It is set according to merit ratings. Employers who provide steady employment are rewarded with reduced rates, whereas those that are primarily responsible for a drain on the unemployment funds receive higher rates. Each employer's unemployment record is reviewed periodically, and its merit rating adjusts the tax rate. Although the range is anywhere from 0% to 6%, the most common tax rate is 3.4%.

Example 14:   Carolyn had cumulative earnings prior to this pay period of $2,415.20. Her gross pay this week was $85. Based on her pay, what SUT would be owed by her employer? Assume a rate of 3.4%.

$6,000.00	Maximum amount	$85.00	Weekly gross earnings
− 2,415.20	Cumulative earnings	× 3.4%	
$3,584.80*	Taxable earnings	$ 2.89	SUT

(*The same rule applies here. If the taxable earnings exceed the gross earnings, the rate is multiplied by the weekly gross earnings.)

Example 15:   David had cumulative earnings prior to this pay period of $5,827.80. His gross pay this week was $425. Based on his pay, what SUT would be owed by his employer? Assume a rate of 3.1%.

$6,000.00	Maximum amount
− 5,827.80	Cumulative earnings
$ 172.20	Taxable earnings
× 3.1%	
$ 5.34	SUT

Jan owns and operates a day-care center. Compute her SUT and FUT, assuming rates of 3.4% and .7%, respectively, on the first $6,000.

18. Cumulative earnings = $5,840
This week's earnings = $ 208

19. Cumulative earnings = $5,917
This week's earnings = $ 189

20. Cumulative earnings = $5,838.71
This week's earnings = $ 315.24

21. Cumulative earnings = $5,801.50
This week's earnings = $ 265.50

THE PAYROLL REGISTER

The Payroll Register is designed to summarize the payroll information at the end of each pay period. The number of hours worked and earnings and deductions are transferred from each of the employees' earnings records. The information on the Payroll Register will be the same as that in the employees' earnings records with the exception of the two columns under the heading "Taxable Earnings." These two columns are used to summarize and total the amounts needed to compute the employer's payroll taxes. There are only two columns needed, one for the FICA taxable earnings, and one for unemployment taxable earnings. (Since the FUT and SUT use the same $6,000 maximum, the single total for the "Unemployment Compensation" column can be used for both the federal rate and the state rate.) A sample Payroll Register is provided below.

Payroll Register

For Week Ending December 12

| Name | Hrs. | Earnings | | | Taxable Earnings | | Deductions | | | | | Net Amount |
		Regular	Over-time	Gross	Unemp. Compen.	FICA	FICA Tax	Inc. Tax	Bonds	Misc.	Total	
Adams, William	30	130.00		130.00		130.00	8.71	9.40	1.50		19.61	110.39
Brown, Samantha	44	140.00	21.00	161.00		161.00	10.79	19.40	1.75	22	53.94	107.06
Cox, Henry	24	69.60		69.60	69.60	69.60	4.66	3.10			7.76	61.84
Davis, David	42	210.00	15.75	225.75		100.00	6.70	20.40			27.10	198.65
Erving, Judith	45	200.00	37.50	237.50				29.80	7.50		37.30	200.20
Fox, Donna	40	170.00		170.00		170.00	11.39	24.90	2.00		38.29	131.71
Garner, Wayne	48	208.00	62.40	270.40	60.00	270.40	18.12	19.20	1.75		39.07	231.33
Totals		1,127.60	136.65	1,264.25	129.60	901.00	60.37	126.20	14.50	22	223.07	1,041.18

Complete the Payroll Register from the information given. A blank form is provided below.

22. The payroll information for eight employees for the week ending November 8 is given as follows. Complete a Payroll Register for that week. The FICA rate is 6.7% on the first $31,800, and overtime is at 1.5 times the regular rate. The FICA rate for the employer is 7.0%.

Name	Cumulative Earnings	Hourly Rate	Hours	Marital Status*	Dependents	Deductions Bonds	Deductions Misc.
Arnold, Jim	$25,615.30	$5.25	47	M	4	$15	$ 3
Benson, Earl	31,530.70	7.20	43	M	6	15	7
Carson, Francis	5,834.11	6.00	40	S	1	10	—
Dawson, Brad	31,788.94	4.50	44½	M	2	12	5
Erskine, George	32,611.17	6.33	41	S	1	—	20
Fulton, Harold	5,935.80	3.75	34	S	0	5	5
Gomez, Carol	13,417.82	4.80	39½	M	5	—	—
Henson, Donna	39,500.00	8.50	40	M	3	20	35

*M = Married, S = Single

Payroll Register

For Week Ending _____

Name	Hrs.	Earnings Regular	Earnings Over-time	Gross	Taxable Earnings Unemp. Compen.	Taxable Earnings FICA	Deductions FICA Tax	Deductions Inc. Tax	Deductions Bonds	Deductions Misc.	Deductions Total	Net Amount
Total												

COMMISSIONS

Salespeople are frequently paid on a commission of their sales rather than by the hour. The commission is determined by multiplying the amount of sales by the commission rate.

Example 16:   Beth has sales of $15,000 and receives a 12% commission. What will be her earnings?

$$\$15,000 \times .12 = \$1,800$$

As an incentive to sell more, a company may use a *sliding scale* of commissions. Be careful to compute each commission, one at a time, and then add them together to get the total commission.

Example 17:   The company used a sliding scale of commissions as follows:

5% on all sales, plus
3% on sales over $10,000, plus
1% on sales over $15,000

If Henry has sales of $18,500, what will be his commission?

$18,500 × 5% =	$925	1st commission
8,500 × 3% =	255	2nd commission
3,500 × 1% =	+ 35	3rd commission
	$1,215	

Salespeople may also be paid a base salary plus commissions. Under this type of pay structure, a certain amount of sales at the beginning of each month is usually not eligible for a commission. After that minimum amount is achieved, the commission rate goes into effect. Any earnings below the minimum level of sales would be due solely to the base salary.

Example 18:   Kathy is paid a base salary of $400 per month and a commission of 5% on all sales over $5,000 each month. This month her sales were $3,950. What were her earnings?

$400	Base salary
+ –0–	Commission (did not reach the minimum amount needed)
$400	Total earnings

Example 19:   Joe Knight is paid a base salary of $600 per month and receives a sliding commission scale as follows:

1% on all sales, plus
3% on sales over $15,000, plus
5% on sales over $25,000

What were his total earnings in a month when his sales totaled $32,000?

$600	Base salary
320	1% commission (1% × $32,000)
510	3% commission (3% × $17,000)
+ 350	5% commission (5% × $ 7,000)
$1,780	Total earnings

Practice Problems

Find the total earnings in each case.

23. The commission scale is:

> 1% on all sales, plus
> 5% on sales over $6,000, plus
> 10% on sales over $12,000

What would be the earnings for sales of $9,800?

24. The commission scale is:

> 3½% on first $2,000, plus
> 7 % on second $2,000, plus
> 10½% on all sales over $8,000

What would be the earnings if sales totaled $10,500?

25. The company pays a base salary of $300 and uses a sliding scale of commissions as follows:

> 3% on the first $12,000, plus
> 5% on sales over $12,000

If Tom has sales of $18,500, what will his earnings be?

The 1977 Social Security Law established the following scheduled tax rates and bases. This law is, of course, subject to change at any time by act of Congress. The 1982 tax base of $31,800 and tax rate of 6.7% was selected for use in this book. Other tax bases and rates may now be in effect. However, the principles and concepts for the computations are the same.

		Employed		Self-employed	
Year	Taxable Wage	Tax Rate	Maximum Tax	Tax Rate	Maximum Tax
1978	$17,700	6.05%	$1,070.85	8.10%	$1,433.70
1979	22,900	6.13	1,403.77	8.10	1,854.90
1980	25,900	6.13	1,587.67	8.10	2,097.90
1981	29,700	6.65	1,975.05	9.30	2,762.10
1982	31,800	6.70	2,130.60	9.35	2,973.30
1983	33,900	6.70	2,271.30	9.35	3,169.65
1984	36,000	6.70	2,412.00	9.35	3,366.00
1985	38,100	7.05	2,686.05	9.90	3,771.90
1986	40,200	7.15	2,874.30	10.00	4,020.00

Practice Problem Solutions

1. 16 + 22 + 24 + 18 + 20 = 100 × $1.16 = $116

2. 312 × 85¢ = $265.20   **3.** $4.24 × 38 = $161.12

4. Regular pay = $5.10 × 40 = $204.00
    Overtime pay = $7.65* ×   6 = <u>+ 45.90</u>   *$5.10 × 1.5 = $7.65
                Gross pay = $249.90

5. $222.74   Straight-time cost (43 hours × $5.18 per hr)
   <u>+   7.77</u>   Overtime bonus [(5.18 × .5) × 3 hours]
   $230.51   Gross pay

6. $244.63   Straight-time (47.5 hours × $5.15)
   <u>+ 19.31</u>   Overtime bonus [($5.15 × .5) × 7.5 hours]
   $263.94   Gross pay

7. $82.70   **8.** $15.80

9. $900 − 860 = $40 × 37% = $14.80 + $208.10 = $222.90

10. $327.85 × 6.7% = $21.97

11. $31,800 − 31,628.72 = $171.28 taxable × 6.7% = $11.48

12. No FICA taxes since the maximum of $31,800 had already been exceeded prior to this week's pay

13.

Employee's Earnings Record

Name: Bill Bailey		Hourly Rate: $7.20							Dependents: 2 (married)	
		Earnings			Deductions					
Period Ending	Hrs.	Regular	Over-time	Gross	FICA Tax	Inc. Tax	Hosp.	Total	Net Amount	Cumulative Earnings
Jan. 7	40	288.00		288.00	19.30	31.50	15	65.80	222.20	288.00
Jan. 14	42	288.00	21.60	309.60	20.74	35.50	15	71.24	238.36	597.60
Jan. 21	39	280.80		280.80	18.81	31.50	15	65.31	215.49	878.40

14.

Employee's Earnings Record

Name: Helen Kramer		Hourly Rate: $4.88							Dependents: 1 (single)	
		Earnings			Deductions					
Period Ending	Hrs.	Regular	Over-time	Gross	FICA Tax	Inc. Tax	Hosp.	Total	Net Amount	Cumulative Earnings
Dec. 13	45	195.20	36.60	231.80	15.53	34.40	10	59.93	171.87	31,741.97
Dec. 20	50	195.20	73.20	268.40	3.89	41.60	10	55.49	212.91	32,010.37
Dec. 27	52	195.20	87.84	283.04		46.40	10	56.40	226.64	32,293.41

15. $872.55 \times 7.0\% = \$61.08$

16. $325.17 \times 7.0\% = \$22.76$

17. $31,800 - 31,590 = \$210$ taxable
$210 \times 7.0\% = \$14.70$

18. $6,000 - 5,840 = \$160$ taxable
$160 \times 3.4\% = \$160 \times .034 = \$5.44$ SUT
$160 \times .7\% = \$160 \times .007 = \$1.12$ FUT

19. $6,000 - 5,917 = \$83$ taxable
$83 \times 3.4\% = \$83 \times .034 = \$2.82$ SUT
$83 \times .7\% = \$83 \times .007 = \$ .58$ FUT

20. $6,000 - 5,838.71 = \$161.29$ taxable
$161.29 \times 3.4\% = \$161.29 \times .034 = \$5.48$ SUT
$161.29 \times .7\% = \$161.29 \times .007 = \$1.13$ FUT

21. $6,000 - 5,801.50 = \$198.50$ taxable
$198.50 \times 3.4\% = \$198.50 \times .034 = \$6.75$ SUT
$198.50 \times .7\% = \$198.50 \times .007 = \$1.39$ FUT

22.

Payroll Register

For Week Ending November 8												
		Earnings			Taxable Earnings		Deductions					
Name	Hrs.	Regular	Over-time	Gross	Unemp. Compen.	FICA	FICA Tax	Inc. Tax	Bonds	Misc.	Total	Net Amount
Arnold, Jim	47	210.00	55.13	265.13		265.13	17.76	20.70	15	3	56.46	208.67
Benson, Earl	43	288.00	32.40	320.40		269.30	18.04	24.10	15	7	64.14	256.26
Carson, Francis	40	240.00		240.00	165.89	240.00	16.08	36.80	10		62.88	177.12
Dawson, Brad	44.5	180.00	30.38	210.38		11.06	.74	18.80	12	5	36.54	173.84
Erskine, George	41	253.20	9.50	262.70				41.60		20	61.60	201.10
Fulton, Harold	34	127.50		127.50	64.20	127.50	8.54	15.90	5	5	34.44	93.06
Gomez, Carol	39.5	189.60		189.60		189.60	12.70	6.00			18.70	170.90
Henson, Donna	40	340.00		340.00				34.70	20	35	94.70	245.30
Totals		1828.50	127.41	1955.71	230.09	1102.59	73.86	203.60	77	75	429.46	1526.25

23.
$9,800 \times 1\% = \$ 98$
$3,800 \times 5\% = + \underline{\ \ 190}$
Total commission $= \ \ \$288$

24.
$2,000 \times 3\frac{1}{2}\% = \$ 70.00$
$2,000 \times 7 \ \% = \ \ 140.00$
$2,500 \times 10\frac{1}{2}\% = + \underline{\ 262.50}$
Total commission $= \ \ \$472.50$

25.
$300 \quad Base salary
$\underline{\ \ 360}$ \quad 3\% \times \$12,000
$+ \underline{\ \ 325}$ \quad 5\% \times \$ 6,500
$985 \quad Total earnings

Exercise 31A: Distribution of Overhead

Name _____

Course/Sect. No. _____   Score_____

Instructions: Distribute the amount given in column 1 according to the ratio given in column 2.

Self-Check Exercise (Answers on back)

	Column 1	Column 2		Column 1	Column 2
(1) _____	**(1)** 4.2	3:2:2	**(2)** $86.90		1:4:6
(2) _____					
(3) _____	**(3)** 5,950	4:4:9	**(4)** $5,220		6:1:3
(4) _____					
(5) _____	**(5)** $333	4:5:6	**(6)** $28,000		5:1:2
(6) _____					
(7) _____	**(7)** 35,248	1:1:2	**(8)** $15,000		9:12:4
(8) _____					

(9) _____   **(9)** The administrative salaries of $62,500 are allocated to two offices in the ratio of 3:1. What is the *difference* between the amounts the two departments would be charged with?

(10) _____   **(10)** The company overhead of $15,600 is to be distributed to the Shoe, Clothing, and Jewelry Departments in the ratio of 2:6:7, respectively. How much *more* would the Jewelry Department receive than the Shoe Department?

(1) _____1.8, 1.2, 1.2_____

(2) _____
$ 7.90
$31.60
$47.40
_____

(3) _____
1,400
1,400
3,150
_____

(4) _____
$3,132
$ 522
$1,566
_____

(5) _____
$ 88.80
$111.00
$133.20
_____

(6) _____
$17,500
$ 3,500
$ 7,000
_____

(7) _____
8,812
8,812
17,624
_____

(8) _____
$5,400
$7,200
$2,400
_____

(9) _____$31,250_____

(10) _____$5,200_____

Exercise 31B: Distribution of Overhead

Name _____

Course/Sect. No. _____   Score_____

Instructions: Distribute the amount given in column 1 according to the ratio given in column 2.

	Column 1	Column 2	Column 1	Column 2
(1) _____	**(1)** 240	1:3:2	**(2)** $16.30	2:3:5
(2) _____				
(3) _____	**(3)** $300	5:7:12	**(4)** 195	1:5:7
(4) _____				
(5) _____	**(5)** 9,900	3:5:1	**(6)** 8,125	8:2:3
(6) _____				
(7) _____	**(7)** 384.8	3:5:8	**(8)** $16,100	1:8:1
(8) _____				

(9) _____

(9) The overhead costs this month of $2,745 are to be allocated to the Cutting, Shaping, and Finishing Departments in the ratio of 6:1:3, respectively. How much would the Finishing Department receive?

(10) _____

(10) The electricity charges of $532 are to be divided among Departments X, Y, and Z in the ratio of 1:4:3, respectively. What amount would department Z receive?

Exercise 32A: Piecework Pay

Name_____

Course/Sect. No._____   Score_____

Instructions: Compute the piecework pay on the following.

	Self-Check Exercise (Answers on back)			
	Units Completed	**Pay Per Unit**	**Units Completed**	**Pay Per Unit**
(1)_____	**(1)** 3,012	3 1/2¢	**(2)** 151	$1.25
(2)_____				
(3)_____	**(3)** 305	$.88	**(4)** 197	62 1/2¢
(4)_____				
(5)_____	**(5)** 94.5	$2.20	**(6)** 55	$3.172
(6)_____				
(7)_____	**(7)** 13	$25.50	**(8)** 217 + 349	67.5¢
(8)_____				

(9)_____

(9) The pitcher was to be paid $3,200 for each win through the first 10, $5,000 for each win from 11 through 15, and $7500 for each win over 15. How much did he earn if he won 23 games and lost 9?

(10)_____

(10) In the previous problem, if he faced 1,320 batters, how much did he earn per batter?

(1) _____$105.42_____

(2) _____$188.75_____

(3) _____$268.40_____

(4) _____$123.13_____

(5) _____$207.90_____

(6) _____$174.46_____

(7) _____$331.50_____

(8) _____$382.05_____

(9) _____$117,000_____

(10) _____$88.64_____

Exercise 32B: Piecework Pay

Name _____

Course/Sect. No. _____   Score_____

Instructions: Compute the piecework pay on the following.

	Units Completed	Pay Per Unit	Units Completed	Pay Per Unit
(1) _____	**(1)** 84	$2.55	**(2)** 800	$.75
(2) _____				
(3) _____	**(3)** 2,417	12.5¢	**(4)** 300 + 185 + 216	$.50
(4) _____				
(5) _____	**(5)** 100	$3.145	**(6)** 6,588	1.5¢
(6) _____				
(7) _____	**(7)** 513 + 184	$.885	**(8)** 1,043	$.90
(8) _____				

(9) _____

(9) Homer has the exciting job of tightening nuts to bolts, for which he is paid $1.80 for each engine completed. If each engine has 12 bolts and he is paid $99, how much does he receive per bolt?

(10) _____

(10) If a baseball pitcher makes $117,000 and pitches 240 innings, how much does he earn per inning?

Exercise 33A: Gross Pay

Name_____

Course/Sect. No._____   Score_____

Instructions: Calculate the gross pay. The overtime rate is 1.5 times the regular rate.

(1) _____	

Self-Check Exercise (Answers on back)

(1) $3.55/hour for 40 hours   **(2)** $7.16/hour for 38.5 hours

(2) _____

(3) $4.93/hour for 48 hours   **(4)** $6.15/hour for 45 hours

(3) _____

(4) _____

(5) $4.50/hour for 41.5 hours   **(6)** $5.15/hour for 34.25 hours

(5) _____

(6) _____

(7) $3.80/hour for 42.75 hours **(8)** $4.10/hour for 49 hours

(7) _____

(8) _____

(9) _____

(9) The normal rate for daytime work is $4.30 per hour. The additional shift differential for night work is $.45 per hour. If the shift worker worked 23 daytime hours and 15.5 nighttime hours, what was the gross pay?

(10) _____

(10) If some worker was paid gross earnings of $184 and had a regular rate of $4 an hour, how many overtime hours did he work?

(1) _____$142_____

(2) _____$275.66_____

(3) _____$256.36_____

(4) _____$292.13_____

(5) _____$190.13_____

(6) _____$176.39_____

(7) _____$167.68_____

(8) _____$219.35_____

(9) _____$172.53_____

(10) _____4_____

Exercise 33B: Gross Pay

Name _____

Course/Sect. No. _____   Score _____

Instructions: Calculate the gross pay. The overtime rate is 1.5 times the regular rate unless indicated otherwise.

(1) _____

(2) _____

(3) _____

(4) _____

(5) _____

(6) _____

(7) _____

(8) _____

(9) _____

(10) _____

(1) $3.40/hour for 35 hours   **(2)** $3.62/hour for 43 hours

(3) $5.00/hour for 42.25 hours   **(4)** $3.90/hour for 39.5 hours

(5) $10.12/hour for 40 hours   **(6)** $3.99/hour for 40 hours

(7) $4.75/hour for 44.5 hours   **(8)** $3.62/hour for 27.5 hours

(9) The workers worked 43, 24, 41.5, and 37 hours during the week. If the pay rate was $4.20 per hour and double time was paid for overtime, what *total* gross amount was earned by the workers?

(10) What would be the gross pay for a worker with a regular rate of $5.15 an hour if she worked 54.25 hours during the week?

Exercise 34A: Withholding Taxes

Name _____

Course/Sect. No. _____   Score _____

Instructions: Using the tax tables in your text, find the weekly withholding tax (income tax).

Self-Check Exercise (Answers on back)

	Marital Status	Earnings	Withholding Allowances
(1) _____	(1) Single	$302	2
(2) _____	(2) Married	$140.50	0
(3) _____	(3) Married	$250	1
(4) _____	(4) Single	$318.21	1
(5) _____	(5) Married	$1,200	5
(6) _____	(6) Single	$700	1
(7) _____	(7) Single	$400	0
(8) _____	(8) Married	$731.25	4

(9) _____   **(9)** Assume a gross pay of $200. What percentage of gross pay would be deducted from a single person claiming 1 withholding allowance? (Round to a tenth of 1%.)

(10) _____   **(10)** The married couple with two children earned $300 and $475 together. Find the *combination* of deductions that would yield the *lowest total* income tax for the couple. (Example: $300 with 1 and $475 with 3; $300 with 2 and $475 with 2; etc.) Enter this lowest amount on the answer line.

(1) _____$46.60_____

(2) _____$13.50_____

(3) _____$29.40_____

(4) _____$54.90_____

(5) _____$312.60_____

(6) _____$189.50_____

(7) _____$88.90_____

(8) _____$147.70_____

(9) _____13.6%_____

(10) _____$110.60_____

Name_____

Course/Sect. No._____   Score_____

Instructions: Using the tax tables in your text, find the weekly withholding tax (income tax).

	Marital Status	Earnings	Withholding Allowances
(1) _____	**(1)** Single	$140.50	0
(2) _____	**(2)** Single	$250	1
(3) _____	**(3)** Married	$302	2
(4) _____	**(4)** Single	$425	0
(5) _____	**(5)** Married	$900	2
(6) _____	**(6)** Single	$550	1
(7) _____	**(7)** Single	$1,000	2
(8) _____	**(8)** Married	$425.31	3

(9) _____   **(9)** The single worker earned $200 this week. If no withholding allowances were claimed, what percentage of the gross pay was deducted for income taxes? (Round to a tenth of 1%.)

(10) _____   **(10)** A married worker earning $500 and claiming 4 withholding allowances would pay how much in income taxes per dollar? Round to cents.

Exercise 35A: FICA Taxes

Name _____

Course/Sect. No. _____   Score _____

Instructions: Compute the FICA taxes using 6.7% of the first $31,800.

Self-Check Exercise (Answers on back)

	Cumulative Earnings Prior to This Week	Earnings This Week
(1) _____	**(1)** $2,140.20	$238.17
(2) _____	**(2)** $28,490.30	$417.73
(3) _____	**(3)** $38,400.00	$600.00
(4) _____	**(4)** $16,573.84	$335.00
(5) _____	**(5)** $31,820.92	$100.00
(6) _____	**(6)** $525.71	$310.55
(7) _____	**(7)** $31,713.80	$416.70
(8) _____	**(8)** $28,900.40	$1,000.00

(9) _____

(9) If the payroll bookkeeper mistakenly deducted 7.6% for FICA taxes instead of 6.7%, what amount should be returned to the worker who had FICA taxable earnings of $120?

(10) _____

(10) In the first payroll period of the year you work 4 hours overtime at 1.5 times your regular rate of $6.50 per hour. How much FICA taxes do you pay *on overtime earnings?*

(1) ___$15.96___

(2) ___$27.99___

(3) ___–0–___

(4) ___$22.45___

(5) ___–0–___

(6) ___$20.81___

(7) ___$5.78___

(8) ___$67___

(9) ___$1.08___

(10) ___$2.61___

Exercise 35B: FICA Taxes

Name _____

Course/Sect. No. _____   Score _____

Instructions: Compute the FICA taxes using 6.7% of the first $31,800.

	Cumulative Earnings Prior to This Week	Earnings This Week
(1) _____	(1) – 0 –	$515.30
(2) _____	(2) $10,431.82	$691.80
(3) _____	(3) $31,750.25	$500.00
(4) _____	(4) $2,120.35	$189.45
(5) _____	(5) $31,699.15	$475.80
(6) _____	(6) $30,416.20	$850.00
(7) _____	(7) $16,311.94	$416.70
(8) _____	(8) $31,689.92	$416.70

(9) _____

(9) If the "taxable earnings" for FICA purposes on John Doe were $187.50 and the gross pay was $435.20, what cumulative earnings did he have *before* this week?

(10) _____

(10) Some workers this week paid FICA taxes on their entire earnings, some on only part of their earnings, and some paid no FICA taxes. If the total deducted from everyone was $78.30 and the total FICA taxable earnings were $1,350, what average rate was paid?

Exercise 36A: Employee's Earnings Record

Name _____

Course/Sect. No. _____   Score _____

Instructions: Complete the Employee's Earnings Record for Becky Jones. Assume the cumulative earnings prior to the October 6 week was $31,235.25. Use 6.7% on the first $31,800 earned for FICA taxes. Ms. Jones is married and claims 1 dependent.

Self-Check Exercise (Answer on back)

Employee's Earnings Record

Name: Becky Jones   Hourly Rate: $5.05   Dependents: 1

	Period Ending	Hrs.	Earnings			Deductions				Net Amount	Cumulative Earnings
			Regular	Overtime	Gross	FICA Tax	Inc. Tax	Hosp.	Total		
(1)	Oct. 6	40	202.00	—				10.00			
(2)	Oct. 13	39	196.95	—				10.00			
(3)	Oct. 20	42	202.00	15.15				10.00			
(4)	Oct. 27	40	202.00	—				10.00			
(5)	Monthly Total		802.95	15.15				40.00			
(6)	Nov. 3	45	202.00	37.88				10.00			
(7)	Nov. 10	43	202.00	22.73				10.00			
(8)	Nov. 17	40	202.00	—				10.00			
(9)	Nov. 24	40	202.00	—				10.00			
(10)	Monthly Total		808.00	60.61				40.00			

(1)	202.00	13.53	20.30	43.83	158.17	31,437.25
(2)	196.95	13.20	18.70	41.90	155.05	31,634.20
(3)	217.15	11.11	21.90	43.01	174.14	31,851.35
(4)	202.00	—	20.30	30.30	171.70	32,053.35
(5)	818.10	37.84	81.20	159.04	659.06	
(6)	239.88	—	25.40	35.40	204.48	32,293.23
(7)	224.73	—	23.50	33.50	191.23	32,517.96
(8)	202.00	—	20.30	30.30	171.70	32,719.96
(9)	202.00	—	20.30	30.30	171.70	32,921.96
(10)	868.61	—	89.50	129.50	739.11	

Exercise 36B: Employee's Earnings Record

Name_____

Course/Sect. No. _____   Score_____

Instructions: Complete the Employee's Earnings Record for John Smith. Assume the cumulative earnings prior to the October 6 week was $30,990.62. Use 6.7% on the first $31,800 earned for FICA taxes. Mr. Smith is married and claims 5 dependents.

Employee's Earnings Record

Name: John Smith   Hourly Rate: $4.55   Dependents: 5

	Period Ending	Hrs.	Earnings		Deductions					Net Amount	Cumulative Earnings
			Regular	Overtime	Gross	FICA Tax	Inc. Tax	Hosp.	Total		
(1)	Oct. 6	40	182.00	—				15.00			
(2)	Oct. 13	40	182.00	—				15.00			
(3)	Oct. 20	42	182.00	13.65				15.00			
(4)	Oct. 27	39.5	179.73	—				15.00			
(5)	Monthly Total		725.73	13.65				60.00			
(6)	Nov. 3	44	182.00	27.30				15.00			
(7)	Nov. 10	41.5	182.00	10.24				15.00			
(8)	Nov. 17	38	172.90	—				15.00			
(9)	Nov. 24	40	182.00	—				15.00			
(10)	Monthly Total		718.90	37.54				60.00			

Exercise 37A: Unemployment Taxes

Name_____

Course/Sect. No._____   Score_____

Instructions: Find the correct amounts for FUT (.7% on first $6,000) and SUT (2.4% on first $6,000).

	Self-Check Exercise (Answers on back)	
	Earnings This Week	**Cumulative Earnings Before This Week**
	$620.00	$8,125.60

(1) _____

    (1) FUT
    (2) SUT

(2) _____

(3) _____

    $143.56 $4,000.31

    (3) FUT
    (4) SUT

(4) _____

(5) _____

    $204.77 $5,863.14

    (5) FUT
    (6) SUT

(6) _____

    $525.00 $6,031.92

(7) _____

    (7) FUT
    (8) SUT

(8) _____

(9) _____

(9) If the total unemployment payroll taxes were $32.50 from taxable earnings of $1,250, what was the SUT rate?

(10) _____

(10) If the *total* payroll taxes totaled $396.40, what percentage did the owner pay if the total payroll was $5100? (Round to a tenth of 1%.)

(1) _____ –0– _____

(2) _____ –0– _____

(3) _____ $1.00 _____

(4) _____ $3.45 _____

(5) _____ $.96 _____

(6) _____ $3.28 _____

(7) _____ –0– _____

(8) _____ –0– _____

(9) _____ 1.9% _____

(10) _____ 7.8% _____

Exercise 37B: Unemployment Taxes

Name _____

Course/Sect. No. _____   Score _____

Instructions: Find the correct amounts for FUT (.7% on first $6,000) and SUT (2.4% on first $6,000).

		Earnings This Week	Cumulative Earnings Before This Week
(1) _____	**(1)** FUT **(2)** SUT	$397.70	$5,712.50
(2) _____			
(3) _____	**(3)** FUT **(4)** SUT	$417.80	$5,909.88
(4) _____			
(5) _____	**(5)** FUT **(6)** SUT	$99.50	$3,177.85
(6) _____			
(7) _____	**(7)** FUT **(8)** SUT	$475.80	$5,621.20
(8) _____			

(9) _____

(9) The total for unemployment payroll taxes was $32.50 based on taxable earnings of $1,250. How much of the $32.50 was for SUT?

(10) _____

(10) The company has a SUT rate of 3.1%. The total payroll taxable for FICA this week is $4,300 and for unemployment taxes is $2,850. Using the FICA tax rate used in the chapter, find the total payroll taxes the employer would have to pay.

Exercise 38A: The Payroll Register

Name_____

Course/Sect. No. _____   Score_____

Instructions: Complete the Payroll Register for the week of December 8 using the form provided and the following information. The cumulative earnings were through December 1. Use an FICA tax rate of 6.7% on the first $31,800. Using a rate of .7% on the first $6,000 for FUT and 2.7% on the first $6,000 for SUT, find the FUT and SUT.

Employee No.	Hours	Hourly Rate	Marital Status*	Dependents	Cumulative Earnings	Deductions Hosp.	U.F.	Other
101	43	$3.80	M	3	$31,620.30	15	5	—
103	37	4.00	S	1	5,875.33	—	10	—
105	52	6.69	M	6	31,714.38	10	—	25
107	20	4.40	S	1	3,892.90	—	—	—
109	30	3.90	S	1	5,908.70	5	—	—

*M = married, S = single

Payroll Register

For Week Ending December 8

	Employee	Hrs.	Earnings Regular	Over-time	Gross	Taxable Earnings Unemp. Compen.	FICA	FICA Tax	Inc. Tax	Deductions Hosp.	U.F.	Other	Total	Net Amount
(1)														
(2)														
(3)														
(4)														
(5)														
(6)	Total													

(7) FUT = ____
(8) SUT = ____

(1)	101	43	152.00	17.10	169.10	—	169.10	11.33	8.60	15.00	5.00	—	39.93	129.17
(2)	103	37	148.00	—	148.00	124.67	148.00	9.92	16.00	—	10.00	—	35.92	112.08
(3)	105	52	267.60	120.42	388.02	—	85.62	5.74	36.10	10.00	—	25.00	76.84	311.18
(4)	107	20	88.00	—	88.00	88.00	88.00	5.90	5.90	—	—	—	11.80	76.20
(5)	109	30	117.00	—	117.00	91.30	117.00	7.84	10.40	5.00	—	—	23.24	93.76
(6)			772.60	137.52	910.12	303.97	607.72	40.73	77.00	30.00	15.00	25.00	187.73	722.39

$2.13 303.97 × .7%
$8.21 303.97 × 2.7%

Exercise 38B: The Payroll Register

Name_____

Course/Sect. No._____   Score_____

Instructions: Complete the Payroll Register for the week of December 8 using the form provided and the following information. The cumulative earnings were through December 1. Use an FICA tax rate of 6.7% on the first $31,800. Using a rate of .7% on the first $6,000 for FUT and 2.7% on the first $6,000 for SUT, find the FUT and SUT.

Employee No.	Hourly Hours	Marital Rate	Status*	Cumulative Dependents	Earnings	Deductions Hosp.	U.F.	Other
102	47.5	$3.52	M	4	$33,607.25	18	5	20
104	40	3.65	S	0	12,415.71	5	—	15
106	40	3.88	M	4	31,693.22	5	10	5
108	45.5	5.16	M	5	31,700.40	10	10	10
110	36.5	3.75	M	2	5,884.12	—	5	10
			*M = married, S = single					

Payroll Register

For Week Ending December 8

	Employee	Hrs.	Earnings Regular	Over-time	Gross	Taxable Earnings Unemp. Compen.	FICA	FICA Tax	Inc. Tax	Deductions Hosp.	U.F.	Other	Total	Net Amount
(1)														
(2)														
(3)														
(4)														
(5)														
(6)	Total													

(7) FUT = ____

(8) SUT = ____

Exercise 39A: Commissions

Name _____

Course/Sect. No. _____   Score _____

Instructions: Find the total earnings for each of the following, using the given sliding scales for commissions.

10% on all sales, plus 20% on sales over $15,000, plus 30% on sales over $30,000.

(1) _____

(1) If sales are $17,500:   **(2)** If sales are $50,500:

(2) _____

15% for the first $10,000 in sales, plus 20% for the second $10,000 in sales, plus 30% on sales over $20,000.

(3) _____

(3) If sales are $8,000:   **(4)** If sales are $70,000:

(4) _____

1½% on all sales, plus 3% on sales over $100,000, plus a bonus of $5,000 if sales total $200,000 or more.

(5) _____

(5) If sales are $340,000:   **(6)** If sales are $1,500,000:

(6) _____

(1) _____$2,250_____

(2) _____$18,300_____

(3) _____$1,200_____

(4) _____$18,500_____

(5) _____$17,300_____

(6) _____$69,500_____

Exercise 39B: Commissions

Name_____

Course/Sect. No._____   Score_____

Instructions: Find the total earnings for each of the following, using the given sliding scales for commissions.

---

(1) _____

(2) _____

(3) _____

(4) _____

(5) _____

(6) _____

10% on all sales, plus 20% on sales over $15,000, plus 30% on sales over $30,000.

(1) If sales are $28,000:   **(2)** If sales are $35,000:

15% for the first $10,000 in sales, plus 20% for the second $10,000 in sales, plus 30% on sales over $20,000.

(3) If sales are $15,000:   **(4)** If sales are $25,000:

1½% on all sales, plus 3% on sales over $100,000, plus a bonus of $5,000 if sales total $200,000 or more.

(5) If sales are $525,220:   **(6)** If sales are $250,000:

Test Yourself—Unit 2

(1) Convert the following as shown (supply the correct answer on each blank line).

	Fraction	Decimal	Percent
(A)	17/20	_____	_____
(B)	_____	3.05	_____
(C)	_____	_____	15 1/8%

(2) What percentage of 20,000 is 500?

(3) What number is 105% of $42?

(4) 136 is 85% of what number?

(5) The net profit in 1983 was $125,000, and in 1984, it increased to $177,000. What percentage of increase does that represent? Round to one-tenth of 1%.

(6) Company A rents a car for $20 per day, which is 2% less than the rate for the same type of car rented by Company B. How much does it cost to rent the Company B car for a day?

(7) Find the ratio of $16,500 to $57,500, expressed "to 1." Round to one-tenth.

(8) Distribute the overhead costs of $810 in the ratio of 1:4:2:2.

(9) Compute the gross pay for someone making $4.25 an hour and receiving time and one-half for all hours over 40 if he worked 44 hours this week. Compute it by the overtime rate approach and by the overtime premium approach.

(10) Compute the net pay if a married man claiming 4 dependents is paid $425 a week and his cumulative earnings prior to this week are $31,450. Assume FICA taxes are 6.7% based on the first $31,800 and other deductions were $25 for United Fund and $39 for a credit union payment.

(11) An employee had cumulative earnings prior to this week of $31,772.39. The earnings this week were $175.75. Find the employer's FICA tax on this employee. (Use 7.0% on the first $31,800.)

(12) Compute the FUT and the SUT given cumulative earnings of $5,850.20 prior to this week. The weekly earnings were $300. (Use .7% and 3.4%, respectively, on the first $6,000.)

(13) Calculate the commission on the following scale for sales of $15,575.

8% on the first $10,000, plus

12% on the second $10,000, plus

5% on sales above $15,000

TEST YOURSELF ANSWERS

(1) **(A)** Decimal = .85

Percentage = 85%

(B) Fraction = 3 1/20

Percentage = 305%

(C) Fraction = 121/800

Decimal = .15125

(2) 2.5%   (3) $44.10   (4) 160   (5) 41.6%   (6) $20.41   (7) .3 to 1

(8) $90, $360, $180, $180

(9) Overtime rate:

Regular pay	=	$170.00
Overtime pay	= +	25.50
Gross pay	=	$195.50

Overtime premium:

Straight time	=	$187.00
Overtime bonus	= +	8.50
Gross pay	=	$195.50

(10) $283.15   (11) $1.93   (12) FUT = $1.05   SUT = $5.09

(13) $1,497.75

Unit

3

Interest

A Preview of What's Next

It is going to cost you approximately $600 more than you now have to go to college next year. Would borrowing the money at 10% simple interest for 12 months be better than arranging an installment loan of $600 at 9% to be paid in 12 installments?

Your father asks you to help him determine how much he would have to invest at one time in an account that would guarantee him $20,000 when he is 65 years old. He is now 42, and the best current rate is 8% compounded quarterly. Could you help him?

You average owing $350 on a revolving charge account at 1.5% interest per month on the unpaid balance. Your savings account is earning you 5 3/4% compounded quarterly on $500. Are you gaining or losing money in interest each month?

A young couple decide they can afford the cost of a new home they have been wanting. Assuming they can handle the down payment, insurance, and taxes, will their budgeted amount for house payments of $850 per month be enough for the $88,000 home they want? (The house is $88,000 after the down payment, and the best mortgage interest they can get is 12% for 25 years.)

Objectives

Interest is the price paid for the use of money. The amount paid depends on the amount borrowed or loaned, the rate charged (expressed as a percentage), and the length of time the money is used. Time may be expressed in days, months, or years. The rate is an annual charge unless stated otherwise. When you have successfully completed this unit, you will be able to:

1. Compute the exact time for which a loan was made and determine the due date.

2. Compute the interest and maturity value for both ordinary and exact interest problems by using the formula $I = PRT$.

3. Determine the maturity value.

4. Find the value of the missing component of the formula $I = PRT$ when three of the four components are known.

5. Calculate the discount and proceeds of a discounted interest-bearing note.

6. Compute compound interest and compound amount.

7. Compute present value.

8. Compute the investment balance for an ordinary annuity.

9. Determine the true, or effective, rate of interest on installment loans and purchases.

10. Calculate the rebate of interest on installment loans and purchases.

11. Compute the finance charge (carrying charge) and amount due on charge accounts.

12. Prepare a home-loan repayment schedule (amortization schedule).

Chapter

Simple Interest Loans and Discounted Loans

Preview of Terms

Discount   In the context of interest, the value obtained when the maturity value, discount rate, and time are multiplied together (discount = maturity value × discount rate × time). When a promissory note is discounted at a bank, it is endorsed and "cashed." The bank deducts the interest (called the *discount*) from the maturity value of the note.

Discount Period   The exact number of days between the discount date and the maturity date.

Maturity Date   The date on which a note is due to be paid.

Maturity Value   The sum of the principal and the interest for a note (maturity value = principal + interest).

Proceeds   The difference between the maturity value and the discount (proceeds = maturity value − discount).

Borrowing money has become an integral part of our economy, and each of you will almost certainly at some time find yourself paying interest for the privilege of using someone else's money.

When money is borrowed, a formal I.O.U. is signed to witness the debt. This I.O.U. is called a *promissory note.* It tells the amount borrowed, the date the money was borrowed, the rate of interest being charged, and the length of time the money is borrowed. It must also be signed. An example of a promissory note is shown on the next page, with the parts identified.

Date *February 4, 19xx*   *$2,000⁰⁰*

Sixty Days from the date I promise to pay to the

order of *THE FIRST NATIONAL BANK OF CAMERON*

the sum of *Two Thousand and* ⁿ°⁄₁₀₀ ~~~ dollars

with interest at *10.5%*

                        Signed *Homer Walker*

Promissory Note

Promissory Note

Maker:	Homer Walker
Payee:	The First National Bank of Cameron
Face value:	$2,000
Date of loan:	February 4, 19xx
Term of loan:	Sixty days
Interest rate:	10.5%

DUE DATES

To determine the due date, you must know the number of days in each month. Then, you simply calculate, month by month, the number of days from one date to the next.

Days in Each Month

January = 31	April = 30	July = 31	October = 31
February = 28	May = 31	August = 31	November = 30
March = 31	June = 30	September = 30	December = 31

Note *Since leap year occurs only one year out of four, all of the problems given will assume it is not a leap year.*

Example 1: When would a note for 90 days dated April 3 be due?

Step 1: Find how many days are left in the initial month (subtract the date of the note from the days in the month).

April	=	30	Days in month
Date of note	=	− 3	(April 3)
		27	Days left in April

Step 2: Add the number of days in each succeeding month until you have the desired total.

April =	27	Days left in April
May =	+31	Days in May
	58	Days from April 3
June =	+30	Days in June
	88	Days from April 3
July =	+ 2	Days needed in July (*Note due on July 2*)
	90	Days from April 3

Example 2: When would a note for 60 days dated December 18 be due?

December =   31   Days in month
         −18   Date of note
          13   Days left in December
         +31   Days in January
          44   Days from December 18
         +16   Days in February needed (*Note due on February 16*)
          60   Days from December 18

DAYS BETWEEN DATES

To calculate the number of days between two dates, add the days in each month between the two dates.

Example 3: How many days are there between January 8 and May 8?

Step 1: Find how many days are left in the initial month (the same as Step 1 of Example 1).

January   =   31   Days in month
Date of note = − 8   (January 8)
               23   Days left in January

Step 2: Add the days for each succeeding month to the days that are left in the initial month.

     23   Days left in January
     28   Days in February
     31   Days in March
     30   Days in April
   + 8   Days in May (*Note due on May 8*)
   120   Days in note

Example 4: How many days are there between November 25 and March 25?

November =   30   Days in month
         −25   Date of note
           5   Days left in November
          31   Days in December
          31   Days in January
          28   Days in February
         +25   Days in March needed
         120   Days in note

Example 5: When the time on a note is stated in months, the note falls due on the same day as the date of the note. If there are not enough days in the month, it is due on the *last day* of that month.

(A) Find the due date on a three-month note dated June 15.

The note would be due on the 15th, three months later. In this case, it would be September 15.

(B) Find the due date on a three-month note dated November 30.

Normally, the note would be due on the 30th, three months later, but that would make it fall due on February 30. Since February does not have 30 days, the note would be due on the last day of February—February 28.

USING A TABLE

Due dates and days between two dates are easier to calculate if you have access to a table for the days in a year (see Table 6.1). When you know the date of a note and length of the note, find the day of the year the note is dated, and add the number of days in the note. The sum will be the day of the year the note will become due.

Table 6.1 Days in a Year*

Day of Month	Jan.	Feb.	Mar.	Apr.	May	June	July	Aug.	Sept.	Oct.	Nov.	Dec.	Day of Month
1	1	32	60	91	121	152	182	213	244	274	305	335	1
2	2	33	61	92	122	153	183	214	245	275	306	336	2
3	3	34	62	93	123	154	184	215	246	276	307	337	3
4	4	35	63	94	124	155	185	216	247	277	308	338	4
5	5	36	64	95	125	156	186	217	248	278	309	339	5
6	6	37	65	96	126	157	187	218	249	279	310	340	6
7	7	38	66	97	127	158	188	219	250	280	311	341	7
8	8	39	67	98	128	159	189	220	251	281	312	342	8
9	9	40	68	99	129	160	190	221	252	282	313	343	9
10	10	41	69	100	130	161	191	222	253	283	314	344	10
11	11	42	70	101	131	162	192	223	254	284	315	345	11
12	12	43	71	102	132	163	193	224	255	285	316	346	12
13	13	44	72	103	133	164	194	225	256	286	317	347	13
14	14	45	73	104	134	165	195	226	257	287	318	348	14
15	15	46	74	105	135	166	196	227	258	288	319	349	15
16	16	47	75	106	136	167	197	228	259	289	320	350	16
17	17	48	76	107	137	168	198	229	260	290	321	351	17
18	18	49	77	108	138	169	199	230	261	291	322	352	18
19	19	50	78	109	139	170	200	231	262	292	323	353	19
20	20	51	79	110	140	171	201	232	263	293	324	354	20
21	21	52	80	111	141	172	202	233	264	294	325	355	21
22	22	53	81	112	142	173	203	234	265	295	326	356	22
23	23	54	82	113	143	174	204	235	266	296	327	357	23
24	24	55	83	114	144	175	205	236	267	297	328	358	24
25	25	56	84	115	145	176	206	237	268	298	329	359	25
26	26	57	85	116	146	177	207	238	269	299	330	360	26
27	27	58	86	117	147	178	208	239	270	300	331	361	27
28	28	59	87	118	148	179	209	240	271	301	332	362	28
29	29		88	119	149	180	210	241	272	302	333	363	29
30	30		89	120	150	181	211	242	273	303	334	364	30
31	31		90		151		212	243		304		365	31

*For leap years, add 1 to each number in the table after February 28.

From Michael L. Kovacic, *Business and Consumer Mathematics,* Table 1, Appendix D, page 356. Allyn and Bacon, 1978.

Example 6: A 90-day note is dated May 15. When is the due date?

May 15 is the 135th day of the year, and the note is due 90 days later.

$$
\begin{array}{rcl}
\text{May 15} &=& 135 \\
&& +\ 90 \\
\hline
\text{August 13} &=& 225
\end{array}
$$

(The note falls due on the 225th day, which is August 13.)

When you know the date of the note and the due date, locate the day of the year for each date and subtract the date of the note from the due date. This will give you the number of days in the note.

Example 7: A note dated March 21 is due to be paid September 21. How many days will the note run?

$$
\begin{array}{rcl}
\text{September 21} &=& \text{264th day} \\
\text{March 21} &=& -\ \text{80th day} \\
\hline
&& \text{184-day note}
\end{array}
$$

Practice Problems

Use Table 6.1 to find each of the following.

6. The due date of a note for 120 days dated January 15.

7. The number of days in a note dated April 17 due on October 17.

BASIC SIMPLE INTEREST FORMULA

The basic interest formula is as follows:

$$I = PRT$$

where: I = Interest—The amount paid for the use of money
P = Principal—The amount borrowed
R = Rate—The percent or interest rate
T = Time—The length of time money is borrowed

Note: *All interest problems are computed using time in relation to one year. A year can be stated as 1, as 12 for months, and as 365 for days. The number of years, months, or days in the note is always expressed as a ratio to one year.*

Example 8: Use the interest formula and substitute in the values P = \$2,000, R = 8%, and T = 3 years.

$$
\begin{aligned}
I &= PRT \\
&= \$2,000 \times .08 \times \frac{3}{1} \\
&= \$480
\end{aligned}
$$

Note: *Remember, when there are no signs between letters in an equation (for example, PRT), multiplication is understood.*

Example 9: Use the interest formula and substitute in the following values: $P = \$300$. $R = 12\frac{1}{2}\%$, and $T = 3$ months.

$$I = PRT$$

$$= \$300 \times .125 \times \frac{3}{12}$$

$$= \$9.38$$

Note: *In this example, time is expressed in months; therefore, the ratio is set up over 12, representing the number of months in one year.*

Example 10: Use the interest formula and substitute in the values $P = \$850$, $R = 11\%$, and $T = 60$ days.

$$I = PRT$$

$$= \$850 \times .11 \times \frac{60}{365}$$

$$= \$15.37$$

Note: *Since the time in this example is expressed in days, the ratio is set up over 365, representing the number of days in a year.*

ORDINARY AND EXACT INTEREST

The interest calculated when a year is expressed as 365 days is called *exact interest*. Interest is called *ordinary interest* when it is based on a year arbitrarily set at 360 days. The 360-day year is widely used in business. In all of the remaining problems in this book assume ordinary interest (360 days) unless stated otherwise.

Example 11: How much interest will be charged when $700 is borrowed at 12% for 60 days?

$$I = PRT$$

$$= \$700 \times .12 \times \frac{60}{360}$$

$$= \$14$$

Example 12: How much interest will be charged when $1,500 is borrowed at 13.5% for 90 days? Compute *exact* interest.

$$I = PRT$$

$$= \$1,500 \times .135 \times \frac{90}{365}$$

$$= \$49.93$$

Practice Problems

Find the amount of interest on each note.

8. A 60-day, $400, 15% note.   **9.** An $8,000, 12½%, 120-day note.

10. A $1,550, 11.75%, 90-day note (exact interest).

The money borrowed—the *principal*—and the interest must be repaid on the due date—the *maturity date.* The principal, or amount stated on the note, is also called the *face value* of the note. The principal and interest together are called the *maturity value (MV).*

$$MV = P + I$$

Maturity value = Principal + Interest

Example 13:   Earl borrows $500 for 120 days at 11.5%. How much will he have to pay at the end of 120 days?

$$I = PRT \qquad\qquad MV = P + I$$

$$= \$500 \times .115 \times \frac{120}{360} \qquad = \$500 + \$19.17$$

$$= \$19.17 \qquad\qquad = \$519.17$$

Practice Problems

11. Alice loaned Dorothy $500 at 8.5% for two years. What did Dorothy have to pay at the end of two years (what was the maturity value)?

12. Sandy borrowed $2,500 from Lil at 10.5% for six months. Find the maturity value.

OTHER INTEREST FORMULAS

Rather than memorize four formulas to handle interest problems, simply solve for the unknown and substitute into the equation the known values.

Example 14:   How much was borrowed at 9% for 60 days if the interest was $15? (Here, everything is known except the principal, *P.* Using the basic simple interest formula, you can therefore solve for *P.*)

$$I = PRT$$

$$\frac{I}{RT} = \frac{PRT}{RT}$$

$$P = \frac{I}{RT}$$

Note: *You now have the formula for P, and can substitute into this formula to solve for P.*

$$P = \frac{I}{RT}$$

$$= \frac{\$15}{.09 \times \frac{60}{360}}$$

$$= \frac{\$15}{.015}$$

$$= \$1,000$$

Example 15:   What interest rate would it take for $1,000 to earn $15 in 60 days? The rate, *R*, is unknown in this problem.

$$I = PRT$$

$$\frac{I}{PT} = \frac{PRT}{PT}$$

$$R = \frac{I}{PT}$$

Note: *You now have the formula for R, and can substitute into this formula to solve for R.*

$$R = \frac{I}{PT}$$

$$= \frac{\$15}{\$1,000 \times \dfrac{60}{360}}$$

$$= \frac{\$15}{\$166.66666*}$$

$$= .09, \text{ or } 9\%$$

(*In your calculator, when you multiplied $1,000 by 60 and divided by 360, you got 166.66666. Add that to "memory" in your calculator and divide it into $15.)

Example 16:   How long (in days) would it take $1,000 to earn interest of $15 at 9%? The time, *T*, is unknown in this problem.

$$I = PRT$$

$$\frac{I}{PR} = \frac{PRT}{PR}$$

$$T = \frac{I}{PR}$$

Note: *You now have the formula for T, and can substitute into this formula to solve for T.*

$$T = \frac{\$15}{\$1,000 \times .09}$$

$$= \frac{\$15}{\$90}$$

$$= .1666666*$$

*The solution for time in interest problems will be in a ratio to one year. To convert that to days, multiply by 360 days (365 days for exact interest). Always round your final answer to an even day.

$$T = .1666666 \times 360 \text{ days}$$

$$= 59.999976 \text{ days}$$

$$= 60 \text{ days}$$

DISCOUNTS AND PROCEEDS

A note is a negotiable instrument like a check. By endorsing it, you can transfer it to another person. The original amount owed (the maturity value) can therefore be paid to anyone who holds the note on the maturity date. The payer of the note is not affected by a change in ownership since he has agreed to pay a certain amount, and that amount is not altered with changes in ownership.

In some instances, the person due to receive payment on a note would rather not wait for the maturity date to receive his money. Since the note is negotiable, he can sell (*discount*) it to a commercial bank for partial payment of the maturity value. The payer would then make payment to the bank on the maturity date after the bank had paid a sum called the *proceeds* on the discount date. The formula for computing the discount is very similar to the simple interest formula.

$$D = (MV)(DR)(T)$$

Discount = (Maturity value)(Discount rate)(Time)

where: D = The amount the bank receives for discounting the note
MV = The amount the note is worth at maturity
DR = The interest rate the bank charges for discounting
T = The length of time the bank holds the note before maturity

Interest is earned only for the length of time the note is held. When a note is discounted, the original time is shared by the payee (the first owner) and the bank. The portion of the maturity value that the payee receives from the bank is called the *proceeds.* The following is the formula for the proceeds:

$$p = MV - D$$

Proceeds = Maturity value − Discount

In computing the discount and proceeds, the first two steps are the same as the steps for finding the maturity value. The complete process requires four steps:

Step 1: Find the interest.

$$I = PRT$$

Step 2: Find the maturity value.

$$MV = P + I$$

Step 3: Find the discount.

$$D = (MV)(DR)(T)$$

Step 4: Find the proceeds.

$$p = MV - D$$

Example 17: Alice borrows $800 for 60 days at 12% from Jim. After 40 days, Jim takes the note to the bank and discounts it at 14%. Find the discount and the proceeds.

Step 1: $I = PRT$

$$= \$800 \times .12 \times \frac{60}{360}$$

$$= \$16$$

Step 2: $MV = P + I$

$$= \$800 + \$16$$

$$= \$816$$

Step 3: $D = (MV)(DR)(T)$

$$= \$816 \times .14 \times \frac{20}{360}$$

$$= \$6.35$$

Step 4: $p = MV - D$

$$= \$816 - \$6.35$$

$$= \$809.65$$

Note: *In Step 3, you must be careful to use the maturity value rather than the original principal and the bank's discount rate rather than the original rate stated on the note. In addition, the number of days over 360 for time should always be the number of days between the discount date and the maturity date—the number of days the bank holds the note. In Step 4, you find the proceeds (the amount Jim receives from the bank on the discount date) by subtracting the discount (the bank's share) from the maturity value. On a time line, the information can be illustrated as follows:*

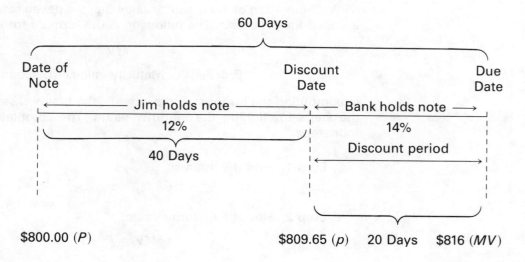

Summary

For Alice		For Jim		For the Bank	
Pays (MV) =	$816	Receives (p) =	$809.65	Receives (MV) =	$816.00
Borrows (P) =	− 800	Loans (P) =	− 800.00	Pays (p) =	− 809.65
Interest (I) =	$16	Earns =	$9.65	Earns (D) =	$6.35

Example 18:

A $5,000 note for three months at 7.5% is dated June 17. It is discounted on July 17 at 8%. Find the discount and proceeds.

Step 1: $I = PRT$

$$= \$5{,}000 \times .075 \times \frac{3}{12}$$

$$= \$93.75$$

Step 2: $MV = P + I$

$$= \$5{,}000 + \$93.75$$

$$= \$5{,}093.75$$

Step 3: $D = (MV)(DR)(T)$

$$= \$5{,}093.75 \times .08 \times \frac{62}{360}$$

$$= \$70.18$$

Step 4: $p = MV - D$

$$= \$5{,}093.75 - \$70.18$$

$$= \$5{,}023.57$$

Note:

In Step 3, you must find the exact number of days in the discount period—the number of days between the discount date and its maturity. In this example, the note dated June 17 for three months would be due on September 17. The exact number of days in the discount period would then be the number of days between July 17 (the discount date) and September 17 (the maturity date).

Practice Problems

Find the interest, discount, maturity value, and proceeds for each.

16. Velma loaned $12,000 for 90 days at 7%. After 30 days, she took the note to the bank to discount it at 8.5%.

17. Chuck loaned out his life savings of $420 for 120 days at 9%. After 40 days, he discounted the note at the bank at 13%.

18. Ray loaned his friend Don $1,000 for three months at 10.5%. The note was dated May 5, and Ray discounted it on June 13 at 12%.

Practice Problems Solutions

1. Step 1:

July	=	31	Days in month
		− 17	
		14	Days left in July

Step 2:

August	=	+ 31	
		45	
September	=	+ 30	
		75	
October	=	+ 31	
		106	
November	=	+ 14	Days in November *(Note due on November 14)*
		120	Days from July 17

2. Step 1:

September	=	30	Days in month
		−29	
		1	

Step 2:

October	=	+31	
		32	
November	=	+28	Days in November *(Note due on November 28)*
		60	Days from September 30

3. Step 1:

March	=	31	Days in month
		−12	
		19	Days left in March

Step 2:

	30	April
	31	May
	30	June
	31	July
	31	August
	+12	Days in September
	184	Days in note

4. Step 1:

July	=	31	Days in month
		− 1	
		30	Days left in July

Step 2:

	31	August
	+ 1	September
	62	Days in note

5. It would be due on September 30, the last day of the month.

6.

January 15	=	15
		+120
		135th day is May 15

7.

October	=	290
April 17	=	−107
		183 days

8. $I = PRT$

$$= \$400 \times .15 \times \frac{60}{360}$$

$$= \$10$$

9. $I = PRT$

$$= \$8,000 \times .125 \times \frac{120}{360}$$

$$= \$333.33$$

10. $I = PRT$

$\quad = \$1,550 \times .1175 \times \dfrac{90}{365}$

$\quad = \$44.91$

11. $\quad I = PRT$

$\qquad = \$500 \times .085 \times 2$

$\qquad = \$85$

$\quad MV = P + I$

$\qquad = \$500 + \$85$

$\qquad = \$585$

12. $\quad I = PRT$

$\qquad = \$2,500 \times .105 \times \dfrac{6}{12}$

$\qquad = \$131.25$

$\quad MV = P + I$

$\qquad = \$2,500 + \$131.25$

$\qquad = \$2,631.25$

13. $\quad I = PRT$

$\quad \dfrac{I}{RT} = \dfrac{PRT}{RT}$

$\quad P = \dfrac{I}{RT}$

$\qquad = \dfrac{\$17.01}{.10 \times \dfrac{50}{360}}$

$\qquad = \$1,225$

14. $\quad I = PRT$

$\quad \dfrac{I}{PT} = \dfrac{PRT}{PT}$

$\quad R = \dfrac{I}{PT}$

$\qquad = \dfrac{\$14}{\$700 \times \dfrac{60}{360}}$

$\qquad = .12$

$\qquad = 12\%$

15. $\quad I = PRT$

$\quad \dfrac{I}{PR} = \dfrac{PRT}{PR}$

$\quad T = \dfrac{I}{PR}$

$\qquad = \dfrac{\$291.81}{\$10,150 \times .115}$

$\qquad = .2499978$

$\qquad = .2499978 \times 360 \text{ days} = 90 \text{ days}$

16. $\quad I = PRT$

$\qquad = \$12,000 \times .07 \times \dfrac{90}{360}$

$\qquad = \$210$

$\quad MV = P + I$

$\qquad = \$12,000 + \$210$

$\qquad = \$12,210$

$\quad D = (MV)(DR)(T)$

$\qquad = \$12,210 \times .085 \times \dfrac{60}{360}$

$\qquad = \$172.98$

$\quad p = MV - D$

$\qquad = \$12,210 - \$172.98$

$\qquad = \$12.037.02$

17. $\quad I = PRT$

$\qquad = \$420 \times .09 \times \dfrac{120}{360}$

$\qquad = \$12.60$

$\quad MV = P + I$

$\qquad = \$420 + \$12.60$

$\qquad = \$432.60$

$\quad D = (MV)(DR)(T)$

$\qquad = \$432.60 \times .13 \times \dfrac{80}{360}$

$\qquad = \$12.50$

$\qquad = \$432.60$

$\quad p = MV - D$

$\qquad = \$432.60 - \$12.50$

$\qquad = \$420.10$

18. $I = PRT$

$\quad\quad = \$1000 \times .105 \times 3/12$

$\quad\quad = \$26.25$

$\quad MV = P + I$

$\quad\quad = \$1000 + \$26.25$

$\quad\quad = \$1,026.25$

$\quad\quad D = (MV)(DR)(T)$

$\quad\quad = \$1,026.25 \times .12 \times 53*/360$

$\quad\quad = \$18.13$

*August 5 (due date) =   217 th day in table

June 13 $= \underline{-164}$ th day in table

$\quad\quad\quad\quad\quad\quad\quad\quad\quad$ 53 day discount period

$\quad\quad p = MV - D$

$\quad\quad = \$1,026.25 - \$18.13$

$\quad\quad = \$1,008.12$

Exercise 40A: Maturity Date

Name _____

Course/Sect. No. _____   Score _____

Instructions: Find the maturity date for each note. (Assume no year is a leap year.)

(1) _____	Self-Check Exercise (Answers on back)

(1) _____

(2) _____

(3) _____

(4) _____

(5) _____

(6) _____

(7) _____

(8) _____

(9) _____

(10) _____

Self-Check Exercise (Answers on back)

(1) Mar. 20 for 60 days   **(2)** Dec. 20 for 120 days

(3) Sep. 20 for 2 months   **(4)** Oct. 31 for 4 months

(5) Feb. 1 for 30 days   **(6)** Jul. 23 for 3 months

(7) Jan. 19 for 150 days   **(8)** Aug. 13 for 3 months

(9) The note dated Jan. 20, 1982 for three years was paid off six months early. When was it paid off?

(10) If the 120-day note was paid on Aug. 24, when was it dated?

(1) _May 19_

(2) _Apr. 19_

(3) _Nov. 20_

(4) _Feb. 28_

(5) _Mar. 3_

(6) _Oct. 23_

(7) _Jun. 18_

(8) _Nov. 13_

(9) _July 20, 1984_

(10) _Apr. 26_

Exercise 40B: Maturity Dates

Name_____

Course/Sect. No._____   Score_____

Instructions: Find the maturity date for each note. (Assume no year is a leap year.)

(1) _____	**(1)** Jan. 8 for 30 days **(2)** Aug. 27 for 90 days
(2) _____	
(3) _____	**(3)** Jun. 15 for 3 months **(4)** May 31 for 4 months
(4) _____	
(5) _____	**(5)** Nov. 10 for 60 days **(6)** Apr. 19 for 180 days
(6) _____	
(7) _____	**(7)** Jul. 23 for 90 days **(8)** May 28 for 120 days
(8) _____	
(9) _____	**(9)** The note dated March 10 for 90 days was extended another 60 days. When was the last due date?
(10) _____	**(10)** The 90-day note was paid off 15 days early. If it was dated on September 10, when was it paid off?

Exercise 41A: Time in Days

Name_____

Course/Sect. No._____   Score_____

Instructions: Find the length of each note in days. (Assume no leap years.)

Self-Check Exercise (Answers on back)

(1) _____

(1) Mar. 20 to Jun. 20   **(2)** Jan. 17 to Apr. 17

(2) _____

(3) _____

(3) Dec. 5 to Feb. 5   **(4)** Jun. 10 to Oct. 8

(4) _____

(5) _____

(5) Apr. 18 to Jun. 18   **(6)** Oct. 31 to Dec. 31

(6) _____

(7) _____

(7) Jul. 25 to Jan. 15   **(8)** Apr. 15 to Jul. 9

(8) _____

(9) _____

(9) The three-month note dated October 8 had how many days between the date of the note and the due date?

(10) _____

(10) A note dated November 17 was due on May 17. How many days in the note?

(1) _____92 days_____

(2) _____90 days_____

(3) _____62 days_____

(4) _____120 days_____

(5) _____61 days_____

(6) _____61 days_____

(7) _____174 days_____

(8) _____85 days_____

(9) _____92 days_____

(10) _____181 days_____

Exercise 41B: Time in Days

Name _____

Course/Sect. No. _____   Score _____

Instructions: Find the length of each note in days. (Assume no leap years.)

---

(1) _____   **(1)** Nov. 27 to Dec. 27   **(2)** Sep. 3 to Dec. 3

(2) _____

(3) _____   **(3)** Aug. 4 to Nov. 30   **(4)** May 18 to Jul. 18

(4) _____

(5) _____   **(5)** Feb. 22 to Jun. 24   **(6)** Nov. 20 to Mar. 20

(6) _____

(7) _____   **(7)** Aug. 7 to Nov. 29   **(8)** Mar. 1 to Jun. 1

(8) _____

(9) _____   **(9)** The note was dated February 7 for 60 days and then renewed. The final maturity date was July 10. How many total days were in the note?

(10) _____   **(10)** A note due on Oct. 20 for three months was therefore dated on July 20. How many days are there between those dates?

Exercise 42A: Simple Interest

Name_____

Course/Sect. No._____   Score_____

Instructions: Compute the amount of interest each note earns. (Use a 365-day year for "exact interest" problems.)

---

(1) _____

(2) _____

(3) _____

(4) _____

(5) _____

(6) _____

(7) _____

(8) _____

(9) _____

(10) _____

Self-Check Exercise (Answers on back)

P = principal R = rate T = time

(1) P = $1,050
R = 8.5%
T = 90 days

(2) P = $900
R = 11.25%
T = 3 months

(3) P = $2,250 (exact interest)
R = 9.75%
T = 120 days

(4) P = $750
R = 10.5%
T = 90 days

(5) P = $7,250
R = 9%
T = 12 months

(6) P = $1,225 (exact interest)
R = 17.2%
T = 30 days

(7) P = $12,000
R = 10.25%
T = 3 years

(8) P = $9,500 (exact interest)
R = 15%
T = 94 days

(9) What is the interest on $5,425.32 at $11\frac{1}{2}$% for 92 days (exact interest)?

(10) The long-term note was $3,500 at $12\frac{1}{4}$% for $2\frac{1}{2}$ years. What amount of interest would be due?

(1) _____$22.31_____

(2) _____$25.31_____

(3) _____$72.12_____

(4) _____$19.69_____

(5) _____$652.50_____

(6) _____$17.32_____

(7) _____$3,690_____

(8) _____$366.99_____

(9) _____$157.26_____

(10) _____$1,071.88_____

Exercise 42B: Simple Interest

Name _____

Course/Sect. No. _____   Score _____

Instructions: Compute the amount of interest each note earns. (Use a 365-day year for "exact interest" problems.)

---

(1) _____

(2) _____

(3) _____

(4) _____

(5) _____

(6) _____

(7) _____

(8) _____

(9) _____

(10) _____

P = principal R = rate T = time

(1) $P = \$800$
$R = 12\%$
$T = 180$ days

(2) $P = \$2000$ (exact interest)
$R = 16\%$
$T = 60$ days

(3) $P = \$5,000$
$R = 10\%$
$T = 30$ days

(4) $P = \$100$
$R = 13\frac{1}{4}\%$
$T = 3$ months

(5) $P = \$10,000$
$R = 14\%$
$T = 150$ days

(6) $P = \$3,000$
$R = 12.25\%$
$T = 83$ days

(7) $P = \$4,800$
$R = 9\frac{3}{4}\%$
$T = 6$ months

(8) $P = \$6,100$
$R = 13\%$
$T = 60$ days

(9) The \$4,000 note at 10% for 90 days was renewed after the interest was paid. The new rate was 12% for 60 days. What was the total interest paid?

(10) The terms were to pay interest on \$3,500 at 9.75% for 15 months. How much interest was paid?

Exercise 43A: Maturity Value

Name _____

Course/Sect. No. _____   Score _____

Instructions: Compute the maturity value on each note. (Use a 365-day year for "exact interest" problems.)

(1) _____

(2) _____

(3) _____

(4) _____

(5) _____

(6) _____

(7) _____

(8) _____

(9) _____

(10) _____

Self-Check Exercise (Answers on back)

P = principal R = rate T = time

(1) P = \$343.22 (exact interest)
R = 10.5%
T = 93 days

(2) P = \$2,138.72
R = 14.15%
T = 64 days

(3) P = \$2,844.60 (exact interest)
R = 11.22%
T = 185 days

(4) P = \$450.75
R = 18.5%
T = 91 days

(5) P = \$1,235.66 (exact interest)
R = $10\frac{1}{4}$%
T = 124 days

(6) P = \$4,000
R = 13.125%
T = $1\frac{1}{2}$ years

(7) P = \$2,307.19 (exact interest)
R = 15.6%
T = 38 days

(8) P = \$1,600.95
R = 13.225%
T = 283 days

(9) Which maturity value would be the largest? Give the amount of the largest. Note 1: \$2,412 at 10.5% for 93 days. Note 2: \$2,360 at 12.375% for 133 days.

(10) The note was \$3,800 at 10% for 3 months. When it came due the *maturity value* was renewed at 9.75% for 90 days. What was the final maturity value on the renewed note?

(1) $352.40

(2) $2,192.52

(3) $3,006.37

(4) $471.83

(5) $1,278.69

(6) $4,787.50

(7) $2,344.66

(8) $1,767.39

(9) $2,477.43

(10) $3,989.94

Exercise 43B: Maturity Value

Name_____

Course/Sect. No._____   Score_____

Instructions: Compute the maturity value on each note.

---

(1) _____

(2) _____

(3) _____

(4) _____

(5) _____

(6) _____

(7) _____

(8) _____

(9) _____

(10) _____

P = principal R = rate T = time

(1) P = $343.22
R = 9.75%
T = 125 days

(2) P = $9,140
R = 8.67%
T = 2 months

(3) P = $1,500.79
R = 12.3%
T = 3 months

(4) P = $3,471.92
R = 9.985%
T = $2\frac{1}{2}$ years

(5) P = $803.05
R = 16.75%
T = 42 months

(6) P = $714.92
R = 12.67%
T = 6 months

(7) P = $6,417.88
R = 10.05%
T = 104 days

(8) P = $355.73
R = $8\frac{3}{4}$%
T = $9\frac{1}{2}$ months

(9) Would the maturity value on $3,505 at $12\frac{1}{4}$% for 6 months be more than $4,000? (yes/no)

(10) What would be the maturity value (rounded to a whole dollar) for $2,525,000 at 13.2% for one-half year?

Exercise 44A: Finding the Unknown

Name_____

Course/Sect. No._____   Score_____

Instructions: Use the basic interest formula ($I = PRT$) and solve for the missing component. (Round to whole dollars and rates to a tenth of 1%.)

Self-Check Exercise (Answers on back)

(1) _____

(2) _____

(3) _____

(4) _____

(5) _____

(6) _____

(7) _____

(8) _____

(9) _____

(10) _____

(1) I = $300
 P = _____
 R = 12%
 T = 5 months

(2) I = $39
 P = $1300
 R = 9%
 T = _____ days

(3) I = $192.16 (exact interest)
 P = _____
 R = 14%
 T = 120 days

(4) I = $312.50
 P = $5,000
 R = _____%
 T = 180 days

(5) I = $51
 P = $2400
 R = 8.5%
 T = _____
  months

(6) I = $83.98
 P = _____
 R = $10\frac{3}{4}$%
 T = 90 days

(7) I = $90.74 (exact interest)
 P = _____
 R = $11\frac{1}{2}$%
 T = 60 days

(8) I = $2,047.50
 P = $6,500
 R = 12.6%
 T = _____ years

(9) How many months will it take $500 to earn interest of $250 if the rate is 10%?

(10) What rate would it take $1,000 to earn interest of $600 for 48 months?

(1) _____$6,000_____

(2) _____120 days_____

(3) _____$4,175_____

(4) _____12.5%_____

(5) _____3 months_____

(6) _____$3,125_____

(7) _____$4,800_____

(8) _____$2\frac{1}{2}$ years_____

(9) _____60 months_____

(10) _____15%_____

Exercise 44B: Finding the Unknown

Name _____

Course/Sect. No. _____   Score _____

Instructions: Use the basic interest formula ($I = PRT$) and solve for the missing component. (Round to whole dollars and rates to a tenth of 1%.)

---

(1) _____

(2) _____

(3) _____

(4) _____

(5) _____

(6) _____

(7) _____

(8) _____

(9) _____

(10) _____

(1) $I =$ _____
  $P = \$3,250$
  $R = 10\%$
  $T = 60$ days

(2) $I = \$81$
  $P = \$2,400$
  $R =$ _____ %
  $T = 90$ days

(3) $I = \$277.50$
  $P = \$925$
  $R =$ _____ %
  $T = 2$ years

(4) $I =$ _____
  $P = \$750$
  $R = 13\frac{1}{4}\%$
  $T = 3$ months

(5) $I = \$183.33$
  $P =$ _____
  $R = 11\%$
  $T = 60$ days

(6) $I = \$630$
  $P = \$1,500$
  $R =$ _____ %
  $T = 3$ years

(7) $I =$ _____
  $P = \$6,000$
  $R = 12\%$
  $T = 30$ days

(8) $I = \$1,312.50$
  $P = \$25,000$
  $R =$ _____ %
  $T = 7$ months

(9) How many days will it take $1,000 to earn interest of $1,000 if the rate is 10%?

(10) How many years will it take $1,000 to earn interest of $2,000 if the rate is 10%?

Exercise 45A: Discount and Proceeds

Name _____

Course/Sect. No. _____   Score _____

Instructions: Find the discount and proceeds on each.

Self-Check Exercise (Answers on back)

Maturity Value	Discount Period	Discount Rate
$ 4,500	120 days	14.5%

(1) _____   **(1)** Discount =

(2) _____   **(2)** Proceeds =

$3,174.58	63 days	$11\frac{3}{4}$%

(3) _____   **(3)** Discount =

(4) _____   **(4)** Proceeds =

Principal	Rate	Date of Note	Time	Discount Date	Discount Rate
$6,750	14%	Dec. 12	90 days	Jan. 1	16%

(5) _____   **(5)** Discount =

(6) _____   **(6)** Proceeds =

$5,995	10.5%	Oct. 20	2 months	Oct. 30	14%

(7) _____   **(7)** Discount =

(8) _____   **(8)** Proceeds =

Abe loaned $5,000 at 16% for 120 days. After 30 days it was discounted with Ben at 17%. Ben discounted Abe's note with Chad 30 days after that for 18%.

(9) _____   **(9)** How much did Ben make?

(10) _____   **(10)** How much did Chad make?

(1) $217.50

(2) $4,282.50

(3) $65.28

(4) $3,109.30

(5) $217.35

(6) $6,768.90

(7) $120.98

(8) $5,978.93

(9) $223.83

(10) $5,042.84

Exercise 45B: Discount and Proceeds

Name _____

Course/Sect. No. _____   Score _____

Instructions: Find the discount and proceeds on each.

	Maturity Value	Discount Period	Discount Rate
	$2432.50	94 days	12%

(1) _____   (1) Discount =

(2) _____   (2) Proceeds =

	$850	77 days	$13\frac{1}{4}\%$

(3) _____   (3) Discount =

(4) _____   (4) Proceeds =

Principal	Rate	Date of Note	Time	Discount Date	Discount Rate
$300	9%	May 30	60 days	Jun. 30	9.5%

(5) _____   (5) Discount =

(6) _____   (6) Proceeds =

$1,230	18.5%	Mar. 1	3 months	May 1	19%

(7) _____   (7) Discount =

(8) _____   (8) Proceeds =

For his new business, Carroll borrowed $75,000 from Freddie on Jan. 22 at 10.75% for 180 days. Freddie discounted the note at the State Bank on Mar. 3 at 12%.

(9) _____   (9) How much did the bank earn?

(10) _____   (10) How much did Freddie receive?

Chapter

Compound Interest, Present Value, and Annuities

Preview of Terms

Annuity
A series of payments, usually at equally spaced time intervals.

Compound Amount
The sum of the principal and all interest that has accumulated.

Compound Interest
The interest calculated on principal plus previously undistributed interest.

Ordinary Annuity
An annuity where the deposit is made at the end of a period. For example, a semiannual annuity for two years would have four investments, each at the end of a six-month interval. See the diagram below.

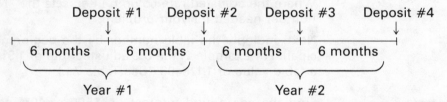

Present Value
Today's value (at the present) of a future amount. It is the amount that must be invested today in order to have a desired sum in the future. The calculation is the opposite of that used to find compound interest amounts from a single investment.

COMPOUND INTEREST

When money is left to earn interest for several interest periods, interest is not only earned on the original principal, it is also earned on interest. This is called *compound interest.* For compound interest problems, you must adjust the rate and time by the number of times interest is to be paid in one year. The most common interest periods are annually (1 time a year), semiannually (2 times a year), quarterly (4 times a year), and monthly (12 times a year). To adjust the rate, *divide* it by the number of times interest is paid in a year. To adjust time, *multiply* it by the number of times interest is paid in a year.

Example 1:　　　　　Adjust for interest compounded semiannually at 10% for three years.

Rate = 10% divided by 2* = 5% per period

Time = 3 years multiplied by 2* = 6 periods

(*The number 2 was used because semiannually means that interest is paid two times in one year.)

Example 2:　　　　　Adjust for interest compounded quarterly at 6% for two years.

Rate = 6% divided by 4 = 1.5% per period

Time = 2 years multiplied by 4 = 8 periods

Once you have adjusted for rate and time, you can complete the computation for determining the compound interest and compound amount. The shortcut method for using a calculator to calculate compound interest follows:

Step 1: Adjust the time and rate to find the number of periods and the interest per period.

Step 2: Convert the adjusted rate to its decimal equivalent and then put a 1 in front of the decimal. (A rate of 2% would be .02; with the 1 added, it would be 1.02. A rate of 3% would be .03; with the 1 added, it would be 1.03.)

Step 3: Multiply the principal (the amount invested) by the adjusted rate. This will give you a new principal, which you should multiply again by the adjusted rate. Continue to multiply the new principal by the adjusted rate until you have multiplied the same number of times as there are interest periods.

To find the compound interest using tables, you make the same adjustments as in examples 1 and 2, and then go to the compound interest tables. The "N" down both sides of the tables stands for the "number" of interest periods. Where the adjusted rate column intersects the "N" row, you have the table number needed for your problem.

Next, multiply the table number by the beginning investment made (principal). For example 2, the 6% adjusted rate of 1.5% for 8 periods would give a table value of 1.126493. (See Table 7.1.)

Example 3:　　　　　Find the compound interest on $500 at 10% compounded quarterly for one year.

By Calculator

Step 1: Rate = 10% divided by 4 = 2.5%
　　　　　Time = 1 year multiplied by 4 = 4 periods

Step 2: 2.5% = .025, then add 1 = 1.025

Step 3: 1.025 × $500.00　　= $512.50
　　　　　1.025 × $512.50　　= $525.3125
　　　　　1.025 × $525.3125　= $538.44531
　　　　　1.025 × $538.44531 = $551.90644
　　　　　　　　　　　　　　= $551.91* rounded

(*This final amount is called the **compound amount**. The original principal of $500 subtracted from the compound amount leaves the difference of $51.91, which is called the **compound interest**. It is best not to round until the last step.)

By Tables

Step 1: Rate = 10% divided by 4 = 2.5%
Time = 1 year multiplied by 4 = 4 periods ($N = 4$)

Step 2: Look in Table 7.1 for 2.5% for 4 periods

Table value = 1.103813

Step 3: Multiply the table value by the principal of $500

1.103813 × $500 = $551.9065
= $551.91 rounded

Note: *Sometimes the table compound amount will not agree exactly with the calculator compound amount. Any difference is due to the number of decimal places used in the table and in the calculator. The differences should be small.*

Note: *In most calculators, the first number entered in a multiplication problem is automatically a constant multiplier. In this problem, the 1.025 would be the constant multiplier. The procedure would be to enter the 1.025 first, then the × sign, then the $500. Then all you would have to do is depress the equal sign eight times—for eight periods.*

EXAMPLE 3 ILLUSTRATED

$500 compounded quarterly at 9% for 1 year

Quarter 1	Quarter 2	Quarter 3	Quarter 4
Interest = $11.25[1]	Interest = $11.50[3]	Interest = $11.76[5]	Interest = $12.03[7]
$500	$511.25[2]	$522.75[4]	$534.51[6]	$546.54[8]

1. $500 × 2.25% = $11.25
2. $500 + $11.25 = $511.25
3. $511.25 × 2.25% = $11.50
4. $511.25 + $11.50 = $522.75

5. $522.75 × 2.25% = $11.76
6. $522.75 + $11.76 = $534.51
7. $534.51 × 2.25% = $12.03
8. $534.51 + $12.03 = $546.54

Example 4: Find the compound amount on $1,200 at 13% compounded semiannually for three years.

By Calculator

Step 1: Rate = 13% divided by 2 = 6.5%
Time = 3 years multiplied by 2 = 6 periods

Step 2: Rate 6.5% = .065, then add 1 = 1.065

Step 3: 1.065 × $1,200.00 = $1,278.00
1.065 × $1,278.00 = $1,361.07
1.065 × $1,361.07 = $1,449.5395
1.065 × $1,449.5395 = $1,543.7595
1.065 × $1,543.7595 = $1,644.1038
1.065 × $1,644.1038 = $1,750.9705
= $1,750.97 rounded

TABLE 7.1
COMPOUND AMOUNT OF $1

N	1 %	1.5 %	2 %	2.5 %	3 %	3.5 %	4 %	N
1	1.010000	1.015000	1.020000	1.025000	1.030000	1.035000	1.040000	1
2	1.020100	1.030225	1.040400	1.050625	1.060900	1.071225	1.081600	2
3	1.030301	1.045678	1.061208	1.076891	1.092727	1.108718	1.124864	3
4	1.040604	1.061364	1.082432	1.103813	1.125509	1.147523	1.169859	4
5	1.051010	1.077284	1.104081	1.131408	1.159274	1.187686	1.216653	5
6	1.061520	1.093443	1.126162	1.159693	1.194052	1.229255	1.265319	6
7	1.072135	1.109845	1.148686	1.188686	1.229874	1.272279	1.315932	7
8	1.082857	1.126493	1.171659	1.218403	1.266770	1.316809	1.368569	8
9	1.093685	1.143390	1.195093	1.248863	1.304773	1.362897	1.423312	9
10	1.104622	1.160541	1.218994	1.280085	1.343916	1.410599	1.480244	10
11	1.115668	1.177949	1.243374	1.312087	1.384234	1.459970	1.539454	11
12	1.126825	1.195618	1.268242	1.344889	1.425761	1.511069	1.601032	12
13	1.138093	1.213552	1.293607	1.378511	1.468534	1.563956	1.665074	13
14	1.149474	1.231756	1.319479	1.412974	1.512590	1.618695	1.731676	14
15	1.160969	1.250232	1.345868	1.448298	1.557967	1.675349	1.800944	15
16	1.172579	1.268986	1.372786	1.484506	1.604706	1.733986	1.872981	16
17	1.184304	1.288020	1.400241	1.521618	1.652848	1.794676	1.947900	17
18	1.196147	1.307341	1.428246	1.559659	1.702433	1.857489	2.025817	18
19	1.208109	1.326951	1.456811	1.598650	1.753506	1.922501	2.106849	19
20	1.220190	1.346855	1.485947	1.638616	1.806111	1.989789	2.191123	20
21	1.232392	1.367058	1.515666	1.679582	1.860295	2.059431	2.278768	21
22	1.244716	1.387564	1.545980	1.721571	1.916103	2.131512	2.369919	22
23	1.257163	1.408377	1.576899	1.764611	1.973587	2.206114	2.464716	23
24	1.269735	1.429503	1.608437	1.808726	2.032794	2.283328	2.563304	24
25	1.282432	1.450945	1.640606	1.853944	2.093778	2.363245	2.665836	25
26	1.295256	1.472710	1.673418	1.900293	2.156591	2.445959	2.772470	26
27	1.308209	1.494800	1.706886	1.947800	2.221289	2.531567	2.883369	27
28	1.321291	1.517222	1.741024	1.996495	2.287928	2.620172	2.998703	28
29	1.334504	1.539981	1.775845	2.046407	2.356566	2.711878	3.118651	29
30	1.347849	1.563080	1.811362	2.097568	2.427262	2.806794	3.243398	30
31	1.361327	1.586526	1.847589	2.150007	2.500080	2.905031	3.373133	31
32	1.374941	1.610324	1.884541	2.203757	2.575083	3.006708	3.508059	32
33	1.388690	1.634479	1.922231	2.258851	2.652335	3.111942	3.648381	33
34	1.402577	1.658996	1.960676	2.315322	2.731905	3.220860	3.794316	34
35	1.416603	1.683881	1.999890	2.373205	2.813862	3.333590	3.946089	35
36	1.430769	1.709140	2.039887	2.432535	2.898278	3.450266	4.103933	36
37	1.445076	1.734777	2.080685	2.493349	2.985227	3.571025	4.268090	37
38	1.459527	1.760798	2.122299	2.555682	3.074783	3.696011	4.438813	38
39	1.474123	1.787210	2.164745	2.619574	3.167027	3.825372	4.616366	39
40	1.488864	1.814018	2.208040	2.685064	3.262038	3.959260	4.801021	40
41	1.503752	1.841229	2.252200	2.752190	3.359899	4.097834	4.993061	41
42	1.518790	1.868847	2.297244	2.820995	3.460696	4.241258	5.192784	42
43	1.533978	1.896880	2.343189	2.891520	3.564517	4.389702	5.400495	43
44	1.549318	1.925333	2.390053	2.963808	3.671452	4.543342	5.616515	44
45	1.564811	1.954213	2.437854	3.037903	3.781596	4.702359	5.841176	45
46	1.580459	1.983526	2.486611	3.113851	3.895044	4.866941	6.074823	46
47	1.596263	2.013279	2.536344	3.191697	4.011895	5.037284	6.317816	47
48	1.612226	2.043478	2.587070	3.271490	4.132252	5.213589	6.570528	48

TABLE 7.1 (Continued)
COMPOUND AMOUNT OF $1

N	4.5 %	5 %	5.5 %	6 %	6.5 %	7 %	7.5 %	N
1	1.045000	1.050000	1.055000	1.060000	1.065000	1.070000	1.075000	1
2	1.092025	1.102500	1.113025	1.123600	1.134225	1.144900	1.155625	2
3	1.141166	1.157625	1.174241	1.191016	1.207950	1.225043	1.242297	3
4	1.192519	1.215506	1.238825	1.262477	1.286466	1.310796	1.335469	4
5	1.246182	1.276282	1.306960	1.338226	1.370087	1.402552	1.435629	5
6	1.302260	1.340096	1.378843	1.418519	1.459142	1.500730	1.543302	6
7	1.360862	1.407100	1.454679	1.503630	1.553987	1.605781	1.659049	7
8	1.422101	1.477455	1.534687	1.593848	1.654996	1.718186	1.783478	8
9	1.486095	1.551328	1.619094	1.689479	1.762570	1.838459	1.917239	9
10	1.552969	1.628895	1.708144	1.790848	1.877137	1.967151	2.061032	10
11	1.622853	1.710339	1.802092	1.898299	1.999151	2.104852	2.215609	11
12	1.695881	1.795856	1.901207	2.012196	2.129096	2.252192	2.381780	12
13	1.772196	1.885649	2.005774	2.132928	2.267487	2.409845	2.560413	13
14	1.851945	1.979932	2.116091	2.260904	2.414874	2.578534	2.752444	14
15	1.935282	2.078928	2.232476	2.396558	2.571841	2.759032	2.958877	15
16	2.022370	2.182875	2.355263	2.540352	2.739011	2.952164	3.180793	16
17	2.113377	2.292018	2.484802	2.692773	2.917046	3.158815	3.419353	17
18	2.208479	2.406619	2.621466	2.854339	3.106654	3.379932	3.675804	18
19	2.307860	2.526950	2.765647	3.025600	3.308587	3.616528	3.951489	19
20	2.411714	2.653298	2.917757	3.207135	3.523645	3.869684	4.247851	20
21	2.520241	2.785963	3.078234	3.399564	3.752682	4.140562	4.566440	21
22	2.633652	2.925261	3.247537	3.603537	3.996606	4.430402	4.908923	22
23	2.752166	3.071524	3.426152	3.819750	4.256386	4.740530	5.277092	23
24	2.876014	3.225100	3.614590	4.048935	4.533051	5.072367	5.672874	24
25	3.005434	3.386355	3.813392	4.291871	4.827699	5.427433	6.098340	25
26	3.140679	3.555673	4.023129	4.549383	5.141500	5.807353	6.555715	26
27	3.282010	3.733456	4.244401	4.822346	5.475697	6.213868	7.047394	27
28	3.429700	3.920129	4.477843	5.111687	5.831617	6.648838	7.575948	28
29	3.584036	4.116136	4.724124	5.418388	6.210672	7.114257	8.144144	29
30	3.745318	4.321942	4.983951	5.743491	6.614366	7.612255	8.754955	30
31	3.913857	4.538039	5.258069	6.088101	7.044300	8.145113	9.411577	31
32	4.089981	4.764941	5.547262	6.453387	7.502179	8.715271	10.117445	32
33	4.274030	5.003189	5.852362	6.840590	7.989821	9.325340	10.876253	33
34	4.466362	5.253348	6.174242	7.251025	8.509159	9.978114	11.691972	34
35	4.667348	5.516015	6.513825	7.686087	9.062255	10.676581	12.568870	35
36	4.877378	5.791816	6.872085	8.147252	9.651301	11.423942	13.511536	36
37	5.096860	6.081407	7.250050	8.636087	10.278636	12.223618	14.524901	37
38	5.326219	6.385477	7.648803	9.154252	10.946747	13.079271	15.614268	38
39	5.565899	6.704751	8.069487	9.703507	11.658286	13.994820	16.785339	39
40	5.816365	7.039989	8.513309	10.285718	12.416075	14.974458	18.044239	40
41	6.078101	7.391988	8.981541	10.902861	13.223119	16.022670	19.397557	41
42	6.351615	7.761588	9.475525	11.557033	14.082622	17.144257	20.852374	42
43	6.637438	8.149667	9.996679	12.250455	14.997993	18.344355	22.416302	43
44	6.936123	8.557150	10.546497	12.985482	15.972862	19.628460	24.097524	44
45	7.248248	8.985008	11.126554	13.764611	17.011098	21.002452	25.904839	45
46	7.574420	9.434258	11.738515	14.590487	18.116820	22.472623	27.847702	46
47	7.915268	9.905971	12.384133	15.465917	19.294413	24.045707	29.936279	47
48	8.271456	10.401270	13.065260	16.393872	20.548550	25.728907	32.181500	48

TABLE 7.1 (Continued)
COMPOUND AMOUNT OF $1

N	8 %	9 %	10 %	11 %	12 %	13 %	14 %	N
1	1.080000	1.090000	1.100000	1.110000	1.120000	1.130000	1.140000	1
2	1.166400	1.188100	1.210000	1.232100	1.254400	1.276900	1.299600	2
3	1.259712	1.295029	1.331000	1.367631	1.404928	1.442897	1.481544	3
4	1.360489	1.411582	1.464100	1.518070	1.573519	1.630474	1.688960	4
5	1.469328	1.538624	1.610510	1.685058	1.762342	1.842435	1.925415	5
6	1.586874	1.677100	1.771561	1.870415	1.973823	2.081952	2.194973	6
7	1.713824	1.828039	1.948717	2.076160	2.210681	2.352605	2.502269	7
8	1.850930	1.992563	2.143589	2.304538	2.475963	2.658444	2.852586	8
9	1.999005	2.171893	2.357948	2.558037	2.773079	3.004042	3.251949	9
10	2.158925	2.367364	2.593742	2.839421	3.105848	3.394567	3.707221	10
11	2.331639	2.580426	2.853117	3.151757	3.478550	3.835861	4.226232	11
12	2.518170	2.812665	3.138428	3.498451	3.895976	4.334523	4.817905	12
13	2.719624	3.065805	3.452271	3.883280	4.363493	4.898011	5.492411	13
14	2.937194	3.341727	3.797498	4.310441	4.887112	5.534753	6.261349	14
15	3.172169	3.642482	4.177248	4.784589	5.473566	6.254270	7.137938	15
16	3.425943	3.970306	4.594973	5.310894	6.130394	7.067326	8.137249	16
17	3.700018	4.327633	5.054470	5.895093	6.866041	7.986078	9.276464	17
18	3.996019	4.717120	5.559917	6.543553	7.689966	9.024268	10.575169	18
19	4.315701	5.141661	6.115909	7.263344	8.612762	10.197423	12.055693	19
20	4.660957	5.604411	6.727500	8.062312	9.646293	11.523088	13.743490	20
21	5.033834	6.108808	7.400250	8.949166	10.803848	13.021089	15.667578	21
22	5.436540	6.658600	8.140275	9.933574	12.100310	14.713831	17.861039	22
23	5.871464	7.257874	8.954302	11.026267	13.552347	16.626629	20.361585	23
24	6.341181	7.911083	9.849733	12.239157	15.178629	18.788091	23.212207	24
25	6.848475	8.623081	10.834706	13.585464	17.000064	21.230542	26.461916	25
26	7.396353	9.399158	11.918177	15.079865	19.040072	23.990513	30.166584	26
27	7.988061	10.245082	13.109994	16.738650	21.324881	27.109279	34.389906	27
28	8.627106	11.167140	14.420994	18.579901	23.883866	30.633486	39.204493	28
29	9.317275	12.172182	15.863093	20.623691	26.749930	34.615839	44.693122	29
30	10.062657	13.267678	17.449402	22.892297	29.959922	39.115898	50.950159	30
31	10.867669	14.461770	19.194342	25.410449	33.555113	44.200965	58.083181	31
32	11.737083	15.763329	21.113777	28.205599	37.581726	49.947090	66.214826	32
33	12.676050	17.182028	23.225154	31.308214	42.091533	56.440212	75.484902	33
34	13.690134	18.728411	25.547670	34.752118	47.142517	63.777439	86.052788	34
35	14.785344	20.413968	28.102437	38.574851	52.799620	72.068506	98.100178	35
36	15.968172	22.251225	30.912681	42.818085	59.135574	81.437412	111.834203	36
37	17.245626	24.253835	34.003949	47.528074	66.231843	92.024276	127.490992	37
38	18.625276	26.436680	37.404343	52.756162	74.179664	103.987432	145.339731	38
39	20.115298	28.815982	41.144778	58.559340	83.081224	117.505798	165.687293	39
40	21.724521	31.409420	45.259256	65.000867	93.050970	132.781552	188.883514	40
41	23.462483	34.236268	49.785181	72.150963	104.217087	150.043153	215.327206	41
42	25.339482	37.317532	54.763699	80.087569	116.723137	169.548763	245.473015	42
43	27.366640	40.676110	60.240069	88.897201	130.729914	191.590103	279.839237	43
44	29.555972	44.336960	66.264076	98.675893	146.417503	216.496816	319.016730	44
45	31.920449	48.327286	72.890484	109.530242	163.987604	244.641402	363.679072	45
46	34.474085	52.676742	80.179532	121.578568	183.666116	276.444784	414.594142	46
47	37.232012	57.417649	88.197485	134.952211	205.706050	312.382606	472.637322	47
48	40.210573	62.585237	97.017234	149.796954	230.390776	352.992345	538.806547	48

By Table

Step 1: Rate = 13% divided by 2 = 6.5%

Time = 3 years multiplied by 2 = 6 periods (N = 6)

Step 2: Look in Table 7.1 for 6.5% for 6 periods

Table value = 1.459142

Step 3: Multiply the table value by the principal of $1,200

1.459142 × $1,200 = $1,750.9704

= $1,750.97 rounded

In both methods the compound amount of $1,750.97 less the principal of $1,200 equals the compound interest of $550.97.

Once you have learned to manipulate your calculator, it is more flexible than using tables because often the available tables will not have the exact interest rates you need. The calculator, however, can be used for any rate and any time.

Practice Problems

Find the compound amount. Use the tables for #1 and #2 and your calculator for #3 and #4.

1. Gayle invested $2,000 of her teaching money at 11% compounded semiannually for 5 years.

2. Janet invested $1,500 for three years at 8% compounded quarterly.

3. Monica invested $800 for four years to be compounded semiannually at 10.5%.

4. Karen invested $800 for four years to be compounded monthly at 10 1/2%.

PRESENT VALUE

The compound interest section showed that money is worth more as time passes because it earns interest. A sum of money now (in the present) is worth more in the future because interest compounds.

1982	Compound interest	1984
$1,000	(Compounded quarterly at 6%)	(?????)*
Present value		Future value

(*The compound amount, or *future value,* would be $1,126.49.)

In compound interest problems, you know the sum of money you are investing (the present value), and you calculate the amount it will be worth (the future value).

Present value problems are the opposite of compound interest problems. In present value problems, you know how much you want to end up with (the future value), and you want to find out how much you must have in the beginning (the present value) to achieve your goal.

1982	Present value	1984
(?????)*	(Compounded quarterly at 6%)	$2,000
Present value		Future value

(*The present value would be $1,775.42.)

You can find the present value by making the same adjustment for rate and time that you make for compound interest. You then divide the future value by the adjusted rate. Repeat the divisions the same number of times as you have periods.

Example 5:   You want to have $3,000 three years from now. You know you could invest money at 8% compounded semiannually. What single sum would you have to invest now (the present value) so you would have a compound amount of $3,000 in three years?

By Calculator

Step 1: Rate = 8% divided by 2 = 4%
Time = 3 years multiplied by 2 = 6 periods

Step 2: 4% = .04, then add 1 = 1.04

Step 3:
$3,000.00   ÷ 1.04 = $2,884.6153
$2,884.6153 ÷ 1.04 = $2,773.6685
$2,773.6685 ÷ 1.04 = $2,666.9889
$2,666.9889 ÷ 1.04 = $2,564.4124
$2,564.4124 ÷ 1.04 = $2,465.7811
$2,465.7811 ÷ 1.04 = $2,370.9433
= $2,370.94 rounded

The present value would be $2,370.94. (You would have to invest that much now—at the present—for it to grow to be $3,000 at the end of three years at 8% compounded semiannually.)

Note: *In most calculators, the second number entered in a division problem automatically becomes a constant divisor. In this problem, the 1.04 would be the constant divisor. The procedure would be to enter the $3,000 first, the ÷ sign second, and then depress the equal sign six times—for six periods.*

By Table

Step 1: Rate = 8% divided by 2 = 4%
Time = 3 years multiplied by 2 = 6 periods ($N = 6$)

Step 2: Look in the Present Value Table (Table 7.2) for 4% for 6 periods.

Table value = .790315

Step 3: Multiply the table value by the future value needed, $3,000.

.790315 × $3,000 = $2,370.945
= $2,370.95 rounded

Practice Problems

Find the present value in each of the following. Use the tables for #5 and #6 and your calculator for #7 and #8.

5. Barbara needs to have $2,000 two years in the future. If she could invest money at 12% compounded quarterly, what single amount would she need to invest now?

6. Mary wants to know the present value of $40,000 compounded semi-annually at 9% for 10 years.

7. Mike figures the last year at college will cost him $5,250. If interest is compounded annually at 14%, what amount will he need to invest now so he will have $5,250 two years from now?

8. Kelly talked Carroll into paying all but $3,000 on a new car for graduation. She has three years left to accumulate that much. How much must she put in now if it is earning 10.5% compounded monthly?

ANNUITIES

An annuity is a series of equal investments at regular intervals. With an annuity, in addition to the interest compounding, the principal is increasing periodically also.

A $500 ordinary annuity made quarterly for one year means $2,000 would be invested, $500 at the end of every 3-month period. The first $500 investment would earn interest for 9 months (3 quarterly periods). The second $500 investment would earn interest for 6 months (2 quarterly periods). The third $500 investment would earn interest for 3 months (1 quarterly period). The final $500 investment would earn no interest that year because it was made on the last day of the year (it will begin to earn interest in future periods).

A deposit made at the beginning of a period is called an *annuity due*. A deposit made at the end of a period is called an *ordinary annuity*. The difference between the two is one period of interest earned.

$1,000 Ordinary Annuity at 5% for 3 Years

		Year 1 December	Year 2 December	Year 3 December
Principal invested	=	$1,000	$1,000	$1,000
Interest earned	=	–0–	$50*	$102.50**
End-of-year balance	=	$1,000	$2,050	$3,152.50

Computations

First year:	$1,000.00	1st year investment (at end of year)
Second year:	$1,000.00	Investment balance
	+ 50.00	Interest (5% × $1,000)*
	$1,050.00	
	+ 1,000.00	2nd year investment (at end of year)
	$2,050.00	Investment balance
Third year:	$2,050.00	Investment balance
	+ 102.50	Interest (5% × $2,050)**
	$2,152.50	
	+ 1,000.00	3rd year investment (at end of year)
	$3,152.50	Investment balance

Compound Interest, Present Value, and Annuities **277**

TABLE 7.2
PRESENT VALUE OF $1 AT COMPOUND INTEREST

N	1 %	1.5 %	2 %	2.5 %	3 %	3.5 %	4 %	N
1	0.990099	0.985222	0.980392	0.975610	0.970874	0.966184	0.961539	1
2	0.980296	0.970662	0.961169	0.951815	0.942596	0.933511	0.924556	2
3	0.970590	0.956317	0.942322	0.928600	0.915142	0.901943	0.888997	3
4	0.960980	0.942184	0.923846	0.905951	0.888487	0.871442	0.854804	4
5	0.951466	0.928261	0.905731	0.883854	0.862609	0.841973	0.821927	5
6	0.942045	0.914542	0.887972	0.862297	0.837484	0.813501	0.790315	6
7	0.932718	0.901027	0.870560	0.841266	0.813092	0.785991	0.759918	7
8	0.923483	0.887711	0.853491	0.820747	0.789409	0.759412	0.730690	8
9	0.914340	0.874592	0.836755	0.800729	0.766417	0.733731	0.702587	9
10	0.905287	0.861667	0.820348	0.781199	0.744094	0.708919	0.675564	10
11	0.896324	0.848933	0.804263	0.762145	0.722422	0.684946	0.649581	11
12	0.887449	0.836388	0.788493	0.743556	0.701380	0.661784	0.624597	12
13	0.878663	0.824027	0.773033	0.725421	0.680952	0.639404	0.600574	13
14	0.869963	0.811849	0.757875	0.707727	0.661118	0.617782	0.577475	14
15	0.861350	0.799852	0.743015	0.690466	0.641862	0.596891	0.555265	15
16	0.852821	0.788031	0.728446	0.673625	0.623167	0.576706	0.533908	16
17	0.844378	0.776386	0.714163	0.657195	0.605017	0.557204	0.513373	17
18	0.836018	0.764912	0.700160	0.641166	0.587395	0.538362	0.493628	18
19	0.827740	0.753608	0.686431	0.625528	0.570286	0.520156	0.474643	19
20	0.819545	0.742471	0.672972	0.610271	0.553676	0.502566	0.456387	20
21	0.811430	0.731498	0.659776	0.595387	0.537550	0.485571	0.438834	21
22	0.803396	0.720688	0.646839	0.580865	0.521893	0.469151	0.421956	22
23	0.795442	0.710037	0.634156	0.566698	0.506692	0.453286	0.405727	23
24	0.787566	0.699544	0.621722	0.552876	0.491934	0.437957	0.390122	24
25	0.779769	0.689206	0.609531	0.539391	0.477606	0.423147	0.375117	25
26	0.772048	0.679021	0.597580	0.526235	0.463695	0.408838	0.360690	26
27	0.764404	0.668986	0.585862	0.513400	0.450190	0.395013	0.346817	27
28	0.756836	0.659100	0.574375	0.500878	0.437077	0.381655	0.333478	28
29	0.749342	0.649359	0.563113	0.488662	0.424347	0.368748	0.320652	29
30	0.741923	0.639763	0.552071	0.476743	0.411987	0.356279	0.308319	30
31	0.734577	0.630308	0.541246	0.465115	0.399988	0.344231	0.296461	31
32	0.727304	0.620993	0.530634	0.453771	0.388338	0.332590	0.285058	32
33	0.720103	0.611816	0.520229	0.442703	0.377027	0.321343	0.274095	33
34	0.712974	0.602774	0.510029	0.431906	0.366045	0.310476	0.263552	34
35	0.705914	0.593866	0.500028	0.421371	0.355384	0.299977	0.253416	35
36	0.698925	0.585090	0.490224	0.411094	0.345033	0.289833	0.243669	36
37	0.692005	0.576443	0.480611	0.401068	0.334983	0.280032	0.234297	37
38	0.685154	0.567925	0.471188	0.391285	0.325227	0.270562	0.225286	38
39	0.678370	0.559532	0.461949	0.381742	0.315754	0.261413	0.216621	39
40	0.671653	0.551263	0.452891	0.372431	0.306557	0.252573	0.208289	40
41	0.665003	0.543116	0.444011	0.363347	0.297629	0.244032	0.200278	41
42	0.658419	0.535090	0.435305	0.354485	0.288960	0.235779	0.192575	42
43	0.651900	0.527182	0.426769	0.345839	0.280543	0.227806	0.185168	43
44	0.645446	0.519391	0.418401	0.337404	0.272372	0.220103	0.178047	44
45	0.639055	0.511715	0.410197	0.329175	0.264439	0.212660	0.171199	45
46	0.632728	0.504153	0.402154	0.321146	0.256737	0.205468	0.164614	46
47	0.626463	0.496702	0.394269	0.313313	0.249259	0.198520	0.158283	47
48	0.620261	0.489362	0.386538	0.305672	0.241999	0.191807	0.152195	48

TABLE 7.2 (Continued)
PRESENT VALUE OF $1 AT COMPOUND INTEREST

N	4.5 %	5 %	5.5 %	6 %	6.5 %	7 %	7.5 %	N
1	0.956938	0.952381	0.947867	0.943396	0.938967	0.934579	0.930233	1
2	0.915730	0.907030	0.898453	0.889997	0.881659	0.873439	0.865333	2
3	0.876297	0.863838	0.851614	0.839619	0.827849	0.816298	0.804961	3
4	0.838562	0.822703	0.807217	0.792094	0.777323	0.762895	0.748801	4
5	0.802451	0.783526	0.765135	0.747258	0.729881	0.712986	0.696559	5
6	0.767896	0.746216	0.725246	0.704961	0.685334	0.666342	0.647961	6
7	0.734829	0.710682	0.687437	0.665057	0.643506	0.622750	0.602755	7
8	0.703185	0.676840	0.651599	0.627413	0.604231	0.582009	0.560702	8
9	0.672905	0.644609	0.617630	0.591899	0.567353	0.543934	0.521583	9
10	0.643928	0.613914	0.585431	0.558395	0.532726	0.508349	0.485194	10
11	0.616199	0.584680	0.554911	0.526788	0.500212	0.475093	0.451343	11
12	0.589664	0.556838	0.525982	0.496970	0.469683	0.444012	0.419854	12
13	0.564272	0.530322	0.498561	0.468839	0.441017	0.414964	0.390562	13
14	0.539973	0.505068	0.472570	0.442301	0.414100	0.387817	0.363313	14
15	0.516721	0.481017	0.447933	0.417265	0.388826	0.362446	0.337966	15
16	0.494470	0.458112	0.424581	0.393647	0.365095	0.338734	0.314387	16
17	0.473177	0.436297	0.402447	0.371365	0.342812	0.316574	0.292453	17
18	0.452801	0.415521	0.381466	0.350344	0.321889	0.295864	0.272049	18
19	0.433302	0.395734	0.361579	0.330513	0.302244	0.276508	0.253069	19
20	0.414643	0.376890	0.342729	0.311805	0.283797	0.258419	0.235413	20
21	0.396788	0.358943	0.324862	0.294156	0.266476	0.241513	0.218989	21
22	0.379701	0.341850	0.307926	0.277505	0.250212	0.225713	0.203711	22
23	0.363350	0.325572	0.291873	0.261798	0.234941	0.210947	0.189498	23
24	0.347704	0.310068	0.276657	0.246979	0.220602	0.197146	0.176277	24
25	0.332731	0.295303	0.262234	0.232999	0.207138	0.184249	0.163979	25
26	0.318403	0.281241	0.248563	0.219810	0.194496	0.172195	0.152539	26
27	0.304692	0.267849	0.235605	0.207368	0.182625	0.160930	0.141896	27
28	0.291571	0.255094	0.223322	0.195630	0.171479	0.150402	0.131997	28
29	0.279015	0.242947	0.211680	0.184557	0.161013	0.140563	0.122787	29
30	0.267000	0.231378	0.200644	0.174110	0.151186	0.131367	0.114221	30
31	0.255503	0.220360	0.190184	0.164255	0.141959	0.122773	0.106252	31
32	0.244500	0.209866	0.180269	0.154958	0.133294	0.114741	0.098839	32
33	0.233971	0.199873	0.170872	0.146187	0.125159	0.107235	0.091943	33
34	0.223896	0.190355	0.161964	0.137912	0.117520	0.100219	0.085529	34
35	0.214255	0.181291	0.153520	0.130105	0.110348	0.093663	0.079562	35
36	0.205028	0.172658	0.145517	0.122741	0.103613	0.087535	0.074011	36
37	0.196199	0.164436	0.137930	0.115793	0.097289	0.081809	0.068847	37
38	0.187751	0.156606	0.130740	0.109239	0.091351	0.076457	0.064044	38
39	0.179666	0.149148	0.123924	0.103056	0.085776	0.071455	0.059576	39
40	0.171929	0.142046	0.117463	0.097222	0.080541	0.066780	0.055419	40
41	0.164525	0.135282	0.111340	0.091719	0.075625	0.062411	0.051553	41
42	0.157440	0.128840	0.105535	0.086528	0.071009	0.058328	0.047956	42
43	0.150661	0.122705	0.100033	0.081630	0.066675	0.054513	0.044610	43
44	0.144173	0.116862	0.094818	0.077009	0.062606	0.050946	0.041498	44
45	0.137965	0.111297	0.089875	0.072650	0.058785	0.047613	0.038603	45
46	0.132024	0.105997	0.085190	0.068538	0.055197	0.044499	0.035910	46
47	0.126338	0.100949	0.080749	0.064658	0.051828	0.041587	0.033404	47
48	0.120898	0.096142	0.076539	0.060999	0.048665	0.038867	0.031074	48

TABLE 7.2 (Continued)
PRESENT VALUE OF $1 AT COMPOUND INTEREST

N	8 %	9 %	10 %	11 %	12 %	13 %	14 %	N
1	0.925926	0.917431	0.909091	0.900901	0.892857	0.884956	0.877193	1
2	0.857339	0.841680	0.826446	0.811623	0.797194	0.783147	0.769468	2
3	0.793832	0.772184	0.751315	0.731191	0.711780	0.693050	0.674972	3
4	0.735030	0.708425	0.683013	0.658731	0.635518	0.613319	0.592080	4
5	0.680583	0.649931	0.620921	0.593451	0.567427	0.542760	0.519369	5
6	0.630169	0.596267	0.564474	0.534641	0.506631	0.480319	0.455587	6
7	0.583490	0.547034	0.513158	0.481658	0.452349	0.425061	0.399637	7
8	0.540269	0.501866	0.466507	0.433927	0.403883	0.376160	0.350559	8
9	0.500249	0.460428	0.424098	0.390925	0.360610	0.332885	0.307508	9
10	0.463193	0.422411	0.385543	0.352184	0.321973	0.294588	0.269744	10
11	0.428883	0.387533	0.350494	0.317283	0.287476	0.260698	0.236617	11
12	0.397114	0.355535	0.318631	0.285841	0.256675	0.230706	0.207559	12
13	0.367698	0.326179	0.289664	0.257514	0.229174	0.204165	0.182069	13
14	0.340461	0.299246	0.263331	0.231995	0.204620	0.180677	0.159710	14
15	0.315242	0.274538	0.239392	0.209004	0.182696	0.159891	0.140097	15
16	0.291890	0.251870	0.217629	0.188292	0.163122	0.141496	0.122892	16
17	0.270269	0.231073	0.197845	0.169633	0.145644	0.125218	0.107800	17
18	0.250249	0.211994	0.179859	0.152822	0.130040	0.110812	0.094561	18
19	0.231712	0.194490	0.163508	0.137678	0.116107	0.098064	0.082948	19
20	0.214548	0.178431	0.148644	0.124034	0.103667	0.086782	0.072762	20
21	0.198656	0.163698	0.135131	0.111742	0.092560	0.076799	0.063826	21
22	0.183940	0.150182	0.122846	0.100669	0.082643	0.067963	0.055988	22
23	0.170315	0.137781	0.111678	0.090692	0.073788	0.060144	0.049112	23
24	0.157699	0.126405	0.101526	0.081705	0.065882	0.053225	0.043081	24
25	0.146018	0.115968	0.092296	0.073608	0.058823	0.047102	0.037790	25
26	0.135202	0.106392	0.083905	0.066314	0.052521	0.041683	0.033149	26
27	0.125187	0.097608	0.076278	0.059742	0.046894	0.036888	0.029078	27
28	0.115914	0.089548	0.069343	0.053822	0.041869	0.032644	0.025507	28
29	0.107327	0.082154	0.063039	0.048488	0.037383	0.028889	0.022375	29
30	0.099377	0.075371	0.057309	0.043683	0.033378	0.025565	0.019627	30
31	0.092016	0.069148	0.052099	0.039354	0.029802	0.022624	0.017217	31
32	0.085200	0.063438	0.047362	0.035454	0.026609	0.020021	0.015102	32
33	0.078889	0.058200	0.043057	0.031940	0.023758	0.017718	0.013248	33
34	0.073045	0.053395	0.039142	0.028775	0.021212	0.015680	0.011621	34
35	0.067634	0.048986	0.035584	0.025924	0.018940	0.013876	0.010194	35
36	0.062624	0.044941	0.032349	0.023355	0.016910	0.012279	0.008942	36
37	0.057986	0.041231	0.029408	0.021040	0.015098	0.010867	0.007844	37
38	0.053690	0.037826	0.026735	0.018955	0.013481	0.009617	0.006880	38
39	0.049713	0.034703	0.024304	0.017077	0.012036	0.008510	0.006035	39
40	0.046031	0.031838	0.022095	0.015384	0.010747	0.007531	0.005294	40
41	0.042621	0.029209	0.020086	0.013860	0.009595	0.006665	0.004644	41
42	0.039464	0.026797	0.018260	0.012486	0.008567	0.005898	0.004074	42
43	0.036541	0.024584	0.016600	0.011249	0.007649	0.005219	0.003573	43
44	0.033834	0.022555	0.015091	0.010134	0.006830	0.004619	0.003135	44
45	0.031328	0.020692	0.013719	0.009130	0.006098	0.004088	0.002750	45
46	0.029007	0.018984	0.012472	0.008225	0.005445	0.003617	0.002412	46
47	0.026859	0.017416	0.011338	0.007410	0.004861	0.003201	0.002116	47
48	0.024869	0.015978	0.010307	0.006676	0.004340	0.002833	0.001856	48

Annuity calculations such as these can be performed by calculators. In instances involving a considerable number of periods, however, these calculations become quite time-consuming. For this reason, tables are generally used for annuity computations.

HOW TO USE TABLE 7.3 FOR AN ORDINARY ANNUITY

Assume an ordinary annuity of $500 made quarterly at 6% interest for two years.

Step 1: Adjust the interest rate and the time.

Rate: 6% ÷ 4 = 1.5%
Time: 2 years × 4 = 8 periods

Step 2: In the table, go across the correct number of periods (found in Step 1) until you reach the correct interest rate column (found in Step 1).

1.5% for 8 periods = Table value of 8.4328 3911

Step 3: Multiply the table value by the amount of the investments being made into the annuity and round off.

8.4328 3911 × $500 = $4,216.42

Practice Problems

Determine the ending investment balance for each of the following. Use the annuity tables (Table 7.3).

9. At the end of each year, Max invests $2,500 at an annual interest rate of 8%. He plans to trade in his old car and buy a new one with the investment balance after 4 years. How much will he have?

10. Jerry invests $500 at the end of each 6-month period into a college education fund for his son. If the annual interest rate is 8%, and he does this for 8 years, how much will his son have for a college education?

TABLE 7.3 ORDINARY ANNUITY TABLE

n	1%	1.5%	2%	2.5%	3%	3.5%	n
1	1.0000 0000	1.0000 0000	1.0000 0000	1.0000 0000	1.0000 0000	1.0000 0000	1
2	2.0100 0000	2.0150 0000	2.0200 0000	2.0250 0000	2.0300 0000	2.0350 0000	2
3	3.0301 0000	3.0452 2500	3.0604 0000	3.0756 2500	3.0909 0000	3.1062 2500	3
4	4.0604 0100	4.0909 0338	4.1216 0800	4.1525 1562	4.1836 2700	4.2149 4288	4
5	5.1010 0501	5.1522 6693	5.2040 4016	5.2563 2852	5.3091 3581	5.3624 6588	5
6	6.1520 1506	6.2295 5093	6.3081 2096	6.3877 3673	6.4684 0988	6.5501 5218	6
7	7.2135 3521	7.3229 9419	7.4342 8338	7.5474 3015	7.6624 6218	7.7794 0751	7
8	8.2856 7056	8.4328 3911	8.5829 6905	8.7361 1590	8.8923 3605	9.0516 8677	8
9	9.3685 2727	9.5593 3169	9.7546 2843	9.9545 1880	10.1591 0613	10.3684 9581	9
10	10.4622 1254	10.7027 2167	10.9497 2100	11.2033 8177	11.4638 7931	11.7313 9316	10
11	11.5668 3467	11.8632 6249	12.1687 1542	12.4834 6631	12.8077 9569	13.1419 9192	11
12	12.6825 0301	13.0412 1143	13.4120 8973	13.7955 5297	14.1920 2956	14.6019 6164	12
13	13.8093 2804	14.2368 2960	14.6803 3152	15.1404 4179	15.6177 9045	16.1130 3030	13
14	14.9474 2132	15.4503 8205	15.9739 3815	16.5189 5284	17.0863 2416	17.6769 8636	14
15	16.0968 9554	16.6821 3778	17.2934 1692	17.9319 2666	18.5989 1389	19.2956 8088	15
16	17.2578 6449	17.9323 6984	18.6392 8525	19.3802 2483	20.1568 8130	20.9710 2971	16
17	18.4304 4314	19.2013 5539	20.0120 7096	20.8647 3045	21.7615 8774	22.7050 1575	17
18	19.6147 4757	20.4893 7572	21.4123 1238	22.3863 4871	23.4144 3537	24.4996 9130	18
19	20.8108 9504	21.7967 1636	22.8405 5863	23.9460 0743	25.1168 6844	26.3571 8050	19
20	22.0190 0399	23.1236 6710	24.2973 6980	25.5446 5761	26.8703 7449	28.2796 8181	20
21	23.2391 9403	24.4705 2211	25.7833 1719	27.1832 7405	28.6764 8572	30.2694 7068	21
22	24.4715 8598	25.8375 7994	27.2989 8354	28.8628 5590	30.5367 8030	32.3289 0215	22
23	25.7163 0183	27.2251 4364	28.8449 6321	30.5844 2730	32.4528 8370	34.4604 1373	23
24	26.9734 6485	28.6335 2080	30.4218 6247	32.3490 3798	34.4264 7022	36.6665 2821	24
25	28.2431 9950	30.0630 2361	32.0302 9972	34.1577 6393	36.4592 6432	28.9498 5669	25
30	34.7848 9153	37.5386 8137	40.5680 7921	43.9027 0316	47.5754 1571	51.6226 7728	30
35	41.6602 7560	45.5920 8789	49.9944 7763	54.9282 0744	60.4620 8181	66.6740 1274	35
40	48.8863 7336	54.2678 9391	60.4019 8318	67.4025 5354	75.4012 5973	84.5502 7775	40
45	56.4810 7472	63.6142 0096	71.8927 1027	81.5161 3116	92.7198 6139	105.7816 7290	45
50	64.4631 8218	73.6828 2804	84.5794 0145	97.4843 4879	112.7968 6729	130.9979 1016	50
55	72.8524 5735	84.5295 9893	98.5865 3365	115.5509 2136	136.0716 1972	160.9468 8984	55
60	81.6696 6986	96.2146 5171	114.0515 3942	135.9915 8995	163.0534 368C	196.5168 8288	60
65	90.9366 4882	108.8027 7216	131.1261 5541	159.1183 3027	194.3327 5782	238.7628 7650	65
70	100.6763 3684	122.3637 5295	149.9779 1114	185.2841 1421	230.5940 6374	288.9378 6459	70
75	110.9128 4684	136.9727 8063	170.7917 7276	214.8882 9705	272.6308 5559	348.5300 1083	75
80	121.6715 2172	152.7108 5247	193.7719 5780	248.3827 1265	321.3630 1855	419.3067 8685	80
85	132.9789 9715	169.6652 2551	219.1439 3897	286.2785 6955	377.8569 5165	503.3673 9448	85
90	144.8632 6746	187.9299 0038	247.1566 5632	329.1542 5328	443.3489 0365	603.2050 2701	90
95	157.3537 5501	207.6061 4246	278.0849 5978	377.6641 5398	519.2720 2568	721.7808 1595	95
100	170.4813 8294	228.8030 4330	312.2323 0591	432.5486 5404	607.2877 3270	862.6116 5666	100

n	4%	4.5%	5%	6%	7%	8%	n
1	1.0000 0000	1.0000 0000	1.0000 0000	1.0000 0000	1.0000 0000	1.0000 0000	1
2	2.0400 0000	2.0450 0000	2.0500 0000	2.0600 0000	2.0700 0000	2.0800 0000	2
3	3.1216 0000	3.1370 2500	3.1525 0000	3.1836 0000	3.2149 0000	3.2464.0000	3
4	4.2464 6400	4.2781 9112	4.3101 2500	4.3746 1600	4.4399 4300	4.5061 1200	4
5	5.4163 2256	5.4707 0973	5.5256 3125	5.6370 9296	5.7507 3901	5.8666 0096	5
6	6.6329 7546	6.7168 9166	6.8019 1281	6.9753 1854	7.1532 9074	7.3359 2904	6
7	7.8982 9448	8.0191 5179	8.1420 0845	8.3938 3765	8.6540 2109	8.9228 0336	7
8	9.2142 2626	9.3800 1362	9.5491 0888	9.8974 6791	10.2598 0257	10.6366 2763	8
9	10.5827 9531	10.8021 1423	11.0265 6432	11.4913 1598	11.9779 8875	12.4875 5784	9
10	12.0061 0712	12.2882 0937	12.5778 9254	13.1807 9494	13.8164 4796	14.4865 6247	10
11	13.4863 5141	13.8411 7879	14.2067 8716	14.9716 4264	15.7835 9932	16.6454 8746	11
12	15.0258 0546	15.4640 3184	15.9171 2652	16.8699 4120	17.8884 5127	18.9771 2646	12
13	16.6268 3768	17.1599 1327	17.7129 8285	18.8821 3767	20.1406 4286	21.4952 9658	13
14	18.2919 1119	18.9321 0937	19.5986 3199	21.0150 6593	22.5504 8786	24.2149 2030	14
15	20.0235 8764	20.7840 5429	21.5785 6359	23.2759 6988	25.1290 2201	27.1521 1393	15
16	21.8245 3114	22.7193 3673	23.6574 9177	25.6725 2808	27.8880 5355	30.3242 8304	16
17	23.6975 1239	24.7417 0689	25.8403 6636	28.2128 7976	30.8402 1730	33.7502 2568	17
18	25.6454 1288	26.8550 8370	28.1323 8467	30.9056 5255	33.9990 3251	37.4502 4374	18
19	27.6712 2940	29.0635 6246	30.5390 0391	33.7599 9170	37.3789 6479	41.4462 6324	19
20	29.7780 7858	31.3714 2277	33.0659 5410	36.7855 9120	40.9954 9232	45.7619 6430	20
21	31.9692 0172	33.7831 3680	35.7192 5181	39.9927 2668	44.8651 7678	50.4229 2144	21
22	34.2479 6979	36.3033 7796	38.5052 1440	43.3922 9028	49.0057 3916	55.4567 5516	22
23	36.6178 8858	38.9370 2996	41.4304 7512	46.9958 2769	53.4361 4090	60.8932 8557	23
24	39.0826 0412	41.6891 9631	44.5019 9887	50.8155 7735	58.1766 7076	66.7647 5922	24
25	41.6459 0829	44.5652 1015	47.7270 9882	54.8645 1200	63.2490 3772	73.1059 3995	25
30	56.0849 3775	61.0070 6966	66.4388 4750	79.0581 8622	94.4607 8632	113.2832 1111	30
35	73.6522 2486	81.4966 1800	90.3203 0735	111.4347 7987	138.2368 7835	172.3168 0368	35
40	95.0255 1570	107.0303 2306	120.7997 7424	154.7619 6562	199.6351 1199	259.0565 1871	40
45	121.0293 9204	138.8499 6510	159.7001 5587	212.7435 1379	285.7493 1084	386.5056 1738	45
50	152.6670 8366	178.5030 2828	209.3479 9572	290.3359 0458	406.5289 2947	573.7701 5642	50
55	191.1591 7299	227.9179 5938	272.7126 1833	394.1720 2657	575.9285 9262	848.9232 0140	55
60	237.9906 8520	289.4979 5298	353.5837 1788	533.1281 8089	813.5203 8355	1253.2132 9584	60
65	294.9683 8045	366.2378 3096	456.7980 1118	719.0828 6076	1146.7551 6064	1847.2480 8276	65
70	364.2904 5876	461.8696 7955	588.5285 1071	967.0321 6965	1614.1341 7425	2720.0800 7377	70
75	448.6313 6652	581.0443 6193	756.6537 1848	1300.9486 7977	2269.6574 1869	4002.5566 2449	75
80	551.2449 7675	729.5576 9854	971.2288 2134	1746.5998 9137	3189.0626 7969	5886.9354 2831	80
85	676.0901 2345	914.6323 3612	1245.0870 6889	2342.9817 4142	4478.5761 1972	8655.7061 1209	85
90	827.9833 3354	1145.2690 0659	1594.6073 0098	3141.0751 8718	6287.1854 2679	12723.9386 1598	90
95	1012.7846 4845	1432.6842 5949	2040.6935 2892	4209.1042 4961	8823.8535 4059	18701.5068 5690	95
100	1237.6237 0461	1790.8559 5627	2610.0251 5693	5638.3680 5857	12381.6617 9381	27484.5157 0427	100

This will give the accumulated amount of $1 deposited periodically with interest compounded.

Practice Problem Solutions

1. **Step 1:** Rate = 11% divided by 2 = 5.5%
   Time = 5 years multiplied by 2 = 10 periods (N = 10)
   Step 2: Table value = 1.708144 (from Table 7.1)
   Step 3: 1.708144 × $2,000 = $3,1416.288 = $3,416.29 rounded

2. **Step 1:** Rate = 8% divided by 4 = 2%
   Time = 3 years multiplied by 4 = 12 periods (N = 12)
   Step 2: Table value = 1.268242 (from Table 7.1)
   Step 3: 1.268242 × $1,500 = $1,902.363 = $1,902.36 rounded

3. **Step 1:** Rate = 10.5% divided by 2 = 5.25%
   Time = 4 years multiplied by 2 = 8 periods
   Step 2: Rate = 5.25% = .0525, then add 1 = 1.0525
   Step 3: 1.0525 × $800 = $1,204.6663 = $1,204.67 rounded
   ⌐ depress the "=" sign 8 times

4. **Step 1:** Rate = 10.5% divided by 12 = .875%
   Time = 4 years multiplied by 12 = 48 periods
   Step 2: Rate = .875% = .00875, then add 1 = 1.00875
   Step 3: 1.00875 × $800 = $1,215.3454 = $1,215.35 rounded
   ⌐ depress the "=" sign 48 times

5. **Step 1:** Rate = 12% divided by 4 = 3%
   Time = 2 years multiplied by 4 = 8 periods (N = 8)
   Step 2: Table value = .789409 (from Table 7.2)
   Step 3: .789409 × $2,000 = $1,578.818 = $1,578.82 rounded

6. **Step 1:** Rate = 9% divided by 2 = 4.5%
   Time = 10 years multiplied by 2 = 20 periods (N = 20)
   Step 2: Table value = .414643 (from Table 7.2)
   Step 3: .414643 × $40,000 = $16,585.72

7. **Step 1:** Rate = 14%
   Time = 2 years = 2 periods
   Step 2: Rate = 14% = .14, then add 1 = 1.14
   Step 3: $5,250 ÷ 1.14 = $4,039.7044 = $4,039.70 rounded
   ⌐ depress the "=" sign 2 times

8. **Step 1:** Rate = 10.5% divided by 12 = .875%
   Time = 3 years multiplied by 12 = 36 periods
   Step 2: Rate .875% = .00875, then add 1 = 1.00875
   Step 3: $3,000 ÷ 1.00875 = $2,192.3668 = $2,191.37 rounded
   ⌐ depress the "=" sign 36 times

9. **Step 1:** Rate = 8%
   Time 4 years = 4 periods (N = 4)
   Step 2: Table value = 4.506112 (from Table 7.3)
   Step 3: 4.506112 × $2,500 = $11,265.28

10. **Step 1:** Rate = 8% ÷ 2 = 4%
    Time = 8 years × 2 = 16 periods (N = 16)
    Step 2: Table value = 21.8245 3114 (Table 7.3)
    Step 3: 21.8245 3114 × $500 = $10,912.265 = $10,912.27 rounded

Exercise 46A: Compound Interest

Name_____

Course/Sect. No. _____   Score_____

Instructions: Use the tables to find the compound interest.

Self-Check Exercise (Answers on back)

(1) _____   **(1)** $500 compounded quarterly at 14% for 2 years.

(2) _____   **(2)** $750 compounded semiannually at 11% for 10 years.

(3) _____   **(3)** $5,000 compounded annually at 13% for 15 years.

(4) _____   **(4)** $3,400 compounded quarterly at 18% for 10 years.

(5) _____   **(5)** $431.72 compounded annually at 9% for 30 years.

(6) _____   **(6)** $2,500 compounded annually at 5% for 25 years.

(7) _____   **(7)** $1,250 compounded quarterly at 20% for 12 years.

(8) _____   **(8)** $1,700 compounded semiannually at 13% for 15 years.

(9) _____   **(9)** Find the compound interest on $15,000 compounded semiannually at 15% for 5 years.

(10) _____   **(10)** Karen invested $3,000 compounded semiannually at 9% for 3 years. Monica invested $3,100 compounded quarterly at 8% for 3 years. Which earned the most interest and how much did she earn?

Compound Interest, Present Value, and Annuities   **285**

(1) _____$158.40_____

(2) _____$1,438.32_____

(3) _____$26,271.35_____

(4) _____$16,375.64_____

(5) _____$5,296.20_____

(6) _____$5,965.89_____

(7) _____$11,751.59_____

(8) _____$9,544.42_____

(9) _____$15,914.48_____

(10) _____Karen—__$906.78_____

Exercise 46B: Compound Interest

Name_____

Course/Sect. No._____   Score_____

Instructions: Use the tables to find the compound interest.

---

(1) _____   (1) $2,000 compounded annually at 10% for 4 years.

(2) _____   (2) $1500 compounded semiannually at 16% for 9 years.

(3) _____   (3) $1,000 compounded semiannually at 12% for 8 months.

(4) _____   (4) $825 compounded annually at $7\frac{1}{2}$% for 6 years.

(5) _____   (5) $1,650 compounded semiannually at 15% for 3 years.

(6) _____   (6) $100 compounded quarterly at 6% for 5 years.

(7) _____   (7) $1,342.67 compounded monthly at 12% for 36 months.

(8) _____   (8) $930 compounded annually at 11% for 15 years.

(9) _____   (9) The single investment of $4,000 was left for 18 months to compound monthly at a 12% rate. What would be the compound interest?

(10) _____   (10) If a $1,500 single investment is left for 10 years to compound quarterly at 18%, what compound interest will have been earned?

Exercise 47A: Compound Interest

Name_____

Course/Sect. No._____   Score_____

Instructions: Use your calculator to find the *compound amount*.

Self-Check Exercise (Answers on back)

(1) _____   **(1)** $450 compounded quarterly at 14.25% for 2 years.

(2) _____   **(2)** $700 compounded semiannually at 11.5% for 10 years.

(3) _____   **(3)** $4,000 compounded annually at 13.4% for 15 years.

(4) _____   **(4)** $3,300 compounded quarterly at 18% for 20 years.

(5) _____   **(5)** $527.38 compounded annually at 8.75% for 30 years.

(6) _____   **(6)** $2,400 compounded annually at 7.9% for 25 years.

(7) _____   **(7)** $1,150 compounded quarterly at 18.5% for 7 years.

(8) _____   **(8)** $1,600 compounded semiannually at 12.5% for 5 years.

(9) _____   **(9)** Find the *compound interest* on $13,200 compounded semiannually at 14.04% for 8 years.

(10) _____   **(10)** The investment was for $2,000 compounded quarterly at 9% for 5 years. The compound amount on that investment was then reinvested at 10.5% compounded semiannually for 3 years. What was the final compound amount?

(1) _$595.43_

(2) _$2,141.44_

(3) _$26,378.85_

(4) _$111,639.31_

(5) _$6,531.33_

(6) _$16,060.07_

(7) _$4,078.41_

(8) _$2,933.66_

(9) _$25,885.27_

(10) _$4,242.57_

Exercise 47B: Compound Interest

Name _____

Course/Sect. No. _____   Score _____

Instructions: Use your calculator to find the *compound amount.*

---

(1) _____   **(1)** $1,500 compounded annually at 10.5% for 4 years.

(2) _____   **(2)** $1,400 compounded semiannually at 16% for 9 years.

(3) _____   **(3)** $900 compounded monthly at 12% for 8 months.

(4) _____   **(4)** $775 compounded annually at $7\frac{1}{2}$% for 6 years.

(5) _____   **(5)** $1,625 compounded semiannually at 15.1% for 3 years.

(6) _____   **(6)** $75 compounded quarterly at 6% for 5 years.

(7) _____   **(7)** $1,289.74 compounded monthly at 12% for 36 months.

(8) _____   **(8)** $905 compounded annually at 11% for 15 years.

(9) _____   **(9)** The single investment of $2,500 was left for 10 months to compound monthly at a 12.25% annual rate. What would be the *compound interest?*

(10) _____   **(10)** A $4,322 single investment is left for 6 years to compound annually at 16.7%. What *compound interest* will have been earned?

Exercise 48A: Present Value

Name_____

Course/Sect. No._____   Score_____

Instructions: Use the tables to find the present value.

	Annual Rate	Time	Compounding Method	(Future Value) Amount
(1) _____	**(1)** 12%	2 years	Quarterly	$3,000
(2) _____	**(2)** 12%	16 months	Monthly	$700
(3) _____	**(3)** 11%	6 years	Semiannually	$16,250
(4) _____	**(4)** 20%	20 months	Quarterly	$150.50
(5) _____	**(5)** 12%	4 years	Annually	$1,500
(6) _____	**(6)** 18%	18 months	Monthly	$5,000
(7) _____	**(7)** 8%	3 years	Quarterly	$1,200
(8) _____	**(8)** 10%	11 years	Annually	$8,531.20

(9) _____ **(9)** Jill wanted to save for a new dress. She needed $125 and only had 9 months left. If money was compounding quarterly at 14%, how much should she invest now?

(10) _____ **(10)** Find the present value of $18,000 compounded annually at 11% for 5 years.

(1) _____$2,368.23_____

(2) _____$596.97_____

(3) _____$8,547.21_____

(4) _____$56.72_____

(5) _____$953.28_____

(6) _____$3,824.56_____

(7) _____$946.19_____

(8) _____$2,990.13_____

(9) _____$112.74_____

(10) _____$10,682.12_____

Exercise 48B: Present Value

Name _____

Course/Sect. No. _____   Score _____

Instructions: Use the tables to find the present value.

		Annual Rate	Time	Compounding Method	(Future Value) Amount
(1) _____	(1)	13%	3 years	Annually	$2,500
(2) _____	(2)	15%	5 years	Semiannually	$15,000
(3) _____	(3)	14%	10 years	Annually	$150,000
(4) _____	(4)	10%	20 years	Annually	$540
(5) _____	(5)	9%	12 years	Semiannually	$1,000
(6) _____	(6)	14%	8 years	Semiannually	$30,000
(7) _____	(7)	$7\frac{1}{2}$%	17 years	Annually	$6,000
(8) _____	(8)	16%	5 years	Semiannually	$4,420

(9) _____   **(9)** Dana needed $500 in 2 years. To obtain that amount, what present value must be invested now compounded semiannually at 15%?

(10) _____   **(10)** Sue wanted to set aside enough to buy a new Lincoln Continental. She needed $27,000 and wanted it within 4 years. If she could earn interest compounded monthly at 18%, what amount would she need to invest now?

Exercise 49A: Present Value

Name_____

Course/Sect. No._____ Score_____

Instructions: Use your calculator to find the present value.

Self-Check Exercise (Answers on back)

	Annual Rate	Time	Compounding Method	(Future Value) Amount
(1) _____	**(1)** 11.75%	2 years	Quarterly	$2,000
(2) _____	**(2)** 10.5%	16 months	Monthly	$650
(3) _____	**(3)** 9.04%	6 years	Semiannually	$15,150
(4) _____	**(4)** 20.4%	28 quarters	Quarterly	$317.25
(5) _____	**(5)** $6\frac{1}{2}$%	4 years	Annually	$1,700
(6) _____	**(6)** 15%	18 months	Monthly	$3,225
(7) _____	**(7)** 7%	3 years	Quarterly	$1,050
(8) _____	**(8)** 10%	11 years	Annually	$600

(9) _____ **(9)** $25,000 five years from now would be worth what present value if it could be invested at 10% compounded quarterly?

(10) _____ **(10)** For the first 3 years the investment earned interest semiannually at 12%. That amount then was transferred to another savings account that earned interest for the next 2 years at 14% compounded annually. If the final compounded amount was $15,200, what was the amount 5 years ago?

(1) _$1,586.50_

(2) _$565.43_

(3) _$8,912.92_

(4) _$78.80_

(5) _$1,321.45_

(6) _$2,578.81_

(7) _$852.66_

(8) _$210.30_

(9) _$15,256.77_

(10) _$8,245.15_

Exercise 49B: Present Value

Name _____

Course/Sect. No. _____   Score _____

Instructions: Use your calculator to find the present value.

	Annual Rate	Time	Compounding Method	(Future Value) Amount
(1) _____	**(1)** 12.5%	3 years	Annually	$2,400
(2) _____	**(2)** 14.8%	5 years	Semiannually	$12,000
(3) _____	**(3)** 15%	10 years	Annually	$125,000
(4) _____	**(4)** 10.3%	17 years	Annually	$775
(5) _____	**(5)** $8\frac{3}{4}$%	12 years	Semiannually	$1,000
(6) _____	**(6)** 5%	8 years	Semiannually	$20,000
(7) _____	**(7)** $4\frac{1}{2}$%	12 years	Annually	$4,000
(8) _____	**(8)** 15.25%	5 years	Semiannually	$3,185.15

(9) Mary Alice needed $1,500 in 5 years. What present amount must be invested now if it will earn 8.5% compounded annually?

(9) _____

(10) How much would need to be invested on January 1 to be worth $1,000 at the end of December? Assume it could be compounded monthly at 15%.

(10) _____

Exercise 50A: Ordinary Annuities

Name _____

Course/Sect. No. _____   Score _____

Instructions: Use the table for ordinary annuities to find the investment balance.

Self-Check Exercise (Answers on back)

		Annual Rate	Time	Compounding Method	Investment Amount
(1) _____	**(1)**	10%	8 years	Semiannually	$850
(2) _____	**(2)**	7%	20 years	Semiannually	$100
(3) _____	**(3)**	8%	25 years	Annually	$250
(4) _____	**(4)**	10%	15 years	Quarterly	$100
(5) _____	**(5)**	8%	5 years	Semiannually	$75
(6) _____	**(6)**	8%	50 years	Annually	$1,000
(7) _____	**(7)**	14%	7 years	Semiannually	$225
(8) _____	**(8)**	8%	12 years	Annually	$5,000

(9) _____

(9) Chad deposits his $5.00 allowance at the end of each month. Starting in January and ending in December, what will be the investment balance if the annual rate is 12%?

(10) _____

(10) Harold invests $500 semiannually in an ordinary annuity at 9% for 12 years. What is his investment balance?

(1) _____ $20,108.87 _____

(2) _____ $8,455.03 _____

(3) _____ $18,276.48 _____

(4) _____ $13,599.16 _____

(5) _____ $900.46 _____

(6) _____ $573,770.15 _____

(7) _____ $5,073.86 _____

(8) _____ $94,885.63 _____

(9) _____ $63.41 _____

(10) _____ $20,844.60 _____

Exercise 50B: Ordinary Annuities

Name_____

Course/Sect. No._____   Score_____

Instructions: Use the table for ordinary annuities to find the investment balance.

		Annual Rate	Time	Compounding Method	Investment Amount
(1) _____	(1)	14%	5 years	Quarterly	$2,000
(2) _____	(2)	6%	15 years	Annually	$300
(3) _____	(3)	16%	10 years	Quarterly	$25
(4) _____	(4)	7%	30 years	Semiannually	$125
(5) _____	(5)	5%	10 years	Annually	$750
(6) _____	(6)	10%	15 years	Quarterly	$150
(7) _____	(7)	12%	10 years	Quarterly	$45
(8) _____	(8)	7%	50 years	Annually	$60

(9) _____ (9) At the end of each year, you invest $1,000 at 7%. What will be your investment balance after 10 years?

(10) _____ (10) Carroll deposits $5,000 at the end of each quarter for 60 quarters. If the annual interest rate is 10%, what is the investment balance?

Chapter

Installment Interest and Rebates

Preview of Terms

Effective Interest Rate
The approximation of the annual percentage rate (APR). The formula used compensates for changing amounts of principal due to payment by installments rather than by a single payment.

Installment Interest
The total interest on an installment purchase or installment loan; sometimes called the *carrying,* or *finance, charge.*

Installment Price
In an installment purchase, the amount paid in installments plus any down payment or trade-in. It is the total amount paid.

Rebate of Interest
The amount of interest that does not have to be paid when an installment purchase or loan is paid off early.

Rule of 78
The method of computing the proportion of the rebate of interest; also called the *sum-of-the-digits* method. The sum of the digits for a year uses the digits 1 through 12 $(1 + 2 + 3 + 4 + 5 + 6 + 7 + 8 + 9 + 10 + 11 + 12)$, which equals 78.

INSTALLMENT INTEREST

You can borrow and repay money in installments, which means that you would have a successively smaller amount of money to use each month (the principal would decrease after each payment). For this reason, the *true,* or *effective, rate* of interest is higher than it is for a single-payment loan. To compute this effective rate of interest, you need a formula different from the $I = PRT$ equation. Also, it is important to know how to compute the monthly payment necessary to pay off the loan and to be able to find the amount necessary to pay off the loan early. The formula used for installment loans is also used for installment purchases.

INSTALLMENT PAYMENTS AND CARRYING CHARGES

When you borrow money, you are charged interest. When you purchase goods, you usually incur a *finance,* or *carrying, charge* (regardless of what it is called, it is still interest). To determine the installment payment for goods, the total amount owed is spread evenly over the number of payments. For a loan, the payment is determined by taking the principal plus interest and dividing it by the number of payments.

For an installment purchase, you find the amount to be paid each time by adding the carrying charge to the cash price, subtracting a down payment, if there is one, and dividing the amount left by the number of payments to be made.

Example 1:   The ABC Co. borrows $6,000 at 8.5% for three years. The company agrees to pay it back monthly. What will be the monthly payment?

$$I = PRT$$

$$= \$6,000 \times .085 \times \frac{3}{1}$$

$$= \$1,530$$

$6,000	Cash loan
+ 1,530	Interest
$7,530	Total owed

Installment payment = $7,530 ÷ 36 months (3 years) = $209.17 per month

Example 2:   The Thompsons bought a dining room set. It had a cash price of $998. They agreed to pay for it in one year with weekly payments. If the carrying charge was $54, what was their installment payment?

$ 998	Cash price
+ 54	Carrying charge
$1,052	Installment price

Installment payment = $1,052 ÷ 52 weeks (1 year) = $20.23 per week

Example 3:   The XYZ Co. bought a delivery truck. It had a list price of $8,400. They traded in their old truck for $1,200 and financed the balance at 9% for 42 months. What were their payments?

$$I = PRT$$

$$= \$7,200 \times .09 \times \frac{42}{12}$$

$$= \$2,268$$

$8,400	Cash price
− 1,200	Trade-in
$7,200	Amount financed
+ 2,268	Interest
$9,468	Amount paid in installments

Installment payments = $9,468 ÷ 42 = $225.43 per month

EFFECTIVE INTEREST RATE

As was mentioned earlier, there is a special formula for finding the true, or effective, interest rate on an installment loan or purchase.

$$r = \frac{2mi}{P(n + 1)}$$

where r = The true, or effective, interest rate.

          m = The number of installment payments that could be made in 1 year (12 if monthly, 52 if weekly). It does *not* refer to the length of the loan.

          i = The dollar amount of the finance or carrying charges (the interest). This includes all charges in excess of the principal or cash price regardless of what they are called (for formula purposes, carrying charges are the same as interest). It is the installment price less the cash price.

          P = The principal or amount borrowed on an installment loan. For an installment purchase, it is the cash price less any down payment or trade-in.

          n = The number of installment payments to be made.

To find the effective interest rate, substitute into the formula and solve for r. Unless instructed otherwise, *round all rates to one-tenth of 1%.*

Example 4:   George bought a lot for $6,000. He agreed to pay $1,300 down and the balance in 60 monthly installments of $95 each. Find the effective interest rate.

$6,000	Cash price
− 1,300	Down payment
$4,700	Principal
$ 95	Installment payment
× 60	Number of monthly installments
$5,700	Amount paid in installments
+ 1,300	Down payment
$7,000	Installment price
$7,000	Installment price
− 6,000	Cash price
$1,000	Interest

m = 12   Number of monthly payments in 1 year
i = $1,000
P = $4,700
n = 60   Number of installments

Note: *There are only two ways of buying—with cash or on the installment plan. Find the total amount paid on the installment plan and subtract the cash price to get the amount of interest.*

$$r = \frac{2mi}{P(n + 1)}$$

$$= \frac{(2)(12)(1000)}{4,700(60 + 1)}$$

$$= .0837$$

$$= 8.4\%$$

Example 5:   Elnora bought a dishwasher. The cash price was $570, and she agreed to pay 10% down and the balance in 26 weekly payments of $21 each. Find the effective rate of interest.

$570	Cash price
× .1	10%
$57	Down payment
$570	Cash price
− 57	Down payment
$513	Principal
$ 21	Weekly payment
× 26	Number of weekly payments
$546	Amount paid in installments
+ 57	Down payment
$603	Installment price
$603	Installment price
− 570	Cash price
$33	Interest

$m = 52$ Number of weekly payments in 1 year
$i = \$33$
$P = \$513$
$n = 26$ Number of installments

$$r = \frac{2mi}{P(n + 1)} = \frac{(2)(52)(33)}{513(26 + 1)} = .2477 = 24.8\%$$

Practice Problems

In each case, find the installment payment and the effective interest rate. Round to one-tenth of 1%.

1. Scott paid $200 down and an additional $850 for a used car that had a cash price of $900. He agreed to pay the balance in 24 equal monthly payments.

2. Dale bought a motor bike on the installment plan. The cash price was $450. He paid $50 down and agreed to pay $500 more over the next 18 months.

REBATE OF INTEREST

When an installment loan or purchase is repaid in full before it is scheduled to be paid off, a portion of the interest does not have to be paid. Many individuals believe that the *interest rebate* — the portion that does not have to be paid — is found by dividing the total interest owed by the number of payments, and then multiplying this amount by the number of payments left at the time the installment loan or purchase is paid off early. However, most businesses use the *sum-of-the-digits method* (also called the *rule of 78*) to calculate an interest rebate. The sum-of-the-digits method was set up to recover as much of the interest as quickly as possible.

Example 6:   Jody purchases a C.B. radio for $120 plus $24 in carrying charges. She is to repay the amount in four equal monthly installments. After making two payments, Jody pays in full the balance due. What was the rebate of interest and the amount needed to make the final payment?

                      120   Cash price
                    + 24   Interest
                    $144   Installment price

            $144 ÷ 4 months = $36 per month
    (The $36 is made up of principal and interest payments.)

In the sum-of-the-digits method, the $24 in interest is divided up so that more is repaid earlier. The number representing how many installment payments are to be made is substituted for n in the formula:

$$\frac{n(n + 1)}{2}.$$

In this example, $n = 4$ because there is a total of four monthly installments. This formula will give you the total number of ways the interest will be divided (in the straight-line method, the interest was divided four ways, 1/4 each month).

$$\frac{n(n + 1)}{2} = \frac{4(4 + 1)}{2} = \frac{4(5)}{2} = \frac{20}{2} = 10.$$

You then use the 10 as the denominator of a fraction; the numerator is the number of payments left. You then will have a fraction for each month that represents the proportion of the total interest that is paid that month. The first month, there were four payments left; the second month, three payments left; the third month, two payments left; and the fourth month, one payment left. The interest would be divided as follows:

$$1\text{st month} = \frac{4}{10} \times \$24 \text{ interest} = \quad \$\ 9.60$$

$$2\text{nd month} = \frac{3}{10} \times \$24 \text{ interest} = \quad 7.20$$

$$3\text{rd month} = \frac{2}{10} \times \$24 \text{ interest} = \quad 4.80$$

$$4\text{th month} = \frac{1}{10} \times \$24 \text{ interest} = + \quad 2.40$$
$$\overline{\$24.00}$$

The $36 monthly payment can be broken down as follows using the sum-of-the-digits method:

Sum-of-the-Digits Method

	Interest		Principal*		
Payment 1:	$ 9.60	+	$ 26.40	=	$ 36
Payment 2:	7.20	+	28.80	=	36
Payment 3:	4.80	+	31.20	=	36
Payment 4:	+ 2.40	+	+ 33.60	=	+ 36
	$24.00		$120.00		$144

(*The principal is found by subtracting the interest from $36 each time.)

To arrive at the sum-of-the-digits fraction, you should use the formula

$$\frac{n(n + 1)}{2}$$

both for the sum of the digits for the payments left and for the total sum of the digits. In the example just completed, there were a *total* of four payments to be made, and there were 2 payments *left* when the balance was paid in full. Substituting both the 2 and the 4 into the equation yields the following:

$$\frac{n(n + 1)}{2} = \frac{2(2 + 1)}{2} = \frac{2(3)}{2} = \frac{6}{2} = 3 \qquad \text{Sum of the digits for the payments } left$$

$$\frac{n(n + 1)}{2} = \frac{4(4 + 1)}{2} = \frac{4(5)}{2} = \frac{20}{2} = 10 \qquad \text{Sum of the digits for the } total \text{ payments}$$

You are now ready to use a formula to calculate the rebate of interest without determining the breakdown of interest and principal for each month.

$$\text{Interest rebate} = \text{Interest} \times \frac{\text{Sum of digits for payments left}}{\text{Sum of digits for total payments}}$$

$$= \$24 \times \frac{3}{10}$$

$$= \$7.20$$

Example 7:

Sherrie opened an account and purchased a fur coat for $1,355. She paid nothing down and agreed to pay $140 a month for 12 months. After 10 months, she paid off the balance due.

A. What was the effective rate of interest? Round to a tenth of 1%

$$
\begin{array}{rl}
\$\ 140 & \text{Monthly payment} \\
\times\quad 12 & \\
\hline
\$1,680 & \text{Installment price} \\
-\ 1,355 & \text{Cash price} \\
\hline
\$\ 325 & \text{Interest}
\end{array}
$$

$$m = 12, \ i = \$325, \ P = \$1,355, \ n = 12$$

$$r = \frac{2mi}{P(n + 1)} = \frac{(2)(12)(325)}{1355(12 + 1)} = 44.3\%$$

B. What was the interest rebate?

$$
\begin{array}{rl}
12 & \text{Total number of payments} \\
-10 & \text{Number of payments made} \\
\hline
2 & \text{Payments left}
\end{array}
$$

$$\frac{n(n + 1)}{2} = \frac{2(2 + 1)}{2} = 3 \qquad \text{Sum of the digits for the payments left}$$

$$\frac{n(n + 1)}{2} = \frac{12(12 + 1)}{2} = 78 \qquad \text{Sum of the digits for the total payments}$$

$$\text{Interest rebate} = \text{Interest} \times \frac{\text{Sum of digits for payments left}}{\text{Sum of digits for total payments}}$$

$$= \$325 \times \frac{3}{78}$$

$$= \$12.50$$

C. What was the amount needed to pay off the installment early?

$ 140	Monthly payment
× 10	Number of months paid
$1,400	Paid
$1,680.00	Installment price (less down payment)
− 1,400.00	Paid
$ 280.00	Owed after 10 payments
− 12.50	Rebate
$ 267.50	Needed to pay off balance

Practice Problems

Find the rebate of interest and the amount needed to pay off the installment debt early.

3. Chad was making a $41 monthly installment payment on an original balance of $492. Of the $492, $53 was for carrying charges. Chad paid the note off early after making 5 of the 12 installments.

4. Mark was going to pay off his installment balance after 12 of the 18 monthly installments. His finance charges (interest) were $72, and his monthly payment was $23.

Practice Problem Solutions

1. $850 \div 24 = \$35.42$ per month

$$r = \frac{2mi}{P(n+1)} = \frac{(2)(12)(150)}{700(24+1)} = .2057 = 20.6\%$$

2. $500 \div 18 = \$27.78$ per month

$$r = \frac{2mi}{P(n+1)} = \frac{(2)(12)(100)}{400(18+1)} = .3158 = 31.6\%$$

3. Sum of digits for payments left: $\dfrac{n(n+1)}{2} = \dfrac{7(7+1)}{2} = 28$

Sum of digits for total payments: $\dfrac{n(n+1)}{2} = \dfrac{12(12+1)}{2} = 78$

Rebate of interest $= \$53 \times \dfrac{28}{78}$

$\qquad\qquad\qquad = \$19.03$

$492.00	Installment price
− 205.00	Paid
$287.00	Owed after 5 payments
− 19.03	Rebate
$267.97	Needed to pay off balance early

4. Sum of digits for payments left: $\dfrac{n(n+1)}{2} = \dfrac{6(6+1)}{2} = 21$

Sum of digits for total payments: $\dfrac{n(n+1)}{2} = \dfrac{18(18+1)}{2} = 171$

Rebate of interest $= \$72 \times \dfrac{21}{171}$

$\qquad\qquad\qquad = \$8.84$

$414.00	Installment price
− 276.00	Paid
$138.00	Owed after 12 payments
− 8.84	Rebate
$129.16	Needed to pay off balance early

Exercise 51A: Installment Interest

Name _____

Course/Sect. No. _____   Score_____

Instructions: Find the regular installment payment and the approximate effective interest rate rounded to one-tenth of 1%.

	Self-Check Exercise (Answers on back)		

(1) _____

	Amount of Loan	**Number of Installments**	**Finance Charge**
	$1,500	26 weeks	$150
(1) Payment			

(2) _____

(2) Rate

	$900	20 weeks	$55
(3) Payment			

(3) _____

(4) _____

(4) Rate

	Amount of Purchase	**Down Payment**	**Finance Charge**	**Number of Installments**
	$6,199	$150	$1,200	24 months
(5) Payment				

(5) _____

(6) _____

(6) Rate

	$2,349.95	$200	$457	30 months
(7) Payment				

(7) _____

(8) _____

(8) Rate

Chad bought some real estate for $12,000. He paid 10% down with the balance due in 60 equal monthly payments of $275. Find the approximate effective rate and the total amount paid.

(9) _____

(9) Rate

(10) _____

(10) Total paid

(1) _$63.46/week_

(2) _38.5%_

(3) _$47.75/week_

(4) _30.3%_

(5) _$302.04/month_

(6) _19.0%_

(7) _$86.90/month_

(8) _16.5%_

(9) _20.8%_

(10) _$17,700_

Exercise 51B: Installment Interest

Name _____

Course/Sect. No. _____   Score _____

Instructions: Find the regular installment payment and the approximate effective interest rate rounded to one-tenth of 1%.

	Amount of Loan $3,000	**Number of Installments** 36 months	**Finance Charge** $580
(1) _____	**(1)** Payment		
(2) _____	**(2)** Rate		
	$6,500	42 months	$1,325
(3) _____	**(3)** Payment		
(4) _____	**(4)** Rate		

	Amount of Purchase $8,500	**Down Payment** $300	**Finance Charge** $1,700	**Number of Installments** 36 months
(5) _____	**(5)** Payment			
(6) _____	**(6)** Rate			
	$995	–0–	$65	16 weeks
(7) _____	**(7)** Payment			
(8) _____	**(8)** Rate			

Scott bought a car with a window sticker price of $5,995 for $1,500 down and $6,000 to be paid out monthly in 42 equal installments. Find the approximate effective rate and the monthly payment.

(9) _____ **(9)** Rate

(10) _____ **(10)** Payment

Exercise 52A: Rebate of Interest

Name_____

Course/Sect. No. _____   Score_____

Instructions: Answer each question as instructed.

---

(1) _____

(2) _____

(3) _____

(4) _____

(5) _____

(6) _____

(7) _____

(8) _____

(9) _____

(10) _____

Self-Check Exercise (Answers on back)

Find the sum-of-the-digits for the following problems.

(1) 24 months   **(2)** 12 weeks

(3) 18 weeks   **(4)** 42 weeks

Find the interest rebate on the next problems.

	Interest	Scheduled Payments	Payments Remaining
(5)	$83.50	42	30
(6)	$1,505	60	35
(7)	$649.20	18	12
(8)	1,100	36	24

(9) Harold and Flo bought a video-tape player for $995. The installment agreement was to pay it off in 24 equal payments of $50. After 12 payments they paid it off. What was the balance of the debt after the interest rebate?

(10) The cash price was $5,200 and the trade-in was $1,500. There were to be 42 equal payments of $112.50. After making 28 payments, it was paid off early. What was the balance needed to pay it off?

(1) _____300_____

(2) _____78_____

(3) _____171_____

(4) _____903_____

(5) _____$43_____

(6) _____$518.11_____

(7) _____$296.13_____

(8) _____$495.50_____

(9) _____$546.70_____

(10) _____$1,455.81_____

Exercise 52B: Rebate of Interest

Name_____

Course/Sect. No._____   Score_____

Instructions: Answer each question as instructed.

(1) _____	Find the sum-of-the-digits for the following problems. **(1)** 30 months **(2)** 48 months
(2) _____	
(3) _____	**(3)** 36 months **(4)** 60 months
(4) _____	

Find the interest rebate on the next problems.

	Interest	Scheduled Payments	Payments Remaining
(5) _____	**(5)** $125	30	15
(6) _____	**(6)** $300	24	18
(7) _____	**(7)** $2,000	48	30
(8) _____	**(8)** $38.25	12	8

(9) _____

(9) Dale borrowed $400 to be repaid with 14 weekly payments of $29.50 and one of $29.75. After making 10 of these 15 installments, he paid it off early. What was the balance of the debt after the rebate?

(10) _____

(10) A television set cost $499.98. A down payment of $50 was made, and the balance was to be repaid in 18 equal monthly installments of $31.50. If it was paid off shortly after the 8th payment, what was the balance due?

Chapter

9

Charge Account Interest and Repayment Schedules

Preview of Terms

Amortization Schedule
The summary of the systematic retirement of a debt by installments.

Finance Charges
When a business charges interest on unpaid balances, they usually call it *finance* or *carrying* charges.

CHARGE ACCOUNT INTEREST

Most charge cards use the "adjusted balance method" of determining what amount will be subjected to a carrying or finance charge. Under this method the monthly finance charge rate is multiplied by the adjusted balance. This adjusted balance is found by taking the amount owed at the end of the previous billing cycle and subtracting payments and credits received during the current billing cycle. All new purchases made during the current billing cycle are added to the account balance after the finance charge is computed. This means the customer has thirty days to pay the full amount before any finance charges are levied.

Some stores continue to use the older rate of 1.5% on the unpaid balance (an 18% annual rate), while other stores have increased their rate to 1.75% (a 21% annual rate). Most allow a rate reduction for unpaid amounts in excess of $500, usually 1% (a 12% annual rate). A minimum payment schedule is also common. An example shown on the next page will be used for all problems referring to a minimum amount due.

Minimum Payment Schedule

If your new balance is	$15 to $100	$101 to $250	$251 to $500	$501 to $800	over $800
Your minimum payment is	$15	$35	$65	$130	$\frac{1}{5}$ of balance

Some charge cards, such as Visa and Master Card, also allow the privilege of obtaining cash at a bank by using the card. Since this privilege is used much less frequently than charging merchandise, it will be noted but not included in the examples and problems.

The formats stores use to show the monthly details of customers' accounts differ considerably. However, the following method is usually the way the new balance is computed, regardless of the format of the statement.

$$
\begin{array}{l}
\text{Previous balance} \\
- \text{ Credits (returns)} \\
\underline{- \text{ Payments}} \\
\text{Adjusted balance} \\
+ \text{ New purchases} \\
\underline{+ \text{ Finance charges (interest)}} \\
\text{New balance}
\end{array}
$$

The finance charge or interest is calculated on the adjusted balance each month. After the credits and payments have been subtracted from last month's balance, the rate used is multiplied by the adjusted balance to give the finance charge. The new balance is the sum of the adjusted balance, any new purchases made this month (not 30 days old yet), and the finance charge.

Example 1:

Previous balance = $167.84   Payments = $50.00   Credits = $58.48
New purchases = $ 89.50   Monthly rate = 1.5%

Previous balance =	$167.84
Credits = −	58.48
Payments = −	50.00
Adjusted balance =	$ 59.36 × 1.5% = $.89
New purchases = +	89.50
Finance charges = +	.89
New balance =	$149.75

Example 2:

On November 1, Freddie owed the MNO Co. $158.16. During the month of November, she returned merchandise worth $55.62, made new purchases of $402.15, and paid $25 on account. What was her balance on December 1 (new balance)? Assume a finance charge rate of 1.75% on the unpaid balance.

Previous balance =	$158.16
Credits (return) = −	55.62
Payments = −	25.00
Adjusted balance =	$ 77.54 × 1.75% = $1.36
New purchases = +	402.15
Finance charge = +	1.36
New balance =	$481.05

Example 3:

The unpaid balance from last month was $752.03. During this month the minimum payment was made (using the payment schedule provided), a return for credit of $36.77 was made, and new purchases of $84.20 were recorded. If the finance charge is 1.75% on unpaid balances through $500 and 1% on unpaid balances over $500, what would be the finance charge and the new balance for the current billing?

$$
\begin{aligned}
\text{Previous balance} &= \$752.03 \\
\text{Credit (returns)} &= - 36.77 \\
\text{Payment} &= - 130.00 \\
\hline
\text{Adjusted balance} &= \$585.26 \\
\text{New purchases} &= + 84.20 \\
\text{Finance charge} &= + 9.60* \\
\hline
\text{New balance} &= \$679.06
\end{aligned}
$$

*(1.75% × $500 = $8.75) plus (1% × $85.26 = $.85) = $9.60. The $85.26 represents the unpaid balance that is greater than $500.

Practice Problems

1. Find the finance charge and new balance on the following. The carrying charge is 1.5% per month.
   Previous balance = $226.14   Returns   = $115.35
   New purchases = $226.79   Payments = $100.00

2. Previous balance = $947.13   Payments   = based on minimum
   New purchases = $103.51   schedule on page 322
   Returns   = $ 87.99   Monthly rate = 1.75% on 1st $500
                                            and 1% on amounts
                                            over $500

AMORTIZATION SCHEDULES

When you purchase a home, or a business, you put some money down and make arrangements to pay the balance in monthly installments over the life of the loan. This long-term loan is called a *mortgage.* In a typical situation, the amount paid monthly to the savings and loan company includes a portion for annual property taxes and homeowner's fire and insurance protection. These tax and insurance payments are held by the savings and loan company in an escrow account, which means that the amount is earmarked for use later in paying your semiannual or annual insurance premium and property tax liabilities.

Recently, because of extremely high interest rates that have been maintained in an effort to dampen inflation, home mortgage rates have been too high for many borrowers to afford. In an effort to compensate, long-term lenders have recently devised several new and creative ways of financing home mortgages. There are now "adjustable rate mortgages," "variable rate mortgages," and "renegotiable rate mortgages," to name just a few. The primary difference in these new methods is that the interest rate is not fixed for the life of the loan but can be adjusted somehow according to prevailing market rates of interest. At the time of printing, these methods were very new and were not acceptable to many lenders. Therefore the examples in this chapter still use the older, more traditional way of financing home loans.

With a traditional mortgage, monthly payments for principal and interest over the life of the loan will be constant. Since the interest is calculated on the

TABLE 9.1

AMOUNT	5 YEARS	10 YEARS	15 YEARS	20 YEARS	25 YEARS	30 YEARS	35 YEARS	40 YEARS
40,500	880.58	557.90	460.33	418.05	396.96	385.70	379.48	375.97
41,000	891.45	564.78	466.01	423.21	401.86	390.46	384.16	380.61
41,500	902.32	571.67	471.70	428.37	406.76	395.22	388.85	385.25
42,000	913.19	578.56	477.38	433.53	411.66	399.99	393.53	389.89
42,500	924.06	585.45	483.06	438.69	416.56	404.75	398.22	394.54
43,000	934.93	592.33	488.75	443.85	421.46	409.51	402.90	399.18
43,500	945.80	599.22	494.43	449.01	426.36	414.27	407.59	403.82
44,000	956.67	606.11	500.11	454.17	431.26	419.03	412.27	408.46
44,500	967.54	613.00	505.79	459.33	436.16	423.79	416.96	413.10
45,000	978.42	619.88	511.48	464.49	441.06	428.56	421.64	417.74
45,500	989.29	626.77	517.16	469.66	445.96	433.32	426.33	422.38
46,000	1,000.16	633.66	522.84	474.82	450.86	438.08	431.01	427.03
46,500	1,011.03	640.55	528.53	479.98	455.76	442.84	435.70	431.67
47,000	1,021.90	647.43	534.21	485.14	460.66	447.60	440.38	436.31
47,500	1,032.77	654.32	539.89	490.30	465.56	452.36	445.06	440.95
48,000	1,043.64	661.21	545.58	495.46	470.46	457.13	449.75	445.59
48,500	1,054.51	668.10	551.26	500.62	475.36	461.89	454.43	450.23
49,000	1,065.38	674.98	556.94	505.78	480.27	466.65	459.12	454.87
49,500	1,076.26	681.87	562.62	510.94	485.17	471.41	463.80	459.52
50,000	1,087.13	688.76	568.31	516.10	490.07	476.17	468.49	464.16
50,500	1,098.00	695.65	573.99	521.26	494.97	480.93	473.17	468.80

AMOUNT	5 YEARS	10 YEARS	15 YEARS	20 YEARS	25 YEARS	30 YEARS	35 YEARS	40 YEARS
60,000	1,304.54	826.50	681.96	619.31	588.07	571.39	562.17	556.98
60,500	1,315.41	833.39	687.64	624.47	592.97	576.16	566.86	561.62
61,000	1,326.28	840.27	693.32	629.63	597.87	580.92	571.54	566.26
61,500	1,337.15	847.16	699.01	634.80	602.77	585.68	576.23	570.90
62,000	1,348.03	854.05	704.69	639.96	607.67	590.44	580.91	575.54
62,500	1,358.90	860.94	710.37	645.12	612.57	595.20	585.60	580.18
63,000	1,369.77	867.82	716.05	650.28	617.47	599.96	590.28	584.83
63,500	1,380.64	874.71	721.74	655.44	622.37	604.73	594.97	589.47
64,000	1,391.51	881.60	727.42	660.60	627.27	609.49	599.65	594.11
64,500	1,402.38	888.49	733.10	665.76	632.17	614.25	604.34	598.75
65,000	1,413.25	895.37	738.79	670.92	637.07	619.01	609.02	603.39
65,500	1,424.12	902.26	744.47	676.08	641.97	623.77	613.71	608.03
66,000	1,434.99	909.15	750.15	681.24	646.87	628.53	618.39	612.67
66,500	1,445.87	916.04	755.84	686.40	651.77	633.29	623.08	617.32
67,000	1,456.74	922.92	761.52	691.57	656.68	638.06	627.76	621.96
67,500	1,467.61	929.81	767.20	696.73	661.58	642.82	632.45	626.60
68,000	1,478.48	936.70	772.88	701.89	666.48	647.58	637.13	631.24
68,500	1,489.35	943.59	778.57	707.05	671.38	652.34	641.82	635.88
69,000	1,500.22	950.47	784.25	712.21	676.28	657.10	646.50	640.52
69,500	1,511.09	957.36	789.93	717.37	681.18	661.86	651.19	645.16
70,000	1,521.96	964.25	795.62	722.53	686.08	666.63	655.87	649.81

TABLE 9.1 (cont.)

MONTHLY PAYMENT
NECESSARY TO AMORTIZE A LOAN

11.75 %   11.75 %

AMOUNT	5 YEARS	10 YEARS	15 YEARS	20 YEARS	25 YEARS	30 YEARS	35 YEARS	40 YEARS
20,000	442.38	284.07	236.84	216.75	206.97	201.89	199.17	197.68
20,500	453.44	291.17	242.76	222.17	212.14	206.94	204.15	202.62
21,000	464.50	298.27	248.68	227.59	217.32	211.99	209.13	207.57
21,500	475.56	305.37	254.60	233.01	222.49	217.03	214.11	212.51
22,000	486.62	312.48	260.52	238.43	227.67	222.08	219.08	217.45
22,500	497.67	319.58	266.44	243.84	232.84	227.13	224.06	222.39
23,000	508.73	326.68	272.36	249.26	238.01	232.17	229.04	227.33
23,500	519.79	333.78	278.28	254.68	243.19	237.22	234.02	232.28
24,000	530.85	340.88	284.20	260.10	248.36	242.27	239.00	237.22
24,500	541.91	347.98	290.12	265.52	253.54	247.32	243.98	242.16
25,000	552.97	355.08	296.04	270.94	258.71	252.36	248.96	247.10
25,500	564.03	362.19	301.96	276.36	263.88	257.41	253.94	252.04
26,000	575.09	369.29	307.88	281.77	269.06	262.46	258.92	256.98
26,500	586.15	376.39	313.81	287.19	274.23	267.50	263.90	261.93
27,000	597.21	383.49	319.73	292.61	279.41	272.55	268.87	266.87
27,500	608.27	390.59	325.65	298.03	284.58	277.60	273.85	271.81
28,000	619.33	397.69	331.57	303.45	289.75	282.64	278.83	276.75
28,500	630.39	404.80	337.49	308.87	294.93	287.69	283.81	281.69
29,000	641.44	411.90	343.41	314.29	300.10	292.74	288.79	286.64
29,500	652.50	419.00	349.33	319.70	305.28	297.79	293.77	291.58
30,000	663.56	426.10	355.25	325.12	310.45	302.83	298.75	296.52
30,500	674.62	433.20	361.17	330.54	315.62	307.88	303.73	301.46
31,000	685.68	440.30	367.09	335.96	320.80	312.93	308.71	306.40
31,500	696.74	447.40	373.01	341.38	325.97	317.97	313.69	311.34
32,000	707.80	454.51	378.93	346.80	331.15	323.02	318.66	316.29
32,500	718.86	461.61	384.85	352.22	336.32	328.07	323.64	321.23
33,000	729.92	468.71	390.77	357.63	341.49	333.12	328.62	326.17
33,500	740.98	475.81	396.70	363.05	346.67	338.16	333.60	331.11
34,000	752.04	482.91	402.62	368.47	351.84	343.21	338.58	336.05
34,500	763.10	490.01	408.54	373.89	357.02	348.26	343.56	341.00
35,000	774.16	497.11	414.46	379.31	362.19	353.30	348.54	345.94
35,500	785.21	504.22	420.38	384.73	367.36	358.35	353.52	350.88
36,000	796.27	511.32	426.30	390.15	372.54	363.40	358.50	355.82
36,500	807.33	518.42	432.22	395.56	377.71	368.44	363.47	360.76
37,000	818.39	525.52	438.14	400.98	382.89	373.49	368.45	365.70
37,500	829.45	532.62	444.06	406.40	388.06	378.54	373.43	370.65
38,000	840.51	539.72	449.98	411.82	393.23	383.59	378.41	375.59
38,500	851.57	546.83	455.90	417.24	398.41	388.63	383.39	380.53
39,000	862.63	553.93	461.82	422.66	403.58	393.68	388.37	385.47
39,500	873.69	561.03	467.74	428.08	408.76	398.73	393.35	390.41
40,000	884.75	568.13	473.66	433.49	413.93	403.77	398.33	395.36

unpaid balance, the payments for the first several years are made up mostly of interest; very little is payment on the principal owed. Since interest is deductible on your income taxes, you should know the total annual amount of interest paid. Some savings and loan companies will provide you with that information; others will not. You can calculate the yearly interest amount

TABLE 9.1 (cont.)

MONTHLY PAYMENT
11.75 % NECESSARY TO AMORTIZE A LOAN 11.75 %

AMOUNT	5 YEARS	10 YEARS	15 YEARS	20 YEARS	25 YEARS	30 YEARS	35 YEARS	40 YEARS
65,000	1,437.70	923.19	769.69	704.41	672.62	656.12	647.27	642.44
65,500	1,448.76	930.30	775.61	709.83	677.79	661.16	652.25	647.38
66,000	1,459.82	937.40	781.53	715.25	682.97	666.21	657.22	652.32
66,500	1,470.88	944.50	787.45	720.67	688.14	671.26	662.20	657.26
67,000	1,481.94	951.60	793.37	726.09	693.32	676.31	667.18	662.20
67,500	1,492.99	958.70	799.29	731.50	698.49	681.35	672.16	667.15
68,000	1,504.05	965.80	805.21	736.92	703.66	686.40	677.14	672.09
68,500	1,515.11	972.91	811.13	742.34	708.84	691.45	682.12	677.03
69,000	1,526.17	980.01	817.05	747.76	714.01	696.49	687.10	681.97
69,500	1,537.23	987.11	822.97	753.18	719.19	701.54	692.08	686.91
70,000	1,548.29	994.21	828.89	758.60	724.36	706.59	697.06	691.86
70,500	1,559.35	1,001.31	834.81	764.01	729.53	711.63	702.04	696.80
71,000	1,570.41	1,008.41	840.74	769.43	734.71	716.68	707.01	701.74
71,500	1,581.47	1,015.51	846.66	774.85	739.88	721.73	711.99	706.68
72,000	1,592.53	1,022.62	852.58	780.27	745.06	726.78	716.97	711.62
72,500	1,603.59	1,029.72	858.50	785.69	750.23	731.82	721.95	716.56
73,000	1,614.65	1,036.82	864.42	791.11	755.40	736.87	726.93	721.51
73,500	1,625.71	1,043.92	870.34	796.53	760.58	741.92	731.91	726.45
74,000	1,636.76	1,051.02	876.26	801.94	765.75	746.96	736.89	731.39
74,500	1,647.82	1,058.12	882.18	807.36	770.93	752.01	741.87	736.33
75,000	1,658.88	1,065.23	888.10	812.78	776.10	757.06	746.85	741.27
75,500	1,669.94	1,072.33	894.02	818.20	781.27	762.10	751.82	746.22
76,000	1,681.00	1,079.43	899.94	823.62	786.45	767.15	756.80	751.16
76,500	1,692.06	1,086.53	905.86	829.04	791.62	772.20	761.78	756.10
77,000	1,703.12	1,093.63	911.78	834.46	796.80	777.25	766.76	761.04
77,500	1,714.18	1,100.73	917.70	839.87	801.97	782.29	771.74	765.98
78,000	1,725.24	1,107.83	923.62	845.29	807.14	787.34	776.72	770.92
78,500	1,736.30	1,114.94	929.55	850.71	812.32	792.39	781.70	775.87
79,000	1,747.36	1,122.04	935.47	856.13	817.49	797.43	786.68	780.81
79,500	1,758.42	1,129.14	941.39	861.55	822.67	802.48	791.66	785.75
80,000	1,769.48	1,136.24	947.31	866.97	827.84	807.53	796.64	790.69
80,500	1,780.53	1,143.34	953.23	872.39	833.01	812.58	801.61	795.63
81,000	1,791.59	1,150.44	959.15	877.80	838.19	817.62	806.59	800.58
81,500	1,802.65	1,157.54	965.07	883.22	843.36	822.67	811.57	805.52
82,000	1,813.71	1,164.65	970.99	888.64	848.54	827.72	816.55	810.46
82,500	1,824.77	1,171.75	976.91	894.06	853.71	832.76	821.53	815.40
83,000	1,835.83	1,178.85	982.83	899.48	858.88	837.81	826.51	820.34
83,500	1,846.89	1,185.95	988.75	904.90	864.06	842.86	831.49	825.28
84,000	1,857.95	1,193.05	994.67	910.32	869.23	847.90	836.47	830.23
84,500	1,869.01	1,200.15	1,000.59	915.73	874.41	852.95	841.45	835.17
85,000	1,880.07	1,207.25	1,006.51	921.15	879.58	858.00	846.43	840.11

yourself by preparing an amortization schedule (home-loan repayment schedule). The size of the monthly payment depends on the size of the loan, the interest rate, and the length of the loan. A table such as Table 9.1 is used to determine the monthly payment.

TABLE 9.1 (cont.)

MONTHLY PAYMENT

				NECESSARY TO AMORTIZE A LOAN			

12 %   12 %

AMOUNT	5 YEARS	10 YEARS	15 YEARS	20 YEARS	25 YEARS	30 YEARS	35 YEARS	40 YEARS
60,000	1,334.69	860.84	720.11	660.66	631.95	617.18	609.34	605.11
62,000	1,379.18	889.54	744.12	682.69	653.01	637.75	629.65	625.28
64,000	1,423.67	918.23	768.12	704.71	674.07	658.32	649.96	645.45
66,000	1,468.16	946.92	792.12	726.73	695.14	678.90	670.27	665.62
68,000	1,512.65	975.62	816.13	748.75	716.20	699.47	690.58	685.79
70,000	1,557.13	1,004.31	840.13	770.77	737.27	720.04	710.90	705.96
72,000	1,601.62	1,033.01	864.13	792.79	758.33	740.61	731.21	726.13
74,000	1,646.11	1,061.70	888.14	814.82	779.40	761.18	751.52	746.30
76,000	1,690.60	1,090.40	912.14	836.84	800.46	781.76	771.83	766.47
78,000	1,735.09	1,119.09	936.15	858.86	821.53	802.33	792.14	786.64
80,000	1,779.58	1,147.78	960.15	880.88	842.59	822.90	812.45	806.81
82,000	1,824.07	1,176.48	984.15	902.90	863.66	843.47	832.76	826.98
84,000	1,868.56	1,205.17	1,008.16	924.93	884.72	864.05	853.07	847.15
86,000	1,913.05	1,233.87	1,032.16	946.95	905.78	884.62	873.38	867.32
88,000	1,957.54	1,262.56	1,056.16	968.97	926.85	905.19	893.69	887.49
90,000	2,002.03	1,291.26	1,080.17	990.99	947.91	925.76	914.01	907.66
92,000	2,046.52	1,319.95	1,104.17	1,013.01	968.98	946.33	934.32	927.83
94,000	2,091.01	1,348.65	1,128.17	1,035.03	990.04	966.91	954.63	948.00
96,000	2,135.50	1,377.34	1,152.18	1,057.06	1,011.11	987.48	974.94	968.17
98,000	2,179.99	1,406.03	1,176.18	1,079.08	1,032.17	1,008.05	995.25	988.34
100,000	2,224.47	1,434.73	1,200.18	1,101.10	1,053.24	1,028.62	1,015.56	1,008.51

MONTHLY PAYMENT

NECESSARY TO AMORTIZE A LOAN

14 %   14 %

AMOUNT	5 YEARS	10 YEARS	15 YEARS	20 YEARS	25 YEARS	30 YEARS	35 YEARS	40 YEARS
70,000	1,628.80	1,086.88	932.23	870.48	842.64	829.42	822.98	819.81
70,500	1,640.44	1,094.65	938.89	876.69	848.66	835.35	828.86	825.66
71,000	1,652.07	1,102.41	945.55	882.91	854.68	841.27	834.74	831.52
71,500	1,663.71	1,110.17	952.21	889.13	860.70	847.19	840.62	837.38
72,000	1,675.34	1,117.94	958.87	895.35	866.72	853.12	846.50	843.23
72,500	1,686.98	1,125.70	965.53	901.57	872.74	859.04	852.37	849.09
73,000	1,698.61	1,133.46	972.19	907.78	878.76	864.97	858.25	854.94
73,500	1,710.24	1,141.23	978.84	914.00	884.78	870.89	864.13	860.80
74,000	1,721.88	1,148.99	985.50	920.22	890.79	876.82	870.01	866.65
74,500	1,733.51	1,156.75	992.16	926.44	896.81	882.74	875.89	872.51
75,000	1,745.15	1,164.52	998.82	932.65	902.83	888.66	881.77	878.37
75,500	1,756.78	1,172.28	1,005.48	938.87	908.85	894.59	887.64	884.22
76,000	1,768.42	1,180.04	1,012.14	945.09	914.87	900.51	893.52	890.08
76,500	1,780.05	1,187.81	1,018.80	951.31	920.89	906.44	899.40	895.93
77,000	1,791.68	1,195.57	1,025.46	957.52	926.91	912.36	905.28	901.79
77,500	1,803.32	1,203.33	1,032.11	963.74	932.93	918.29	911.16	907.64
78,000	1,814.95	1,211.10	1,038.77	969.96	938.95	924.21	917.04	913.50
78,500	1,826.59	1,218.86	1,045.43	976.18	944.96	930.14	922.91	919.36
79,000	1,838.22	1,226.62	1,052.09	982.39	950.98	936.06	928.79	925.21
79,500	1,849.85	1,234.39	1,058.75	988.61	957.00	941.98	934.67	931.07
80,000	1,861.49	1,242.15	1,065.41	994.83	963.02	947.91	940.55	936.92

TABLE 9.1 (cont.)

```
                                   MONTHLY PAYMENT
15 %                       NECESSARY TO AMORTIZE A LOAN                    15 %
```

AMOUNT	5 YEARS	10 YEARS	15 YEARS	20 YEARS	25 YEARS	30 YEARS	35 YEARS	40 YEARS
50,000	1,189.51	806.69	699.80	658.41	640.43	632.23	628.42	626.62
51,000	1,213.30	822.82	713.80	671.57	653.23	644.88	640.98	639.15
52,000	1,237.09	838.95	727.80	684.74	666.04	657.52	653.55	651.69
53,000	1,260.88	855.09	741.79	697.91	678.85	670.17	666.12	664.22
54,000	1,284.67	871.22	755.79	711.08	691.66	682.81	678.69	676.75
55,000	1,308.46	887.35	769.78	724.24	704.47	695.45	691.26	689.28
56,000	1,332.25	903.49	783.78	737.41	717.28	708.10	703.83	701.82
57,000	1,356.04	919.62	797.78	750.58	730.08	720.74	716.39	714.35
58,000	1,379.83	935.76	811.77	763.75	742.89	733.39	728.96	726.88
59,000	1,403.62	951.89	825.77	776.92	755.70	746.03	741.53	739.41
60,000	1,427.41	968.02	839.76	790.08	768.51	758.68	754.10	751.94
61,000	1,451.20	984.16	853.76	803.25	781.32	771.32	766.67	764.48
62,000	1,474.99	1,000.29	867.76	816.42	794.13	783.97	779.23	777.01
63,000	1,498.78	1,016.42	881.75	829.59	806.93	796.61	791.80	789.54
64,000	1,522.57	1,032.56	895.75	842.76	819.74	809.25	804.37	802.07
65,000	1,546.36	1,048.69	909.74	855.92	832.55	821.90	816.94	814.61
66,000	1,570.15	1,064.82	923.74	869.09	845.36	834.54	829.51	827.14
67,000	1,593.94	1,080.96	937.73	882.26	858.17	847.19	842.08	839.67
68,000	1,617.73	1,097.09	951.73	895.43	870.98	859.83	854.64	852.20
69,000	1,641.52	1,113.22	965.73	908.60	883.78	872.48	867.21	864.73
70,000	1,665.31	1,129.36	979.72	921.76	896.59	885.12	879.78	877.27
71,000	1,689.10	1,145.49	993.72	934.93	909.40	897.77	892.35	889.80
72,000	1,712.89	1,161.62	1,007.71	948.10	922.21	910.41	904.92	902.33
73,000	1,736.68	1,177.76	1,021.71	961.27	935.02	923.05	917.48	914.86
74,000	1,760.47	1,193.89	1,035.71	974.44	947.83	935.70	930.05	927.40
75,000	1,784.26	1,210.03	1,049.70	987.60	960.63	948.34	942.62	939.93

Example 4: A $70,000 loan at 14% for 20 years would yield a monthly payment of $870.48.

Example 5: A $66,000 loan at 11 3/4% for 30 years would yield a monthly payment of $666.21.

Example 6: A $100,000 loan at 12% for 25 years would yield a monthly payment of $1,053.24.

Once you have determined the monthly payment, you can construct a loan repayment schedule. A one-year schedule is shown in Table 9.2.

The interest each month is on the unpaid balance from the previous month. This monthly installment loan uses the following formula to determine the amount of interest each month:

$$\text{Interest} = \text{Loan balance} \times \text{Rate} \times \frac{1}{12}$$

TABLE 9.2

Loan Repayment Schedule

$10,000 at 10% for 1 Year

Payment No.	Monthly Payment	Payment on Interest	Payment on Principal	Loan Balance after Payment
1	$879.16	$83.33	$795.83	$9,204.17
2	879.16	76.70	802.46	8,401.71
3	879.16	70.01	809.15	7,592.56
4	879.16	63.27	815.89	6,776.67
5	879.16	56.47	822.69	5,953.98
6	879.16	49.62	829.54	5,124.44
7	879.16	42.70	836.46	4,287.98
8	879.16	35.73	843.43	3,444.55
9	879.16	28.70	850.46	2,594.09
10	879.16	21.62	857.54	1,736.55
11	879.16	14.47	864.69	871.86
12	879.16	7.27	871.89	(.03)

To determine the amount of the monthly payment that applies to the principal, subtract the interest each month from the monthly payment. The payment on the interest plus the payment on the principal *must* always equal the amount of the monthly payment.

To find the loan balance after payment, subtract the amount of principal payment from the previous month's loan balance.

Example 7: Find the loan balance after three months on a $75,000 loan at 15% for 25 years.

Step 1: Find the monthly payment in the table.

$75,000 at 15% for 25 years = $960.63

Step 2: Find the interest on the first month's payment.

Interest = Loan balance × Rate × 1/12
= $75,000 × 15% × 1/12
= $937.50 Payment on interest

Step 3: Subtract the payment on interest from the monthly payment.

$960.63	Monthly payment
− 937.50	Payment on interest
$23.13	Payment on principal

Step 4: Subtract the payment on the principal from the last loan balance.

$75,000.00	Loan balance
− 23.13	Payment on principal
$74,976.87	Loan balance after payment

Repeat steps 2–4 using the new loan balance each time.

The loan repayment schedule—amortization schedule—for the first three months with the interest computations for Step 2 shown for each month is shown in Table 9.3.

TABLE 9.3

Loan Repayment Schedule

$75,000 at 15% for 25 Years

Payment No.	Monthly Payment	Payment on Interest	Payment on Principal	Loan Balance After Payment
1	$960.63	$937.50	$23.13	$74,976.87
2	960.63	937.21	23.42	74,953.45
3	960.63	936.92	23.71	74,929.74

Payment 1: $75,000 × 15% × 1/12 = $937.50
Payment 2: $74,976.87 × 15% × 1/12 = $937.21
Payment 3: $74,953.45 × 15% × 1/12 = $936.92

As you can see, a home loan requires that a fantastic amount of interest be paid over the life of a loan. In this example, the total interest would be $213,189 ($960.63 per month × 300 payments in 25 years = $288,189 − $75,000 principal borrowed = $213,189).

Practice Problems

Complete a loan repayment schedule for the first three months in each of the following.

3. $47,500 at 11% for 30 years   **4.** $80,000 at 11 3/4% for 25 years

5. $70,000 at 11% for 30 years

3.

Payment No.	Monthly Payment	Payment on Interest	Payment on Principal	Loan Balance After Payment
1				
2				
3				

4.

Payment No.	Monthly Payment	Payment on Interest	Payment on Principal	Loan Balance After Payment
1				
2				
3				

5.

Payment No.	Monthly Payment	Payment on Interest	Payment on Principal	Loan Balance After Payment
1				
2				
3				

Practice Problem Solutions

1. Previous balance = $226.14
   Returns = − 115.35
   Payments = − 100.00
   Subtotal = $10.79 × 1.5% = 16¢
   New purchases = + 226.79
   Finance charges = + .16
   New balance = $237.74

2. Previous balance = $947.13
   Credits (returns) = − 87.99
   Payment = − 189.43   (by the schedule, 1/5 × $947.13)
   Adjusted balance = $669.71
   New purchases = + 103.51
   Finance charge = + 10.45   (1.75% × $500) plus (1% × $169.71)
   New balance = $783.67

3.

Payment No.	Monthly Payment	Payment on Interest	Payment on Principal	Loan Balance After Payment
1	$452.36*	$435.42	$16.94	$47,483.06
2	452.36	435.26	17.10	47,465.96
3	452.36	435.10	17.26	47,448.70

*$47,500 at 11% for 30 years = $452.36

4.

Payment No.	Monthly Payment	Payment on Interest	Payment on Principal	Loan Balance After Payment
1	$827.84*	$783.33	$44.51	$79,955.49
2	827.84	782.90	44.94	79,910.55
3	827.84	782.46	45.38	79,865.17

*80,000 at 11.75% for 25 years = $827.84

5.

Payment No.	Monthly Payment	Payment on Interest	Payment on Principal	Loan Balance After Payment
1	$666.63*	$641.67	$24.96	$69,975.04
2	666.63	641.44	25.19	69,949.85
3	666.63	641.21	25.42	69,924.43

*$70,000 at 11% for 30 years = $666.63

Exercise 53A: Charge Account Interest

Name _____

Course/Sect. No. _____   Score _____

Instructions: Answer the following as required.

(1) _____	Self-Check Exercise (Answers on back) Using a monthly rate charge of 1.75% on the first $500 that is unpaid and 1% on balances over $500, find the *finance charge*. **(1)** Unpaid balance = $1,251.13 **(2)** Unpaid balance = $432.66
(2) _____	
(3) _____	**(3)** Unpaid balance = $755.83 **(4)** Unpaid balance = $943.60
(4) _____	
(5) _____	**(5)** Unpaid balance = $507.94 **(6)** Unpaid balance = $767.88
(6) _____	For the next problems find the *new balance.* Use the minimum payment schedule in the chapter to determine the payment amount. Assume a monthly rate charge of 1.75% on the first $50 unpaid and 1.5% on amounts unpaid over $500.

	Previous Balance	New Purchases	Returns
(7) _____	**(7)** $582.30	$149.99	–0–
(8) _____	**(8)** $982.31	$72.79	–0–

(9) _____ **(9)** New purchases of $85.13 were to be added to a previous balance of $105.17. During the month there were also returns of $25.19. What would be the new balance?

(10) _____ **(10)** The previous balance was $761.35, and there were new purchases of $104.60 and returns of $15.99. Find the new balance.

(1) _____$16.26_____

(2) _____$7.57_____

(3) _____$11.31_____

(4) _____$13.19_____

(5) _____$8.83_____

(6) _____$11.43_____

(7) _____$610.21_____

(8) _____$871.68_____

(9) _____$130.90_____

(10) _____$730.44_____

Exercise 53B: Charge Account Interest

Name_____

Course/Sect. No._____   Score_____

Instructions: Answer the following as required.

(1) _____

(2) _____

(3) _____

(4) _____

(5) _____

(6) _____

(7) _____

(8) _____

(9) _____

(10) _____

Using a monthly rate charge of 1.75% on the first $500 that is unpaid and 1% on balances over $500, find the *finance charge.*

(1) Unpaid balance = $384.74   **(2)** Unpaid balance = $687.90

(3) Unpaid balance = $187.91   **(4)** Unpaid balance = $1,104.79

(5) Unpaid balance = $807.51   **(6)** Unpaid balance = $498.13

For the next problems find the *new balance.* Use the minimum payment schedule in the chapter to determine the payment amount. Assume a monthly rate charge of 1.75% on the first $500 unpaid and 1.5% on amounts over $500.

Previous Balance	New Purchases	Returns
(7) $641.50	$35.50	$50
(8) $615.94	–0–	$37.12

(9) The previous balance was $417.31, there were new purchases of $15.90, and no returns were made. What is the new balance?

(10) New purchases   = –0–
Previous balance = $1,004.83
Returns   = –0–
Find the new balance.

Exercise 54A: Amortization

Name _____

Course/Sect. No. _____   Score _____

Instructions: Find the correct monthly payment on the first 6 problems, complete the repayment schedules, and answer the other questions.

(1) _____	Self-Check Exercise (Answers on back)

(1) _____

(2) _____

(3) _____

(4) _____

(5) _____

(6) _____

(7) _____

(8) _____

(9) _____

(10) _____

Self-Check Exercise (Answers on back)

(1) $71,000 at 15% for 20 years.

(2) $68,000 at 11.75% for 20 years.

(3) $65,000 at 11% for 40 years.

(4) $53,000 at 15% for 40 years.

(5) $85,000 at $11\frac{3}{4}$% for 30 years.

(6) $67,000 at 15% for 30 years.

(7) $64,000 at 12% for 30 years (put last loan balance on answer line).

Payment No.	Monthly Payment	Payment on Interest	Payment on Principal	Loan Balance after Payment

(8) $50,000 at 11% for 30 years (put last loan balance on answer line).

Payment No.	Monthly Payment	Payment on Interest	Payment on Principal	Loan Balance after Payment

(9) The Stewarts financed $58,000 for 20 years at 15%. What part of their *second* payment will be *interest*?

(10) Kendi and Terrell bought a house for $65,500. They made a down payment of $22,000, financing the rest at 11% for 30 years. If taxes and insurance included in the monthly payment are $146.35, how much is the total *monthly payment*?

(1)   _$934.93_

(2)   _$736.92_

(3)   _$603.39_

(4)   _$664.22_

(5)   _$858_

(6)   _$847.19_

(7)   _$63,944.49_

(8)   _$49,945.99_

(9)   _$724.52_

(10)   _$560.62_

Exercise 54B: Amortization

Name _____

Course/Sect. No. _____   Score _____

Instructions: Find the correct monthly payment on the first 6 problems, complete the repayment schedules, and answer the other questions.

---

(1) _____   **(1)** $82,000 at 12% for 20 years.

(2) _____   **(2)** $77,500 at 14% for 5 years.

(3) _____   **(3)** $96,000 at 12% for 25 years.

(4) _____   **(4)** $71,000 at 14% for 25 years.

(5) _____   **(5)** $64,000 at 12% for 30 years.

(6) _____   **(6)** $75,000 at 14% for 35 years.

(7) _____   **(7)** $25,000 at 11.75% for 30 years (put last loan balance on answer line).

Payment No.	Monthly Payment	Payment on Interest	Payment on Principal	Loan Balance after Payment

(8) _____   **(8)** $72,500 at 14% for 30 years (put last loan balance on answer line).

Payment No.	Monthly Payment	Payment on Interest	Payment on Principal	Loan Balance after Payment

(9) _____   **(9)** Ted and Mary Ann bought a new house for $80,000, paying 10% down and financing the rest at 14% for 30 years. What is the amount paid on the *principal* the first month?

(10) _____   **(10)** Roy bought a ranch for $125,000, paying $50,000 down and financing the balance at 11.75% for 35 years. What is the *balance of the loan* after the second payment?

Test Yourself — Unit 3

(1) Find the interest and maturity value on each.
(A) $5,050 at 7.75% for 120 days
(B) $425 at 9% for 52 days (exact interest)
(C) $1,572.38 at 8.25% for 90 days

(2) What interest rate would earn interest of $8.67 on a principal of $650 for 60 days?

(3) How many days would it take to earn $37.50 interest on $1,000 at 7.5%?

(4) $2,000 is invested to earn interest at 7% compounded quarterly for 4 years. What would be the compound amount at the end of 4 years?

(5) What would be the present value of $8,000 if money earned interest of 8% compounded semiannually for 3 years?

(6) How much would someone have at the end of 5 years if they invested $500 at the end of each six-month period when the annual interest was 10%?

(7) Find the discount and proceeds for each.

	Principal	Rate	Date of Note	Time	Discount Rate	Discount Date
(A)	$800	8%	April 10	60 days	9%	May 20
(B)	$350	6%	January 2	90 days	10%	January 30

(8) You borrowed $2,000 and agreed to pay $95 monthly for 24 months. You paid off the loan early, shortly after the 18th payment. Find the effective interest rate (round the rate to one-tenth of 1%), the rebate of interest, and the amount needed to pay off the loan.

(9) If you owed $214.20 last month and this month paid $75 on your revolving charge account, returned merchandise worth $32.50, and made new purchases of $51.95, what would be your carrying charge at 1.5% per month?

(10) Find the loan balance after the second monthly payment on a $78,000 home loan at 12% for 30 years.

(1) (A) $I = \$130.46$ **(2)** 8%   **(3)** 180 days
          $MV = \$5,180.46$
     (B) $I = \$5.45$
          $MV = \$430.45$
     (C) $I = \$32.43$
          $MV = \$1,604.81$
(4) \$2,639.86   **(5)** \$6,322.52   **(6)** \$6,288.95
(7) (A) Discount = \$4.05
          Proceeds = \$806.62
     (B) Discount = \$6.12
          Proceeds = \$349.13
(8) Rate   = 13.4%   **(9)** \$1.60   **(10)** \$77,955.12
     Rebate   = \$19.60
     Balance due = \$550.40

Unit
4

Merchandise Discounts and Markups

A Preview of What's Next

The wholesaler said her trade discounts of 15%, 10%, and 5% were the same as a single discount of 30%. Was she right?

If you had to borrow the money at 12%, would it still pay you to take advantage of terms 2/10, n/60 on a purchase of $500?

The terms were 3/20, n/90, and you received a payment of $800 within the discount period. If the original amount owed was $1,000, do they still owe you $200?

Your store received some new merchandise that cost you $8.50 per item. You would like to mark it up 30% above your cost. What would you price it for?

The advertisement claimed that all the merchandise in the store was marked down 20%. If an item is now priced at $15.50 and was originally priced at $18.00, was it marked down 20%?

Objectives

One of the most crucial decisions a business has to make involves pricing its products. The price must cover the cost of purchasing or making the merchandise, the advertising and other normal business expenses, and a desired return on the investment (*profit*). Discounts and markups affect what the price eventually will be. After successfully completing this unit, you will be able to:

1. Distinguish between trade and cash discounts.
2. Calculate the amount of a single trade discount or series of trade discounts.
3. Reduce any chain discount to a single equivalent discount.
4. Calculate the rate of discount when the list and net prices are known.
5. Compute the cash discount for single or multiple discount periods.
6. Determine the proper cash discount and the balance due when partial payments are made within the discount period.
7. Determine when it is profitable to borrow money to take advantage of a cash discount.
8. Compute equivalent markup percentages.
9. Calculate the markup based on sales price or cost.
10. Determine the cost when the sales price and the percentage of markup on sales price are known.
11. Determine the sales price when the cost and percentage of markup on sales price ar known.
12. Determine the sales price when the cost and the markup percentage on cost are known.
13. Determine the cost when the sales price and the percentage of markup on cost are known.
14. Compute the percentage of markup on either sales price or cost.
15. Calculate the amount and percentage of markdowns.

Chapter

10

Merchandise Discounts

Preview of Terms

Cash Discount The discount the seller offers the buyer if payment is made early, usually within 10 days from the date of the invoice. **Example**: The terms 2/10, n/30 are the offer of a 2% cash discount if the invoice is paid within 10 days. The "n/30" means the full amount should be paid within 30 days if the discount is not taken.

FOB Destination Terms where the seller rather than the buyer must pay the freight charges.

FOB Shipping Point Terms where the buyer rather than the seller must pay the freight charges.

List Price The price listed in the seller's catalog (before any possible discounts).

Net Price The list price less any discounts; the price the buyer must pay after trade discounts.

SEDR Abbreviation for *single equivalent discount rate;* the combined discount rate for several trade discounts. **Example**: For discounts of 10% and 5%, the SEDR is 14.5%.

Trade Discount An adjustment in the list price that can be taken regardless of when payment is made.

Discounts are given for volume buying, to sell slow-moving merchandise, and to encourage early payment. Buyers have been conditioned to expect discounts of some type, and contemporary marketing methods usually suggest the seller offer one or more of these discounts.

TRADE DISCOUNTS

A *trade discount* is usually offered between manufacturer and wholesaler or between wholesaler and retailer. It is an agreed reduction in the list price of an item and is given regardless of when the merchandise is paid for. A common use would be in conjunction with merchandise advertised in a catalog. A single catalog with set list prices is issued to all buyers. Separate trade discount sheets are then issued to different classes of buyers (those purchasing more merchandise would receive larger discounts).

Example 1:

The Vetter's Co. sold nursing supplies with a list price of $200 and a trade discount of 20%. What were the discount and the net price?

$$
\begin{array}{rl}
\$200 & \text{List price} \\
\times \quad .2 & \text{20\% discount} \\
\hline
\$40 & \text{Discount} \\
\\
\$200 & \text{List price} \\
- \quad 40 & \text{Discount} \\
\hline
\$160 & \text{Net price}
\end{array}
$$

Example 2:

The McCrary Co. sold car parts with a list price of $380 and trade discounts of 15%, 10%, and 5%. What were the total discount and the net price?

Discount 1: $380.00 List price $380.00 List price
        ×   .15 15% discount   −   57.00 Discount
              $57.00 1st discount $323.00 Adjusted list price

Discount 2: $323.00 Adjusted list price $323.00 Adjusted list price
        ×   .1 10% discount   −   32.30 Discount
              $32.30 2nd discount $290.70 Adjusted list price

Discount 3: $290.70 Adjusted list price $290.70 Adjusted list price
        ×   .05 5% discount   −   14.54 Discount
              $14.54 3rd discount $276.16 Net price

List price   = $380.00   Discount 1   = $57.00
Net price   = − 276.16   or   Discount 2   =   32.30
Total discount   = $103.84   Discount 3   = +   14.54
                                    Total discount = $103.84

Note:

The discounts can be taken in any order, and succeeding discounts are calculated on the adjusted list price—after the previous discount(s) have been subtracted.

Practice Problems

Find the total discount and the net price.

1. List price of $650 with a trade discount of 17.5%

2. List price of $925 with trade discounts of 20% and 12%

3. List price of $1,800 with trade discounts of 15%, 10%, and 5%

SINGLE EQUIVALENT DISCOUNT RATE

Multiple discounts can be converted to a *single equivalent discount rate (SEDR)*, which will give the same total discount as the multiple discounts when taken separately. The use of the SEDR will reduce the number of calculations and therefore the possibility of errors. (*Caution:* You must *not* add the separate discounts; the rate will not be correct.)

$$SEDR = 1 - [(100\% - \text{1st discount rate})(100\% - \text{2nd discount rate})$$
$$(100\% - \text{3rd discount rate})]$$

Note: *Substitute into this formula the multiple discount rates and solve for the SEDR. The solution will be the decimal equivalent of the SEDR, which should be converted to a percentage.*

Example 3: Find the SEDR for trade discounts of 10%, 8%, and 2%.

$$SEDR = 1 - [(100\% - 10\%)(100\% - 8\%)(100\% - 2\%)]$$
$$= 1 - [(90\%)(92\%)(98\%)]$$
$$= 1 - [(.9)(.92)(.98)]$$
$$= 1 - .81144$$
$$= .18856$$
$$= 18.856\%$$

Example 4: Find the SEDR for trade discounts of 20%, 10%, and 1%.

$$SEDR = 1 - [(100\% - 20\%)(100\% - 10\%)(100\% - 1\%)]$$
$$= 1 - [(80\%)(90\%)(99\%)]$$
$$= 1 - [(.8)(.9)(.99)]$$
$$= 1 - .7128$$
$$= .2872$$
$$= 28.72\%$$

Note: *Do not round off the SEDR.*

Example 5: Find (A) the SEDR and (B) the net price if the Newton Manufacturing Co. sold microwave ovens worth $380 at trade discounts of 15%, 10%, and 5%.

A.
$$SEDR = 1 - [(100\% - 15\%)(100\% - 10\%)(100\% - 5\%)]$$
$$= 1 - [(85\%)(90\%)(95\%)]$$
$$= 1 - [(.85)(.9)(.95)]$$
$$= 1 - .72675$$
$$= .27325$$
$$= 27.325\%$$

B.

$380	List price
× .27325	SEDR
$103.84	Total discount
$380.00	List price
− 103.84	Total discount
$276.16	Net price

Note: *This problem uses the same list price and trade discount as Example 2. Notice that if you add the trade discounts of 15%, 10%, and 5%, you get a 30% discount, which is more than the actual SEDR.*

The formula for the SEDR is much easier to apply if you remember that the *complements* of the discounts are multiplied together and subtracted from 1. The complement of 15% is 85%; the complement of 10% is 90%. The complement of a given number is another number which, when added to the given number, sums to 1 (1.00 or 100%).

Also note in the formula that the decimal number you get after you multiply together complements of the discounts (before subtracting from 1 in the formula) represents the net price as a percentage of the list price. In Example 5, the complements of 15%, 10%, and 5% multiplied gave a decimal number of .72675.

100.000%	List price
− 72.675%	Net price as a percentage of the list price
27.325%	SEDR
$380	List price
× .72675	Net price as a percentage of the list price
$276.17	Net price (a 1¢ difference due to rounding)

Therefore, if you do not want to know the SEDR but want only the net price, multiply the complements of the trade discounts together (which gives the net price as a percentage of the list price) and multiply that product by the list price.

Practice Problems

Find the SEDR and the net price.
4. List price of $595 with trade discounts of 20% and 5%.
5. List price of $1,350 with trade discounts of 10%, 7%, and 3%.
6. List price of $2,800 with trade discounts of 18%, 12%, and 6%.

CASH DISCOUNTS

A *cash discount* (or *sales discount*) differs from a trade discount in that it is contingent upon when the merchandise is paid for. A cash discount has a discount period. If the invoice is paid within the discount period, the discount can be subtracted from the invoice price before payment is sent in. The discount applies only to the merchandise kept. Returned merchandise and freight charges must be excluded from the invoice total before computing the discount. A typical cash discount would be written "2/10, n/60."

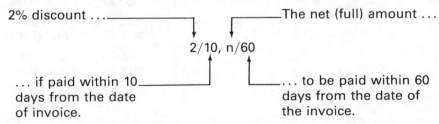

2% discount . . . ———————————— The net (full) amount . . .

2/10, n/60

. . . if paid within 10 days from the date of invoice. ———————————— . . . to be paid within 60 days from the date of the invoice.

The cash discount is stated in the terms of the purchase agreement and would be deducted from the invoice to determine the amount the buyer needs to remit.

Example 6: What should the buyer pay for merchandise that cost $400, with terms 2/10, n/60, if the invoice was dated December 2 and paid on December 12?

$400	Invoice price
× .02	2% Discount
$8	Cash discount
$400	Invoice price
− 8	Cash discount
$392	Amount due

Note: *The seller would accept payment of $392 as full payment for the $400 invoice since it was paid within the 10-day discount period.*

Example 7: What would be the amount due for merchandise that cost $1,200, with terms 3.5/15, n/90, if the invoice was dated January 7 and paid on January 20? Assume $300 of damaged merchandise was returned on January 12.

$1,200.00	Merchandise kept
− 300.00	Returned goods
$900.00	Merchandise kept
× .035	3.5% discount
$31.50	Cash discount
$900.00	Merchandise kept
− 31.50	Cash discount
$868.50	Amount due

TRANSPORTATION CHARGES

When the seller ships merchandise, the freight charges may be the responsibility of either the buyer or the seller, depending on the terms.

1. *FOB Shipping Point.* The abbreviation FOB means *free on board* and refers to the buyer. Under the terms of FOB shipping point, the buyer pays the freight from the seller's place of business (the shipping point).

2. *FOB Destination.* Under these terms, the seller pays the freight charges all the way to the destination (the buyer's home or place of business).

Seller ⊢————————————————————————————⊣ Buyer
(Shipping Point)   (Destination)

Note: *As an aid, think of FOB as "free on buyer" when referring to transportation charges.*

Example 8: The South American Importing Co. sold $5,000 in merchandise, with terms 2/10, n/30. The invoice was dated April 12, the freight charges were $50 FOB shipping point, and there were no returns of merchandise. If the invoice was paid on April 21, how much was paid?

$5,000	Merchandise kept
× .02	2% discount
$100	Cash discount

$5,000	Merchandise kept
− 100	Cash discount
$4,900	
+ 50	Freight charges
$4,950	Amount due

COMPREHENSIVE EXAMPLE

The invoice that follows is a comprehensive example that includes trade discounts, freight charges (F.O.B. shipping point), and cash discount terms. Check each of the amounts in the invoice to verify that you know how to compute them.

INVOICE

SOUTHWEST AUTO SUPPLY COMPANY   Invoice #3580
339 Main Street
Oklahoma City, Oklahoma

Sold To: Discount Auto Parts   Date: March 4, 1984
       101 West Avenue
       Houston, TX 77002   Shipped: Prepaid Freight

Terms: 2/10, n/60

Quantity	Unit	Part Number	Unit Price	Extension
34	ea.	8055	.90	30.60
17	ea.	7413	2.53	43.01
45	ea.	301	.50	22.50
120	ea.	5669	1.25	150.00
400	ea.	6232	.25	100.00
60	ea.	745	17.20	1,032.00
		Total list price		1,378.11
		Less: Trade discounts (40%, 5%)		−592.58
		Total net price		785.53
		Plus: Prepaid freight charges		+ 7.80
				793.33*

*The net amount to pay within the cash discount period would be $777.62.

Practice Problems

Find the cash discount and amount paid.

7. Merchandise valued at $10,200 was bought on terms 2/10, n/90 on January 4, and payment was made on January 14.

8. A piano worth $2,800 was bought, with terms 3/10, n/60, FOB shipping point. Additional freight charges of $75 were added, making the invoice $2,875. The invoice was paid within the discount period.

9. Sue's Perpetual Sale Store, Inc., purchased four dozen new dresses for $1,000, with terms 2/20, n/60, freight of $30, and terms FOB destination. After 5 days, the store returned $150 worth of dresses because they were the wrong color. The payment was made 15 days after the date of the invoice.

PARTIAL PAYMENT WITHIN THE DISCOUNT PERIOD

If a business is going to offer a cash discount on the full amount, there is no reason why the same discount should not be granted on a portion of the full amount. A buyer may choose to make a partial payment within the discount period, and the seller would need to know how to compute the amount to credit to the buyer's account. (If you will check, in each previous cash discount, the amount received in full payment is less than the amount the buyer is given credit for. This means the seller must give the buyer credit for *more* than is sent in.)

$$\text{Credit given for partial payment} = \frac{\text{Amount of partial payment}}{100\% - \% \text{ of cash discount}}$$

Example 9:

Larry buys $700 of merchandise, with terms 3/15, n/60. If he pays $500 within the discount period, (A) how much credit should be given on the partial payment and (B) how much will he still owe?

A. $\text{Credit given for partial payment} = \dfrac{\text{Amount of partial payment}}{100\% - \% \text{ of cash discount}}$

$$= \frac{\$500}{100\% - 3\%}$$

$$= \frac{\$500}{97\%}$$

$$= \$515.46$$

Note:

Here, $500 was sent in, and $515.46 was the credit given; that is, the credit given was more than the amount sent in.

B.

$700.00	Merchandise kept
− 515.46	Credit given
$184.54	Balance owed

Practice Problems

Find the credit given for each partial payment in the discount period and the balance due.

10. Of the $800 owed, $200 was paid within the discount period. The terms were 2/10, n/30.

11. Of the $2,000 owed, $1,250 was paid within the discount period. The terms were 3.5/15, n/60.

MULTIPLE DISCOUNTS

The cash discount terms may provide for more than one discount period and rate. The important point to remember is that *each* discount period starts from the date of the invoice. This means that there are overlaps in discount periods. As an example, consider the terms 3/10, 1/20, n/60.

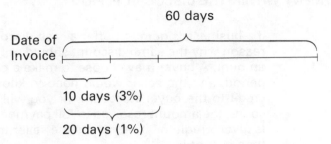

Note here that there is a 10-day overlap in discount periods. Therefore, in reality, the 1% discount period is only from the 11th day to the 20th day from the date of the invoice. If the invoice were paid before the 11th day, the larger 3% discount would be taken.

Example 10:   Find the discount on merchandise worth $500, with terms 3/10, 1/20, n/60, if the invoice is dated March 1 and paid March 7.

$$\begin{array}{ll} \$500 & \text{Merchandise kept} \\ \times \ .03 & \text{3\% discount*} \\ \hline \$ \ 15 & \text{Cash discount} \end{array}$$

*The discount was paid within the first 10 days from the date of the invoice; therefore, the first discount applies.

Example 11:   Find the discount on merchandise worth $850, with terms 3/10, 1.5/20, n/60, if the invoice is dated August 15 and paid August 27.

$$\begin{array}{ll} \$ \ 850 & \text{Merchandise kept} \\ \times \ .015 & \text{1.5\% discount*} \\ \hline \$12.75 & \text{Cash discount} \end{array}$$

*The discount was paid between the 10th and 20th day from the date of the invoice; therefore, the second discount applies.

Practice Problems

Find the cash discount.

12. Merchandise worth $1,800, with terms 2/10, n/60, had an invoice date of October 8. Payment was made on October 18.

13. Merchandise worth $375.58, with terms 4/15, 2.5/30, n/90, had an invoice date of May 15. Payment was made on June 5.

BORROWING TO TAKE ADVANTAGE OF DISCOUNTS

Cash discounts add up over the course of a year. An efficient business will take advantage of as many cash discounts as possible, even if it has to borrow the money to do so.

Example 12:

Merchandise of $1,500 is bought, with terms 2/10, n/30. What will be the *net savings* (cash discount less loan interest) if the money is borrowed at 9% to take advantage of the discount?

Step 1: Find the cash discount.

$1,500	Merchandise kept
× .02	2% discount
$ 30	Cash discount
$1,500	Merchandise kept
− 30	Cash discount
$1,470	Needed in discount period

Step 2: Find the interest charge.

$$I = PRT$$
$$= \$1,470 \times .09 \times \frac{20}{360}$$
$$= \$7.35*$$

*Although the interest here is only $7.35, most banks have a minimum charge on any loan—such as $10.

Note:

Only $1,470, and not $1,500, needs to be borrowed if payment is made within the discount period. Also, it is normal to make the payment on the last day of the discount period—the 10th day in this example. Therefore, if the money is borrowed and paid on the 10th day, there will be 20 days left in the original terms. This is why the number 20 is used in the interest computation for time.

Step 3: Find the net savings.

$30.00	Cash discount
− 7.35	Interest charge
$22.65	Net savings

Practice Problems

Find the net savings through borrowing to take advantage of a discount.

14. $700 at 2/10, n/30 are the merchandise value and terms. Money can be borrowed at 9.25% or $10, whichever is more.

15. You made a $4,800 purchase with terms 3/15, n/60 on June 4. On June 19, you borrowed enough to pay the amount needed at 10.5% interest.

Practice Problem Solutions

1. $650.00 $650.00
     ×   .175   − 113.75
     $113.75 Discount $536.25 Net price

2. $925
     ×   .2
     $185

     $925 $740.00
     − 185   ×   .12
     $740 $ 88.80

     $740.00 $185.00
     − 88.80   + 88.80
     $651.20 Net price $273.80 Discount

3. $1,800 $1,530 $1,377.00
     ×   .15   ×   .1   ×   .05
     $ 270 $ 153 $ 68.85

     $1,800 $1,530 $1,377.00
     −   270   −   153   −   68.85
     $1,530 $1,377 $1,308.15 Net price

     $270.00
      153.00
     + 68.85
     $491.85 Discount

4. SEDR = 1 − [(100% − 20%) (100% − 5%)]
          = 1 − [(80%) (95%)]
          = 1 − .76
          = .24
          = 24%

     $595.00 List price
     ×   .76
     $452.20 Net price

5. SEDR = 1 − [(100% − 10%) (100% − 7%) (100% − 3%)]
          = 1 − [(90%) (93%) (97%)]
          = 1 − .81189
          = .18811
          = 18.811%

     $1,350.00 List price
     ×   .81189
     $1,096.05 Net price

6. SEDR = 1 − [(100% − 18%) (100% − 12%) (100% − 6%)]
　　　= 1 − [(82%) (88%) (94%)]
　　　= 1 − .678304
　　　= .321696
　　　= 32.1696%

　$2,800.00　List price
　× 　.678304 ◄
　$1,899.25　Net price

7.　$10,200
　　× 　　.02
　　$　204　Cash discount

　　$10,200
　　− 　　204
　　$ 9,996　Amount paid

8.　$2,800
　　− 　.03
　　$　84　Cash discount

　　$2,875
　　− 　84
　　$2,791　Amount paid

9.　$1,000　Bought
　　− 　150　Returned
　　$　850　Kept

　　$850
　　× .02
　　$ 17　Cash discount

　　$850
　　− 17
　　$833　Amount paid

10. Credit $= \dfrac{\$200}{100\% - 2\%} = \dfrac{\$200}{98\%} = \$204.08$

　　$800.00
　− 204.08
　$595.92　Balance due

11. Credit $= \dfrac{\$1,250}{100\% - 3.5\%} = \dfrac{\$1,250}{96.5\%} = \$1,295.34$

　　$2,000.00
　− 1,295.34
　$　704.66　Balance due

12.　$1,800
　　× 　.02
　　$　36　Cash discount

13.　$375.58
　　× 　.025
　　$　9.39　Cash discount

14. $700
× .02

$ 14   Cash discount

$700
− 14

$686   Amount needed to pay within discount period

30   Days net amount due
−10   Days in discount period
20   Days for loan period

$I = PRT$

$= \$686 \times 9.25\% \times \dfrac{20}{360}$

$= \$3.53$

$14   Cash discount
− 10   Minimum loan charge (exceeds interest of $3.53)
$ 4   Net savings

15. $4,800
× .03

$ 144   Cash discount

$4,800
− 144

$4,656   Amount needed to pay within discount period

60   Days net amount due
−15   Days in discount period
45   Days for loan period

$I = PRT$

$= \$4,656 \times 10.5\% \times \dfrac{45}{360}$

$= \$61.11$

$144.00   Cash discount
− 61.11   Interest
$ 82.89   Net savings

Exercise 55A: Trade Discount and Net Price

Name _____

Course/Sect. No. _____   Score_____

Instructions: Find the discount or net price as instructed.

Self-Check Exercise (Answers on back)

Find the total discounts.

(1) _____

(1) List price = $99.95   **(2)** List price = $650
       Discount = 12.5%   Discount = 8%, 3.5%

(2) _____

(3) _____

(3) List price = $1,000   **(4)** List price = $149.99
       Discount = 5%, 3%, 1%   Discount = 10%, 5%

(4) _____

Find the net price.

(5) _____

(5) List price = $4,250   **(6)** List price = $89.50
       Discount = 6%, 3%   Discount = 15.5%, 10.5%

(6) _____

(7) _____

(7) List price = $750.35   **(8)** List price = $600
       Discount = 13.5%, 5.5%   Discount = 13%, 12%

(8) _____

(9) _____

(9) For buying in volume, Lewis Grocery received discounts of 8%, 6%, and 4% on a purchase of $52,386.38. What amount was discounted?

(10) _____

(10) Due to a decrease in production costs, Allied discounted all of its prices by 5%. Preferred customers received a second discount of 8%. One of those preferred customers ordered goods worth $5,134.86. What was the net price?

(1) ____$12.49____

(2) ____$72.93____

(3) ____$87.71____

(4) ____$21.75____

(5) ___$3,875.15___

(6) ____$67.69____

(7) ____$613.35____

(8) ____$459.36____

(9) ___$8,894.79___

(10) ___$4,487.87___

Exercise 55B: Trade Discount and Net Price

Name _____

Course/Sect. No. _____   Score _____

Instructions: Find the discount or net price as instructed.

(1) _____

(2) _____

(3) _____

(4) _____

(5) _____

(6) _____

(7) _____

(8) _____

(9) _____

(10) _____

Find the total discounts.

(1) List price = $1,800
Discount = 10%, 7%

(2) List price = $200
Discount = 5%, 2.5%

(3) List price = $3,150
Discount = 15%, 5%

(4) List price = $759.95
Discount = 20%, 10%, 5%

Find the net price.

(5) List price = $515
Discount = 13%, 8%, 4%

(6) List price = $5,000
Discount = 20%, 10%, 2%

(7) List price = $3,495
Discount = 6%, 4%, 2%

(8) List price = $600
Discount = 20%, 5%

(9) ABC Sporting Goods bought $4,200 worth of tennis rackets from Willings Sporting Goods. The company received discounts of 10% and 8%. What was its bill?

(10) A good customer of Erdman's bought a diamond for $6,000 with discounts of 5% and 5%. What was the net price?

Exercise 56A: Single Equivalent Discounts

Name _____

Course/Sect. No. _____   Score _____

Instructions: Find the single equivalent discount rate (SEDR). *Do not round.*

Self-Check Exercise (Answers on back)

(1) _____

(1) List price = $100
Discounts = 5%, 4%

(2) List price = $350
Discounts = 17.5%, 2%

(2) _____

(3) _____

(3) List price = $1,200
Discounts = 10%, 6%, 3%

(4) List price = $459.95
Discounts = 16%, 8.5%

(4) _____

(5) _____

(5) List price = $12,450
Discounts = 11.5%, 5.5%

(6) List price = $1,350.31
Discounts = 4%, 2%, 1%

(6) _____

(7) _____

(7) List price = $689.49
Discounts = 5.5%, 3.5%

(8) List price = $1,820.50
Discounts = 7%, 5%, 3%

(8) _____

(9) _____

(9) Since Marcia was such a good customer, Donna's Dresses agreed to give her discounts of 5% and 3%. If she bought $644.75 worth of dresses at list price, what was her net price?

(10) _____

(10) How much *less* than list price did Drew pay for a case of paint listed at $78.50 if he received discounts of 9% and 6.5%?

(1) _8.8%_

(2) _19.15%_

(3) _17.938%_

(4) _23.14%_

(5) _16.3675%_

(6) _6.8608%_

(7) _8.8075%_

(8) _14.3005%_

(9) _$594.14_

(10) _$11.71_

Exercise 56B: Single Equivalent Discounts

Name _____

Course/Sect. No. _____   Score _____

Instructions: Find the single equivalent discount rate (SEDR). *Do not round.*

---

(1) _____

(2) _____

(3) _____

(4) _____

(5) _____

(6) _____

(7) _____

(8) _____

(9) _____

(10) _____

(1) List price = $40
Discounts = 3%, 1%

(2) List price = $150
Discounts = 10%, 2%

(3) List price = $500
Discounts = 8%, 6%

(4) List price = $5,050
Discounts = 11%, 7%

(5) List price = $739.25
Discounts = 3%, 1%

(6) List price = $1,600
Discounts = 20%, 10%

(7) List price = $513
Discounts = 7.5%, 3.5%

(8) List price = $10,000
Discounts = 13%, 9%

(9) Larry's Linen Store received trade discounts of 10%, and 5% on an order of sheets. The list price was $963.80. What was the SEDR?

(10) Sid's Sport Shop purchased $4,110.38 worth of ski equipment from Skico, which gave discounts of 12% and 4%. What was the SEDR?

Exercise 57A: Cash Discounts

Name_____

Course/Sect. No._____   Score_____

Instructions: Find the cash discount.

	Cost	Terms	Invoice Date	Date Paid
(1) _____	**(1)** $500	2/10, n/60	Feb. 3	Feb. 20
(2) _____	**(2)** $125	2/15, n/45	Jun. 4	Jun. 10
(3) _____	**(3)** $2,000	3.5/15, n/90	Oct. 7	Oct. 20
(4) _____	**(4)** $415.25	1/10, n/60	Nov. 7	Nov. 27
(5) _____	**(5)** $1,000	2/10, n/45	May 3	May 8
(6) _____	**(6)** $16,000	2/15, n/90	Jan. 17	Feb. 1
(7) _____	**(7)** $2,150.75	2/15, n/45	Jun. 10	Jun. 20
(8) _____	**(8)** $4,307.10	3/15, n/90	July 27	Aug. 17

(9) _____

(9) Willie and Louise bought $450 worth of frozen food for their cafe. The invoice date was August 11, and the terms were 4/20, n/60. How much do they owe on August 31?

(10) _____

(10) On September 19, Tick Tock sold $685 worth of clocks to Fenton Furniture, with terms of 3.5/10, n/30. What is the amount of the total discount if they pay within the discount period?

(1) _–0–_

(2) _$2.50_

(3) _$70_

(4) _–0–_

(5) _$20_

(6) _$320_

(7) _$43.02_

(8) _–0–_

(9) _$432_

(10) _$23.98_

Exercise 57B: Cash Discounts

Name _____

Course/Sect. No. _____   Score _____

Instructions: Find the cash discount.

	Cost	Terms	Invoice Date	Date Paid
(1)	**(1)** $2,050	2/10, n/30	May 1	May 8
(2)	**(2)** $3,149	3/15, n/60	Aug. 8	Aug. 25
(3)	**(3)** $182.30	5/20, n/60	Mar. 17	Apr. 4
(4)	**(4)** $1,417	3/10, n/30	Dec. 4	Dec. 14
(5)	**(5)** $8,275	1.5/10, n/30	Jan. 20	Jan. 25
(6)	**(6)** $2,335	3.5/10, n/60	Aug. 17	Sep. 5
(7)	**(7)** $1,510	1/10, n/30	Mar. 24	Mar. 30
(8)	**(8)** $3,505.95	2/10, n/60	Oct. 20	Oct. 30

(9) _____

(9) On a purchase of $3,550 worth of plastic pipe on July 10, the terms were 3/10, n/30. If the invoice is paid on July 20, what is the net price?

(10) _____

(10) Norris Co. sells drop-in ranges to Mr. Burke of Burke Builders, the terms being 5/15, n/30. An invoice dated April 16 is for $530, and one dated April 20 is for $160. Burke wants to wait until the last possible day to pay each. If he pays each within the discount period, what is the total amount he will pay?

Exercise 58A: Partial Payments within Discount Period

Name_____

Course/Sect. No. _____   Score_____

Instructions: Find the credit given for the partial payments and the balance due.

Self-Check Exercise (Answers on back)

	Invoice Price	Terms	Partial Payment
	$600	2/10, n/30	$450

(1) _____   **(1)** Credit

(2) _____   **(2)** Balance

	$1,000	3/10, n/60	$500

(3) _____   **(3)** Credit

(4) _____   **(4)** Balance

	$3,000	2.5/10, n/45	$1,500

(5) _____   **(5)** Credit

(6) _____   **(6)** Balance

	$5,000	3.5/10, n/60	$3,500

(7) _____   **(7)** Credit

(8) _____   **(8)** Balance

(9) _____   **(9)** Because of a cash flow problem, Good Sandwich Shop could pay only $340 on May 3. The bill was for $575. If the terms were 4/10, n/30 and the invoice was dated April 23, how much did they still owe?

(10) _____   **(10)** A car dealer received a shipment of 10 new cars. The invoice, dated September 18, was for $58,500. The terms were 6/10, n/60. The dealer could pay only $18,500 on September 28. How much did he still owe?

(1) _____$459.18_____

(2) _____$140.82_____

(3) _____$515.46_____

(4) _____$484.54_____

(5) _____$1,538.46_____

(6) _____$1,461.54_____

(7) _____$3,626.94_____

(8) _____$1,373.06_____

(9) _____$220.83_____

(10) _____$38,819.15_____

Exercise 58B: Partial Payments within Discount Period

Name _____

Course/Sect. No. _____   Score _____

Instructions: Find the credit given for the partial payments and the balance due.

		Invoice Price	Terms	Partial Payment
		$400	2/10, n/30	$300
(1) _____	**(1)** Credit			
(2) _____	**(2)** Balance			
		$850	3/10, n/30	$500
(3) _____	**(3)** Credit			
(4) _____	**(4)** Balance			
		$2,150	3/10, n/60	$1,500
(5) _____	**(5)** Credit			
(6) _____	**(6)** Balance			
		$3,429	5/15, n/90	$2,000
(7) _____	**(7)** Credit			
(8) _____	**(8)** Balance			

(9) _____   **(9)** Luis bought some lumber for his woodworking shop on March 13, with terms 3/10, n/30. The invoice was for $634.58. On March 23, he paid $300. How much credit did he get?

(10) _____   **(10)** On terms of 3/5, n/20, how much credit does Teddy get for paying $300 on his bill of $850, if he pays within five days?

Exercise 59A: Multiple Cash Discounts

Name_____

Course/Sect. No._____   Score_____

Instructions: Find the cash discount.

Self-Check Exercise (Answers on back)

	Merchandise Value	Terms	Invoice Date	Payment Date
(1) _____	**(1)** $1,350.40	3/10, 1/20, n/60	Jul. 20	Jul. 30
(2) _____	**(2)** $1,250.95	4/10, 2/20, n/90	Jun. 18	Jun. 27
(3) _____	**(3)** $3,000.00	3/10, 1/20, n/60	Mar. 11	Mar. 25
(4) _____	**(4)** $980.75	1.5/10, 1/20, n/45	Nov. 12	Nov. 25
(5) _____	**(5)** $1,710.00	5/10, 3/15, n/90	Sep. 8	Sep. 28
(6) _____	**(6)** $693.17	$3\frac{1}{2}$/15, $1\frac{1}{2}$/25, n/90	Nov. 17	Nov. 27
(7) _____	**(7)** $4,325.64	3/10, 1/15, n/60	Jun. 28	Jul. 30
(8) _____	**(8)** $4,107.61	3/10, 1/20, n/90	May 7	May 13

(9) _____

(9) Goods for $930 from Fry Supply to a local electrician were invoiced on May 16 with terms of 6/10, 4/20, n/30. How much would the electrician save by paying within 10 days instead of between the 11th and 20th days?

(10) _____

(10) Since a supplier of campaign literature wants to get his money from politicians quickly, he offers terms of 6/5, 4/10, n/30. He invoiced one candidate for $1,475.60 on September 28. The candidate paid on October 5. How much did he pay?

(1) _____40.51_____

(2) _____50.04_____

(3) _____30_____

(4) _____9.81_____

(5) _____$-0-$_____

(6) _____24.26_____

(7) _____$-0-$_____

(8) _____123.23_____

(9) _____18.60_____

(10) _____$1,416.58$_____

Exercise 59B: Multiple Cash Discounts

Name _____

Course/Sect. No. _____   Score _____

Instructions: Find the cash discount.

	Merchandise Value	Terms	Invoice Date	Payment Date
(1) _____	**(1)** $5,620.50	3/10, 1/20, n/60	May 13	May 23
(2) _____	**(2)** $9,449.99	2/15, 1/25, n/60	Feb. 5	Feb. 25
(3) _____	**(3)** $3,995.00	3/10, 1/20, n/60	Oct. 7	Nov. 7
(4) _____	**(4)** $2,450.25	2.5/15, .5/30, n/90	Jan. 8	Jan. 22
(5) _____	**(5)** $850.00	$2\frac{1}{2}$/10, $1\frac{1}{2}$/15, n/60	Aug. 17	Aug. 30
(6) _____	**(6)** $555.00	2/10, $\frac{1}{2}$/20, n/60	Apr. 4	Apr. 15
(7) _____	**(7)** $3,100.00	2/10, 1/20, n/45	May 1	May 15
(8) _____	**(8)** $417.82	3/15, 1/25, n/60	Jan. 24	Feb. 10

(9) _____   **(9)** Pew Supply Co. sold First Church some new pews, listed at $3,575. The terms were 5/10, 2/20, n/60. The invoice was dated August 10 and was paid August 21. What amount was paid?

(10) _____   **(10)** Photoco gives terms of 4.5/10, 3/20, n/60 to Phil, who bought $4,350 worth of photographic supplies on April 14. Phil paid on April 29. How much did he pay?

Exercise 60A: Borrow to Take Advantage of Discount

Name _____

Course/Sect. No. _____   Score _____

Instructions: Find the net savings (or net loss) of borrowing to take advantage of a cash discount.

		List Price	Terms	Interest Rate
(1) _____	(1)	$5,100	2/10, n/30	18.5%
(2) _____	(2)	$755	2.5/10, n/60	21.5%
(3) _____	(3)	$1,250	4/15, n/45	12.5%
(4) _____	(4)	$450	4.5/10, n/60	8.5%
(5) _____	(5)	$4,100	2/10, n/60	9%
(6) _____	(6)	$8,500	4/10, n/60	14%
(7) _____	(7)	$4,000	2/15, n/90	18%
(8) _____	(8)	$10,000	2/20, n/90	9%

(9) _____

(9) The Racquet Club bought a supply of tennis outfits for $1,156.38 on terms of 4.5/20, n/60. They borrowed the money to pay the invoice on the 20th day. How much did they save if the interest rate was 11%?

(10) _____

(10) Through sales of the product, a store raises $885.69 within 10 days to help pay the invoice for the goods sold. The invoice is for $3,004.76 with terms of 3.5/10, n/30. If the store borrows the rest necessary to pay on the 10th day at 12%, how much savings do they realize?

(1) _____$50.63 Gain_____

(2) _____$3.10 Loss_____

(3) _____$37.50 Gain_____

(4) _____$15.18 Gain_____

(5) _____$31.77 Gain_____

(6) _____$181.33 Gain_____

(7) _____$67 Loss_____

(8) _____$28.50 Gain_____

(9) _____$38.54 Gain_____

(10) _____$91.74 Gain_____

Exercise 60B: Borrow to Take Advantage of Discount

Name_____

Course/Sect. No._____   Score_____

Instructions: Find the net savings (or net loss) of borrowing to take advantage of a cash discount.

	List Price	Terms	Interest Rate
(1) _____	(1) $820	2/10, n/30	10%
(2) _____	(2) $2,150	3/10, n/30	11%
(3) _____	(3) $1,000	3.5/10, n/45	22%
(4) _____	(4) $1,339	3/15, n/60	13%
(5) _____	(5) $3,000	5/15, n/60	15%
(6) _____	(6) $3,825	2.5/10, n/30	10.5%
(7) _____	(7) $5,000	3.5/10, n/60	12%
(8) _____	(8) $3,500	4.5/20, n/60	10%

(9) _____

(9) Money can be borrowed at 9.75%, with a minimum charge of $10. What are the savings, if any, when Acme, Inc., borrows to take advantage of 5/10, n/30 terms on a bill of $5,310?

(10) _____

(10) Bill's Garden Supply got a shipment of garden hoses. The invoice, dated February 25, was for $885 on terms 3.5/10, n/60. How much savings would there be if Bill borrowed the money and paid on March 5? The interest rate was 13%.

Chapter

11

Merchandise Markups and Markdowns

Preview of Terms

Equivalent Markup Percentages

The rate of markup on selling price that yields the same dollar markup as the rate of markup on cost. **Example:** A 20% markup on a $100 selling price yields a $20 markup. A 25% markup on an $80 cost yields a $20 markup. Therefore, a 20% selling price markup is equivalent to a 25% cost markup.

Markdown

A downward adjustment in the original selling price.

Markup

The difference between an item's selling price and its cost; also known as *gross* profit.

MARKUP

The price of merchandise sold must cover the costs of purchasing it (cost of goods sold) plus the operating expenses of the business and provide some operating profit. The difference between the amount a seller pays for goods and the price he sells them for is called the *markup* or *gross profit*.

$$\text{Selling price} - \text{Cost} = \text{Markup}$$

This equation is usually expressed as:

$$SP = C + M$$
$$\text{Selling price} = \text{Cost} + \text{Markup}$$

Example 1: If an item costs $1.25 and is marked up another $.35, what would the selling price be?

$$SP = C + M$$
$$SP = \$1.25 + \$.35$$
$$SP = \$1.60$$

Example 2: If merchandise costing $80 is sold for $110, what is the markup?

$$SP = C + M$$
$$\$110 = \$80 + M$$
$$\$110 - \$80 = M$$
$$\$30 = M$$

Practice Problems

1. The book cost the store manager $8.95 and she marked it up another $4.50. What was its selling price?

2. The grocer purchases a gallon of milk for $1.25 and sells it for $1.89. What is his gross profit (markup)?

3. The bicycle had a selling price of $119.99. It had been marked up $39.99. What was its cost?

PERCENTAGE OF MARKUP

Since markup can be compared to two values (cost and selling price), the markup is always expressed as a percentage of cost or as a percentage of selling price.

Example 3:

Markup   = $ 15
Cost   = + 85
Selling price = $100

Markup as a percentage of selling price = Markup/Selling price
= M/SP
= $15/$100
= .15, or 15%

Example 4:

Markup   = $ 75
Cost   = + 200
Selling price = $275

Markup as a percentage of cost = Markup/Cost
= M/C
= $75/$200
= .375, or 37.5%

Practice Problems

4. A clock costs $8 and sells for $12. What is the percentage of markup based on selling price?

5. A poker table and four chairs cost $95. If the buyer sells them for $150, what percentage of markup based on selling price will she have?

6. If a raincoat costs $16 and sells for $24, what is the percentage of markup based on cost?

EQUIVALENT MARKUP PERCENTAGES

The same item can have two different markups, depending on the basis of the markup, cost or selling price.

Example 5:

$$
\begin{aligned}
\text{Markup} &= \$\ 20 \\
\text{Cost} &= \underline{+\ 80} \\
\text{Selling price} &= \$100
\end{aligned}
$$

Markup as a % of Selling price	Markup as a % of Cost
Percent Markup $(SP) = M/SP$	Percent Markup $(C) = M/C$
$= \$20/\$100$	$= \$20/\$80$
$= .20$	$= .25$
$= 20\%$	$= 25\%$

To find the equivalent (same) markup for each, you have to apply one of the following rules:

Rule 1:

> If you know the percentage of cost markup, divide it by 100% **plus itself** to find the percentage of selling price markup.

Example 6:

If the percentage of cost markup is 25%, what would the percentage of selling price markup be?

$$
\begin{aligned}
\% \ SP \ \text{Markup} &= 25\%/(100\% + 25\%) \\
&= 25\%/125\% \\
&= .25/1.25 \\
&= .20, \text{ or } 20\%
\end{aligned}
$$

Rule 2:

> If you know the percentage of selling price markup, divide it by 100% **minus itself** to find the percentage of cost markup.

Example 7:

If the percentage of selling price markup is 20%, what would be the percentage of cost markup?

$$
\begin{aligned}
\% \ C \ \text{Markup} &= 20\%/(100\% - 20\%) \\
&= 20\%/80\% \\
&= .20/.80 \\
&= .25, \text{ or } 25\%
\end{aligned}
$$

Practice Problems

From the percentage given, find the missing percentage. (Round to a tenth if necessary.)

7. A 20% cost markup is equivalent to a ____ % markup on selling price.

8. A 40% cost markup is equivalent to a _____% markup on selling price.

9. A 23.1% markup on selling price is equivalent to a _____% cost markup.

10. A 9.1% markup on selling price is equivalent to a _____% cost markup.

For the other markup problems, begin with the basic equation, substitute known values, and solve for the unknown.

$$SP = C + M$$

DETERMINING THE COST WHEN THE SELLING PRICE AND THE PERCENTAGE OF MARKUP ON SELLING PRICE ARE KNOWN

Example 8: Find the cost of an item selling at $40 if the markup is 40% of selling price.

Markup = 40% of Selling price = 40% SP = $.4SP$

$$SP = C + M$$

$$\$40 = C + .4SP$$
$$\$40 = C + .4(\$40)$$
$$\$40 = C + \$16$$
$$\$24 = C$$

Note: *Since markup equals 40% of selling price, then 40% SP, or .4SP, can be substituted for markup. Also, since SP is known to be $40 in this problem, $40 can be substituted for SP. Then .4 × $40 gives a markup of $16.*

DETERMINING THE SELLING PRICE WHEN THE COST AND THE PERCENTAGE OF MARKUP ON SELLING PRICE ARE KNOWN

Example 9: Find the selling price for a typewriter costing $42 and marked up 30% on selling price.

Markup = 30% of Selling price = 30% SP = $.3SP$

$$SP = C + M$$
$$SP = \$42 + .3SP$$
$$SP - .3SP = \$42$$
$$.7SP = \$42$$
$$SP = \frac{\$42}{.7}$$
$$SP = \$60$$

Note: *In this problem, the selling price is the unknown. After substituting .3SP for markup, the unknown SP can be found.*

DETERMINING THE SELLING PRICE WHEN THE COST AND THE PERCENTAGE OF MARKUP ON COST ARE KNOWN

Example 10: A store owner purchases a dress for $20 and marks it up 40% on cost. Find the selling price.

$$Markup = 40\% \text{ of Cost} = 40\% \ C = .4C$$
$$SP = C + M$$

$$= \$20 + .4C$$
$$= \$20 + .4(\$20)$$
$$= \$20 + \$8$$
$$= \$28$$

Note: *Since the value for* C *is given, $20 can be substituted for* C *everywhere in the equation. Then .4 × $20 gives a markup of $8.*

DETERMINING THE COST WHEN THE SELLING PRICE AND THE PERCENTAGE OF MARKUP ON COST ARE KNOWN

Example 11: A sport coat that sells for $50 has a markup of 60% on cost. Find the cost.

$$Markup = 60\% \text{ of cost} = 60\% \ C = .6C$$
$$SP = C + M$$
$$\$50 = C + .6C$$
$$\$50 = 1.6C$$
$$\frac{\$50}{1.6} = C$$
$$\$31.25 = C$$

Practice Problems

Substitute the known values into the basic markup equation ($SP = C + M$) and find the unknown.

11. A surfboard sells for $80. The markup on it is 30% of selling price. Find the cost.

12. Find the selling price for a coat that costs $50 and was marked up 25% on the selling price.

13. A swimsuit costs $15 and is marked up 40% on cost. What is the selling price?

14. If a rug sells for $100 and is marked up 50% on cost, what is the cost?

MARKDOWNS

Merchandise is usually marked down because a store owner needs to sell excess stock to make room for new stock or wants to entice customers into the store in hopes they will also buy some merchandise that is not on sale. A *markdown* is the difference between the original selling price and the reduced selling price.

Original selling price − Reduced selling price = Markdown

Example 12: If all the merchandise, with a selling price of $4,000, is to be sold for $3,400, what is the markdown?

Original selling price − Reduced selling price = Markdown
$4,000 − $3,400 = $600

Merchandise Markups and Markdowns **385**

The percentage of markdown can be found by dividing the dollar markdown amount by the original selling price.

$$\frac{\text{Markdown}}{\text{Original selling price}} = \text{Percentage of markdown}$$

Example 13:   A coat was priced at $100 and marked down to $80. What was the percentage of markdown?

$$\$100 - \$80 = \$20 \text{ Markdown}$$
$$\$20/\$100 = \text{Percentage markdown}$$
$$20\% = \text{Percentage markdown}$$

Practice Problems

Find the dollar markdown and the rate of markdown (percentage of markdown). Round to one-tenth of 1%.

15. A lounge chair was to be sold for $395 but was reduced to sell for $325.

16. The original selling price of the dining room set was $1,575 and the marked down price was $1,450.

17. The used car was reduced from $2,950 to $2,350.

18. The set of golf clubs was originally priced at $199.95 and later was reduced by $25.

Practice Problem Solutions

1. $SP = C + M$
$ = \$8.95 + \$4.50$
$ = \$13.45$

2. $SP = C + M$
$1.89 = \$1.25 + M$
$\$1.89 - \$1.25 = M$
$\$.64 = M$

3. $SP = C + M$
$\$119.99 = C + \$39.99$
$\$119.99 - \$39.99 = C$
$\$80 = C$

4. $SP = C + M$
$\$12 = \$8 + M$
$\$4 = M$

$\dfrac{M}{SP} =$ Markup as a percentage of selling price

$\dfrac{\$4}{\$12} = .33,$ or 33.3%

5. $SP = C + M$
$\$150 = \$95 + M$
$\$55 = M$

$\dfrac{M}{SP} =$ Markup as a percentage of selling price

$\dfrac{\$55}{\$150} = .366,$ or 36.7%

6. $SP = C + M$
$\$24 = \$16 + M$
$\$8 = M$

$\dfrac{M}{C} =$ Markup as a percentage of cost

$\dfrac{\$8}{\$16} = .5$ or 50%

7. $\% \, SPM = \dfrac{20\%}{100\% + 20\%}$

$ = \dfrac{20\%}{120\%}$

$ = .1666,$ or 16.7%

8. $\% \, SPM = \dfrac{40\%}{100\% + 40\%}$

$ = \dfrac{40\%}{140\%}$

$ = .2857,$ or 28.6%

9. $\% \, CM = \dfrac{23.1\%}{100\% - 23.1\%}$

$ = \dfrac{23.1\%}{76.9\%}$

$ = .3003,$ or 30.0%

10. $\% \, CM = \dfrac{9.1\%}{100\% - 9.1\%}$

$ = \dfrac{9.1\%}{90.9\%}$

$ = .1001,$ or 10.0%

11. Markup $= 30\%$ selling price $= 30\% \, SP = .3SP$
$SP = C + M$
$SP = C + .3SP$
$\$80 = C + .3(\$80)$
$\$80 = C + \$24$
$\$56 = C$

12. Markup = 25% selling price = 25% SP = .25SP

$$SP = C + M$$
$$SP = \$50 + .25SP$$
$$SP - .25SP = \$50$$
$$.75SP = \$50$$
$$SP = \frac{\$50}{.75}$$
$$SP = \$66.67$$

13. Markup = 40% cost = 40% C = .4C

$$SP = C + M$$
$$SP = C + .4C$$
$$SP = \$15 + .4(\$15)$$
$$SP = \$15 + \$6$$
$$SP = \$21$$

14. Markup = 50% cost = 50% C = .5C

$$SP = C + M$$
$$\$100 = C + .5C$$
$$\$100 = 1.5C$$
$$\frac{\$100}{1.5} = C$$
$$\$66.67 = C$$

15.

$395	Original price
− 325	Marked down price
$ 70	Markdown

Percentage of markdown = $\dfrac{\$70}{\$395}$

= .1772

= 17.7%

16.

$1,575	Original price
− 1,450	Marked down price
$ 125	Markdown

Percentage of markdown = $\dfrac{\$125}{\$1,575}$

= .0793

= 7.9%

17.

$2,950	Original price
− 2,350	Marked down price
$ 600	Markdown

Percentage of markdown = $\dfrac{\$600}{\$2,950}$

= .2033

= 20.3%

18. $199.95 = Original price
$25.00 = Markdown

Percentage of markdown = $\dfrac{\$25}{\$199.95}$

= .1250

= 12.5%

Exercise 61A: Markup Formula

Name_____

Course/Sect. No._____   Score_____

Instructions: Determine the selling price, cost, or markup.

Self-Check Exercise (Answers on back)

(1) _____

(2) _____

(3) _____

(4) _____

(5) _____

(6) _____

(7) _____

(8) _____

(9) _____

(10) _____

(1) SP = $39.95
C = $29.95
M = ?

(2) SP = $103.50
C = ?
M = $35.00

(3) SP = ?
C = $8.30
M = $1.15

(4) SP = $42.39
C = ?
M = $7.21

(5) SP = $189.19
C = $153.75
M = ?

(6) SP = $.15
C = $.12
M = ?

(7) SP = $160
C = $150
M = ?

(8) SP = ?
C = $105.50
M = $35.95

(9) The store manager paid $1.85 each for the trinkets he sold. He sold them for $25.80 a dozen. What was the amount of markup per trinket?

(10) The three pans sold for $3.95, $45.95, and $8.95. They cost a total of $12.35. What was the total markup for the set?

(1) _____$10_____

(2) _____$68.50_____

(3) _____$9.45_____

(4) _____$35.18_____

(5) _____$35.44_____

(6) _____$.03_____

(7) _____$10_____

(8) _____$141.45_____

(9) _____$.30_____

(10) _____$46.50_____

Exercise 61B: Markup Formula

Name _____

Course/Sect. No. _____   Score _____

Instructions: Determine the selling price, cost, or markup.

(1) _____   (1) $SP = \$50.00$ (2) $SP = \$59.99$
                                   $C = \$24.50$ $C = ?$
(2) _____ $M = ?$ $M = \$15.25$

(3) _____   (3) $SP = ?$ (4) $SP = \$2,350$
                                   $C = \$82.75$ $C = \$1,760$
(4) _____ $M = \$31.25$ $M = ?$

(5) _____   (5) $SP = ?$ (6) $SP = \$8,460$
                                   $C = \$1,520$ $C = ?$
(6) _____ $M = \$430$ $M = \$1,500$

(7) _____   (7) $SP = ?$ (8) $SP = \$5,321.50$
                                   $C = \$24.99$ $C = ?$
(8) _____ $M = \$2.75$ $M = \$975.75$

(9) _____   **(9)** The table tennis set was selling for $150, which was $35.50 more than it cost. What did it cost?

(10) _____   **(10)** A new electronic game cost $39.95 and was marked up $15.95. What did it sell for?

Exercise 62A: Markup Amounts and Percentages

Name_____

Course/Sect. No._____   Score_____

Instructions: Solve for the missing part. Round all percentages to one-tenth of 1%.

Self-Check Exercise (Answers on back)

	Selling Price	Cost	% of Markup on Selling Price	% of Markup on Cost
(1)	?	$400	40%	
(2)	$75.95	?		20%
(3)	$15.25	$13.50		?
(4)	?	$1.45	90%	
(5)	$2,339	$1,950	?	
(6)	$159.95	$139.50		?
(7)	?	$19.99		75%
(8)	?	$13.35	65.5%	

(1) _____

(2) _____

(3) _____

(4) _____

(5) _____

(6) _____

(7) _____

(8) _____

(9) _____

(9) The concession stand at the county fair operates on a markup of 120% based on cost. How much did goods that cost $2,150 sell for?

(10) _____

(10) Sue buys a self-defense chemical spray from Don for $8. She marks it up 10% based on her cost. At what price does Sue sell it?

(1) _$666.67_

(2) _$63.29_

(3) _13.0%_

(4) _$14.50_

(5) _16.6%_

(6) _14.7%_

(7) _$34.98_

(8) _$38.70_

(9) _$4,730_

(10) _$8.80_

Exercise 62B: Markup Amounts and Percentages

Name_____

Course/Sect. No._____   Score_____

Instructions: Solve for the missing part. Round all percentages to one-tenth of 1%.

		Selling Price	Cost	% of Markup on Selling Price	% of Markup on Cost
(1) _____	**(1)**	$1,300	?	40%	
(2) _____	**(2)**	?	$6.50		20%
(3) _____	**(3)**	$109.95	$99.95	?	
(4) _____	**(4)**	$8,000	?	12.5%	
(5) _____	**(5)**	?	$335		18.2%
(6) _____	**(6)**	$5.99	?		35%
(7) _____	**(7)**	$27.50	?	10.8%	
(8) _____	**(8)**	?	$19.99	75%	

(9) _____

(9) First Grocery has a markup of .8% based on cost. If they pay $56,700 for a shipment from lollegg, what will these goods sell for?

(10) _____

(10) Don sells self-defense chemical sprays at a 50% markup based on selling price. If each device costs him $4, what is his selling price for each?

Exercise 63A: Converting Equivalent Markup Rates

Name _____

Course/Sect. No. _____   Score _____

Instructions: From the rate of markup given, find the other rate. Round to one-tenth of 1%.

Self-Check Exercise (Answers on back)

	Percentage of Markup on Cost	Percentage of Markup on Selling Price
(1) _____	(1) ?	22%
(2) _____	(2) 25%	?
(3) _____	(3) 30%	?
(4) _____	(4) ?	50%
(5) _____	(5) 37.5%	?
(6) _____	(6) 3.4%	?
(7) _____	(7) ?	11.56%
(8) _____	(8) ?	60.08%

(9) _____   **(9)** The store manager increased his percentage of markup on selling price from 25% to 30%. What *increase* (not the new percentage) did that cause in the percentage of markup on cost?

(10) _____   **(10)** The owner told the store manager to increase the percentage of markup on cost by 5%. That would make the *new* percentage of markup on selling price 4%. What was the *new* percentage of markup on cost?

(1) _____ 28.2% _____

(2) _____ 20% _____

(3) _____ 23.1% _____

(4) _____ 100% _____

(5) _____ 27.3% _____

(6) _____ 3.3% _____

(7) _____ 13.1% _____

(8) _____ 150.5% _____

(9) _____ 9.6% _____

(10) _____ 66.7% _____

Exercise 63B: Converting Equivalent Markup Rates

Name_____

Course/Sect. No._____   Score_____

Instructions: From the rate of markup given, find the other rate. Round to one-tenth of 1%.

	Percentage of Markup on Cost	Percent of Markup on Selling Price
(1) _____	(1) ?	5%
(2) _____	(2) 10%	?
(3) _____	(3) ?	35%
(4) _____	(4) ?	18.5%
(5) _____	(5) 62.5%	?
(6) _____	(6) ?	3.4%
(7) _____	(7) 80%	?
(8) _____	(8) 30.45%	?

(9) _____ **(9)** If an item is sold at a 60% markup on cost, what is the percentage of markup on selling price for that item?

(10) _____ **(10)** Each item was to be changed from a 40% markup on cost to a 36% markup on cost. What would be the *new* percentage of markup on selling price?

Exercise 64A: Percentage of Markdown

Name_____

Course/Sect. No._____   Score_____

Instructions: Find the percentage of markdown accurate to one-tenth of 1%.

Self-Check Exercise (Answers on back)

	Original Price	Reduced Price	Markdown Amount
(1)	(1) $25.50	$23.85	
(2)	(2) $1.99		$.37
(3)	(3)	$399	$20
(4)	(4) $16.99		$1.45
(5)	(5) $149.95	$139.95	
(6)	(6) $8.58		$.19
(7)	(7) $450.00	$429.95	
(8)	(8) $1,995	$1,750	

(1) _____

(2) _____

(3) _____

(4) _____

(5) _____

(6) _____

(7) _____

(8) _____

(9) _____

(9) Louis has some shirts in his shop that have not sold. He decides to mark them down from $17 to $9. What percentage did he mark them down?

(10) _____

(10) At the end of the baseball season, the sporting goods store reduced baseball gloves to $27.50. Since they were selling for $39.95, what was the rate of markdown?

(1) _____ 6.5% _____

(2) _____ 18.6% _____

(3) _____ 4.8% _____

(4) _____ 8.5% _____

(5) _____ 6.7% _____

(6) _____ 2.2% _____

(7) _____ 4.5% _____

(8) _____ 12.3% _____

(9) _____ 47.1% _____

(10) _____ 31.2% _____

Exercise 64B: Percentage of Markdown

Name _____

Course/Sect. No. _____   Score _____

Instructions: Find the percentage of markdown accurate to one-tenth of 1%.

	Original Price	Reduced Price	Markdown Amount
(1) _____ **(1)** $99.50		$79.50	
(2) _____ **(2)** $8,499			$750
(3) _____ **(3)**		$15.49	$1.15
(4) _____ **(4)** $435		$425	
(5) _____ **(5)**		$119.95	$10.00
(6) _____ **(6)**		$27.40	$2.60
(7) _____ **(7)** 89¢		69¢	
(8) _____ **(8)** $1.39			$.67

(9) _____   **(9)** Donna's dresses, regularly $69.95, are on sale for $49.95. Women are rushing to take advantage of this markdown. What is the percentage of markdown?

(10) _____   **(10)** The bakery marks down day-old loaves of bread from 94¢ to 64¢. What is the rate of markdown?

Test Yourself — Unit 4

(1) Find the net price for merchandise listed at $850 with trade discounts of 15%, 10%, and 5%.

(2) Find the SEDR for trade discounts of 20%, 12%, and 4%.

(3) Merchandise bought on December 3 for $325 had terms of 3/15, n/45. If it was paid for on December 17, what amount should have been paid?

(4) Partial payment of $200 was made within the discount period. If the terms were 1.5/10, n/30, how much credit should have been given for the payment?

(5) The purchase of $4,000 had terms of 3/15, n/60. How much could be saved (net) if the money was borrowed at 10.5% to take advantage of the discount?

(6) If the percentage of markup on cost is 12%, find the equivalent percentage of markup on selling price. Round to a tenth of 1%.

(7) A bicycle costs $35 and sells for $59.95. What is the percentage of markup based on selling price? Round to a tenth of 1%.

(8) A skateboard costs $8.50 and sells for $12.49. What is the percentage of markup based on cost? Round to a tenth of 1%.

(9) A desk sells for $250. The markup is 15% of selling price. Find the cost.

(10) Find the selling price for a boat that costs $3,500 and is marked up 10% on selling price.

(11) A dress costs $49.99 and is marked up 30% on cost. What is the selling price?

(12) If a lamp sells for $75 and is marked up 25% on cost, what is the cost?

(13) A winter coat was to be sold for $135 but was reduced to sell for $119.95. What is the percentage of markdown? Round to a tenth of 1%.

Unit

5

Depreciation and Inventory

A Preview of What's Next

You have been given the responsibility of replacing the equipment when it wears out. The cost of the equipment was $82,500, and there is accumulated depreciation of $51,300. Could you determine the book value and therefore help yourself project the money needs you will have in the near future for replacement of these assets?

There are acceptable ways of accelerating the depreciation expense on plant assets so more of their costs can be recovered earlier in their life. What are these methods? What new method will eventually replace the old traditional ways of computing depreciation?

You expect there are some items on your shelves that do not have a satisfactory turnover when compared to an industry standard. How do you find your merchandise inventory turnover?

When prices are rising (inflation), one method of valuing ending inventory will allow the business to show a lower profit (which is better for tax purposes). What is the method, and how does it work?

Objectives

The loss in value of a business asset is called *depreciation.* The owner or business manager must be able to determine the amount to charge as a business expense due to the depreciation of assets. The merchandise a business has not yet sold is called its *inventory.* Several important business decisions are made concerning these inventories. After successfully completing this unit, you will be able to:

1. Compute the depreciation and book value for any year of an asset's life by each of the following methods:
   a. Straight-line.
   b. Units-of-production.
   c. Declining balance.
   d. Sum-of-the-years'-digits.
   e. ACRS method.

2. Determine if merchandise is selling as fast as it should by calculating the inventory turnover at retail and the inventory turnover at cost.

3. Compute the valuation of ending inventory for each of the following methods:
   a. FIFO (First-in, first-out).
   b. LIFO (Last-in, first-out).
   c. Weighted average.
   d. Retail.

4. Find the cost of goods sold using FIFO, LIFO, and the weighted average methods.

Chapter

12

Depreciation

Preview of Terms

Accumulated Depreciation	The total depreciation of a plant asset since it was purchased.
Asset	Anything of value that a person or business owns.
Book Value	The asset's cost less its accumulated depreciation; its current worth on the books.
Depreciable Amount	The amount a plant asset can be depreciated; the *net cost* of an asset after deducting the residual value.
Depreciation	The process of allocating the cost of a plant asset to the periods it will benefit (spreading its cost over its useful life).
Obsolescence	The decrease in the value of an asset caused by availability of new, improved assets.
Residual Value	The amount that can be recovered for an asset when you are finished with it (trade-in value, scrap value, or sales value).

DEPRECIATION

As objects of value — *assets* — are used, they gradually wear out and become worth less. This process of wearing out is called *depreciation.* When something is worth less than before because it has been replaced by a newer, better asset that can perform better with greater quality, the process is called *obsolescence.* Generally, depreciation and obsolescence are grouped together and simply called *depreciation.*

To be current and to reflect their worth accurately, businesses should annually record the amount their assets have depreciated. This calculation, added to the previous depreciation, gives the *total,* or *accumulated, depreciation.* The accumulated depreciation is shown in a financial statement along with the cost of the asset to reflect the current worth, or *book value,* of the asset.

Asset cost − Accumulated depreciation = Book value

When a business buys office supplies, these supplies are recorded as an expense because they will be used up within one year—the length of time accountants use to define *current.* An asset that has more than one year of expected usefulness or life is not recorded initially as an expense. This type of asset is called a *fixed asset,* or a *plant asset.* It is depreciated over its expected life. This helps balance the income derived from its use by the costs of purchasing and operating it. This balance is called the *matching concept* in accounting.

You must know three things to be able to calculate the annual depreciation expense, that is, the amount of value the asset lost that year. These are the *cost,* the *expected life,* and the anticipated *residual value. Residual value* refers to the amount you can receive for the asset when you are finished with it. It includes such things as trade-in value, sale value, and scrap value.

Of these three components, two are educated guesses (the expected life and the anticipated residual value). For this reason, depreciation is not usually recorded in cents, but is rounded to the nearest dollar. Also, an asset should not be depreciated for any period of time less than a month. An asset purchased during the month would be recorded on the depreciation records as either the first of the current month or the first of the next month, whichever is closer to the purchase date.

The amount you receive for an asset when you are finished with it (the residual value) is, in fact, a return of the original cost to you and therefore not an expense. The net cost, or *depreciable amount,* is the original cost less the residual value, and is the maximum amount that can be depreciated.

Original cost − Residual value = Depreciable amount (Net cost)

The four traditional methods of depreciating assets with a life expectancy of more than one year are the straight-line method, the units-of-production method, the declining balance method, and the sum-of-the-years'-digits method. Of these, the last two are *accelerated* methods because they charge more depreciation expense in the early years of the asset's life than in the later years. The accelerated cost recovery system (ACRS), a relatively new approach, is another accelerated depreciation method.

Note: *Unless instructed otherwise, round each answer in depreciation problems to whole dollars.*

STRAIGHT-LINE METHOD

The *straight-line method* of depreciation, defined by the following equation, is the easiest to calculate and therefore one of the most commonly used.

$$\frac{\text{Cost} - \text{Residual value}}{\text{Estimated life}} = \text{Annual depreciation}$$

Example 1: A copying machine was purchased for $8,200. It was expected to have a useful life of 5 years (or 300,000 copies) and a residual value of $700. It reproduced total copies for each of the 5 years as follows: 42,500; 60,250; 68,000; 66,420; 62,500. Find the amount of straight-line depreciation each year.

$$\frac{\text{Cost} - \text{Residual value}}{\text{Estimated life}} = \frac{\$8,200 - \$700}{5 \text{ years}}$$

$$= \frac{\$7,500}{5 \text{ years}}$$

$$= \$1,500 \text{ per year}$$

Note: *The depreciable amount of $7,500 will be spread evenly over the five-year period. Also, dividing the $1,500 annual depreciation by the depreciable amount of $7,500 is one way of finding the straight-line rate (1,500/ $7,500 = 20%).*

Practice Problems

Find the annual straight-line depreciation expense.

	Cost	Estimated Life	Residual Value
1.	$12,000	5 years	$2,000
2.	$60,000	20 years	-0-

UNITS-OF-PRODUCTION METHOD

The *units-of-production method* is very similar to the straight-line method except it depreciates an asset based on *usage* rather than on time elapsed. If you use an asset a lot, you depreciate it a lot; if you do not use it, you do not depreciate it.

$$\frac{\text{Asset cost} - \text{Residual value}}{\text{Expected units of use}} = \text{Depreciation expense per unit}$$

You then multiply the depreciation expense per unit by the number of units the asset was used that year. You ignore the adjustment for a partial-year depreciation in this method.

Example 2:   A copying machine was purchased for $8,200. It was expected to have a useful life of 5 years, (or 300,000 copies) and a residual value of $700. It reproduced total copies for each of the 5 years as follows: 42,500; 60,250; 68,000; 66,420; 62,500. Find the depreciation expense for the first year, using the units-of-production method.

$$\frac{\text{Asset cost} - \text{Residual value}}{\text{Expected units of use}} = \frac{\$8,200 - \$700}{300,000 \text{ copies}}$$

$$= \frac{\$7,500}{300,000}$$

$$= \$.025 \text{ per copy}$$

42,500 copies × $.025 per copy = $1,062.50 depreciation expense
for year 1

Note:

The units can be any reasonable measurement, such as miles driven, hours of use, or number of items produced. Although the units-of-production method is probably the fairest method of depreciation, it is really only feasible to use on assets that have some type of counter to record the units — like a truck, which has an odometer.

Practice Problems

Compute the depreciation expense using the units-of-production method.

4. A delivery truck cost $10,500 and was expected to be driven 50,000 miles and have a trade-in value of $1,500. What depreciation expense should be recorded for a year when it was driven 11,520 miles? Round to even dollars.

5. The gasoline pumps were installed at a cost of $3,000. They are supposed to pump 500,000 gallons of gas before they need to be replaced. What amount would be depreciated on the pumps if there was no residual value and they pumped 27,000 gallons this year?

DECLINING BALANCE METHOD

Use of an accelerated method such as the *declining balance method* of depreciation is justified on the basis that the earning power of an asset is greater in the early years, and therefore the depreciation expense should be greater in the early years to match it. The declining balance *rate* customarily is twice the straight-line rate. The straight-line rate can be found quickly by dividing 100% by the number of years of expected life of the asset.

$$\text{Straight-line rate (S.L. rate)} = \frac{100\%}{\text{No. years}}$$
$$\text{Declining balance rate} = \text{Straight-line rate} \times 2$$

Example 3:

Find the straight-line rate and the declining balance rate (at twice the straight-line rate) for each of the following time periods: 5 years, 8 years, 20 years.

5 years: 100%/5 =   20% S.L. rate × 2 = 40% declining balance rate
8 years: 100%/8 = 12.5% S.L. rate × 2 = 25% declining balance rate
20 years: 100%/20 =   5% S.L. rate × 2 = 10% declining balance rate

Once you have found the declining balance *rate,* you can calculate the annual depreciation as follows:

Declining balance rate × Book value = Annual depreciation

Example 4:

A copying machine was purchased for $8,200. It was expected to have a useful life of 5 years (or 300,000 copies) and a residual value of $700. It reproduced total copies for each of the 5 years as follows: 42,500; 60,250; 68,000; 66,420; 62,500. Using the declining balance method, find the first year's depreciation.

S.L. rate = 100%/5 = 20% × 2 = 40% Declining balance rate

$8,200	Cost
− -0-	Accumulated depreciation
$8,200	Book value*
× .4	Declining balance rate
$3,280	1st year's depreciation

(*The book value and the cost will always be the same for the first year's depreciation because no previous depreciation has been recorded. Also notice that the residual value is ignored in this method — *except that the book value should never be less than the estimated residual value;* otherwise, you would be depreciating it by more than the depreciable amount allowed.)

Example 5:
Continuing with Example 4, find the depreciation for the second year.

$8,200   Cost
− 3,280   Accumulated depreciation
$4,920   Book value

Book value × Declining balance rate = Annual depreciation
$4,920   ×   40%   = $1,968

The complete depreciation schedule follows:

Declining Balance Illustration

Year	Cost	Annual Depreciation	Accumulated Depreciation	End-of-year Book Value
1	$8,200	$3,280	$3,280	$4,920
2	8,200	1,968	5,248	2,952
3	8,200	1,181	6,429	1,771
4	8,200	708	7,137	1,063
5	8,200	363*	7,500	700

Computations

Year 1: $8,200 × 40% = $3,280
Year 2: $4,920 × 40% = $1,968
Year 3: $2,952 × 40% = $1,181
Year 4: $1,771 × 40% = $ 708
Year 5: $1,063 × 40% = $ 363

(*Since the trade-in value is $700, the book value cannot be less than $700. In the fifth year, the calculated depreciation of $425 would cause the book value to be $638, which is $62 too much. Therefore, $62 must be taken from the $425 ($425 − $62 = $363), so that the depreciable amount of $7,500 is not exceeded.)

Practice Problems

Using the declining balance method at twice the straight-line rate (the double-declining balance method), find the depreciation expense for the second year in each problem. Round to dollars.

5. Cost = $8,450, Estimated life = 10 years, Estimated residual value = $450.

6. Cost = $87,000, Estimated life = 25 years, Estimated residual value = $12,000.

SUM-OF-THE-YEARS'-DIGITS METHOD

In the *sum-of-the-years'-digits method,* you use a successively smaller fraction to determine the depreciation expense each year. The results of this method are similar to those of the declining balance method; more depreciation is charged in the early years and less in the later years (an accelerated method).

The numerator of the fraction represents the remaining years of the asset's useful life. This numerator decreases in value by 1 each year. The denominator of the fraction remains constant and is the sum of the digits in the asset's life. For an asset with an expected life of five years, the sum of the digits is $5 + 4 + 3 + 2 + 1 = 15$. In this case, the fraction for the first year would be 5/15, for the second year 4/15, for the third year 3/15, and so on.

You multiply this fraction by the depreciable amount (cost less residual value) each year to calculate the depreciation for that year.

Fractional value × (Cost − Residual value) = Annual depreciation

Example 6:

A copying machine was purchased for $8,200. It was expected to have a useful life of 5 years (or 300,000 copies) and a residual value of $700. It reproduced total copies for each of the 5 years as follows: 42,500; 60,250; 68,000; 66,420; 62,500. Use the sum-of-the-years'-digits method to find the depreciation expense for each year.

		Annual Depreciation
First year:	5 = 5/15 × ($8,200 − $700) =	$2,500
Second year:	4 = 4/15 × ($8,200 − $700) =	2,000
Third year:	3 = 3/15 × ($8,200 − $700) =	1,500
Fourth year:	2 = 2/15 × ($8,200 − $700) =	1,000
Fifth year:	+1 = 1/15 × ($8,200 − $700) = +	500
	15	$7,500 Depreciable amount

Note:

As a shortcut for finding the sum-of-the-years' digits (the denominator in the fraction), use the formula $\dfrac{n(n + 1)}{2}$, where n is the number of years of useful life. In the example just completed, the expected life is five years, and the use of the shortcut formula would be as follows:

$$\frac{n(n + 1)}{2} = \frac{5(5 + 1)}{2} = \frac{5(6)}{2} = \frac{30}{2} = 15$$

The application of the formula for an asset with an expected life of 20 years would be as follows:

$$\frac{n(n + 1)}{2} = \frac{20(20 + 1)}{2} = \frac{20(21)}{2} = \frac{420}{2} = 210$$

The completed depreciation schedules for each of the four methods are shown on the next page so you can compare them.

Example 7:

A copying machine was purchased for $8,200. It was expected to have a useful life of 5 years (or 300,000 copies) and a residual value of $700. It reproduced total copies for each of the 5 years as follows: 42,500; 60,250; 68,000; 66,420; 62,500.

Practice Problems

Find the depreciation expense for the year indicated using the sum-of-the-years'-digits method.

COMPARISON OF DEPRECIATION SCHEDULES

| | STRAIGHT-LINE METHOD | | | UNITS-OF-PRODUCTION METHOD | |
Year	Annual Depreciation	Accumulated Depreciation	Year	Annual Depreciation	Accumulated Depreciation
1	$1,500	$1,500	1	$1,063	$1,063
2	1,500	3,000	2	1,506	2,569
3	1,500	4,500	3	1,700	4,269
4	1,500	6,000	4	1,661	5,930
5	1,500	7,500	5	1,563	7,493

| | DOUBLE-DECLINING BALANCE METHOD | | | SUM-OF-THE-YEARS'-DIGITS METHOD | |
Year	Annual Depreciation	Accumulated Depreciation	Year	Annual Depreciation	Accumulated Depreciation
1	$3,200	$3,280	1	$2,500	$2,500
2	1,968	5,248	2	2,000	4,500
3	1,181	6,429	3	1,500	6,000
4	708	7,137	4	1,000	7,000
5	363	7,500	5	500	7,500

PARTIAL-YEAR DEPRECIATION

You can find the partial-year depreciation by multiplying the fractional part of the first year by the depreciation expense that first year. Begin depreciating at the nearest *first day of the month,* since depreciation is not calculated for a period of time shorter than a month. That is, a September 8 purchase would include the entire month of September, since September 8 is closer to September 1 than to October 1. A May 18 purchase would not begin depreciating until June 1.

First find the annual depreciation and then adjust that by the appropriate fractional part of a year. All subsequent years will not need to be adjusted except when using the sum-of-the-years'-digits method.

Example 8:   A typewriter that cost $425 and had an expected life of five years and a residual value of $55 was purchased on June 20. What was the first year depreciation, using the straight-line method?

$$\frac{\$425 - \$55}{5 \text{ years}} = \frac{\$370}{5 \text{ years}} = \$74 \text{ per year}$$

Since the depreciation expense is recorded for a year, you must find the partial-year depreciation. June 20 is closer to July 1 than to June 1, so for depreciation purposes, depreciate the asset from July 1 to December 31 that first year (6 months). This would mean that you would adjust the $74 annual depreciation by multiplying it by 6/12 (the number of months the asset was owned that first year placed over 12 months).

$$\$74 \times \frac{6}{12} = \$37 \quad \text{Depreciation expense the 1st year}$$

Practice Problems

Find the partial-year depreciation (round to whole dollars).

9. An adding machine that cost $199 was bought on October 8. It was expected to last four years and have a residual value of $15. Use the straight-line method to find the first year depreciation.

10. A desk that cost $450 was bought on March 18. It was expected to last 10 years and have no residual value. Use the straight-line method to find the first year depreciation.

Example 9: If an asset costs $24,000, has no residual value, and has an expected life of eight years, find the double-declining balance depreciation on December 31 if it was purchased August 25.

$$100\%/8 = 12.5\% \times 2 = 25\% \quad \text{Declining balance rate}$$

$$\$24,000 \times 25\% = \$6,000 \times \frac{4}{12}^* = \$2,000$$

*The 4/12 is used because August 25 is closer to September 1 than to August 1, which leaves four months in the year. After the first year, the computation would not require a partial-year adjustment.

Practice Problems

11. Cost = $9,420, Estimated life = 5 years, Purchase date = February 21. Find the first year depreciation using the double-declining balance method.

12. Cost = $250,000, Estimated life = 40 years, Purchase date = May 8. Find the first year depreciation using the double-declining balance method.

Finding partial-year depreciation for the sum-of-the-years'-digits method is considerably more complex than for the other methods. To charge off the entire amount allowable, each fractional amount must be used completely. This means that the unused portion of a fraction one year must be used the next year.

Example 10: An asset cost $1,450 and had a trade-in value of $250 and an expected useful life of two years. If it was purchased on May 3, 1978, find the depreciation expense on December 31, 1978, and on December 31, 1979, using the sum-of-the-years'-digits method.

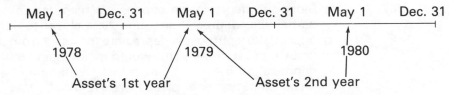

As you can see, the calendar year does not correspond to the first full year of the asset's life. The fractions for an asset with a life of two years would be 2/3

and 1/3 in the sum-of-the-years'-digits method. In 1978, only 8/12 of the asset's first year would be completed (from May 1 to December 31).

$$\frac{2}{3} \times (\$1,450 - \$250) = \$800 \times \frac{8}{12} = \$533 \quad \text{Depreciation expense for 1978}$$

For 1979, the first and second year fractions overlap. There is still 4/12 left of the first fraction (December 31, 1978, to May 1, 1979) plus 8/12 of the second fraction (May 1, 1979, to December 31, 1979). Therefore, the depreciation would be computed as follows:

$$\frac{2}{3} \times (\$1,450 - \$250) \times \frac{4}{12} = \quad \$267 \quad \text{First 4 months of 1979}$$

$$\frac{1}{3} \times (\$1,450 - \$250) \times \frac{8}{12} = \underline{+ \ 267} \quad \text{Last 8 months of 1979}$$

$$\hspace{7cm} \$534 \quad \text{Depreciation expense for 1979}$$

Note: *The asset would still have four months of depreciation left in 1980. The allowable expense would be 1/3 × ($1,450 − $250) × 4/12 = $133. This, added to the 1978 amount of $533 and the 1979 amount of $534, would give $1,200, which is the depreciable amount allowed.*

Practice Problems

13. Cost = $15,700, Life = 10 years, Residual value = $2,700, Purchase date = November 1. Find the depreciation for the first year using the sum-of-the-years'-digits method.

14. Cost = $24,850, Life = 25 years, Residual value = $1,350, Purchase date = June 25. Find the depreciation for the second year using the sum-of-the-years'-digits method.

ACCELERATED COST RECOVERY SYSTEM (ACRS)

The Economic Recovery Tax Act of 1981 made many substantial changes affecting taxes. One of those changes concerning business was the "Accelerated Cost Recovery System" (ACRS), which simplifies and speeds up depreciation of business assets.

Under ACRS the concepts of estimated useful life, salvage value, and partial year depreciation have been eliminated. In addition, the element of choice for the type of depreciation method was basically eliminated. Except for

ACRS APPLICABLE PERCENTAGES

If the Age of the Asset is	(BY COST RECOVERY PERIODS OF 3 AND 5 YEARS) The Percentage of Depreciation Deductible	
	For 3 years	For 5 years
1	25%	15%
2	38%	22%
3	37%	21%
4	—	21%
5	—	21%

unusual instances, the ACRS is a mandatory system for all business assets purchased and put into use anytime during or after 1981. All assets fall into one of four different "cost recovery periods" of either 3, 5, 10, or 15 years.* After the cost recovery period is identified, a table of percentages indicates the percentage of depreciation to be deducted for that year without concern for residual value or when during the year the asset was purchased (no partial-year depreciations).

*See end of chapter.

Example 11:

Assume a machine is purchased on May 8, 1981, at a cost of $22,000 and is determined to be in the 5-year class. Find the depreciation for each year.

$$\text{1st year depreciation: } 15\% \times \$22,000 = \$3,300$$
$$\text{2nd year depreciation: } 22\% \times \$22,000 = \$4,840$$
$$\text{3rd-5th year depreciation: } 21\% \times \$22,000 = \$4,620$$

Practice Problems

Using the ACRS, find the second year depreciation for each of the following.

15. Office furniture costing $5,300 classified as a 5-year asset.

16. Tools costing $900 classified as a 3-year asset.

Almost all business assets will fall into the recovery periods of 3 or 5 years. Therefore only those periods will be used in the text examples.

ACRS ASSET CLASSIFICATION EXAMPLES

3-year property: Automobiles, light trucks, certain special tools
5-year property: Heavy trucks, equipment, most office furniture
10-year property: Primarily public utility property
15-year property: Almost all depreciable realty (exclusive of land cost)

Practice Problem Solutions

1. $\dfrac{\$12,000 - \$2,000}{5 \text{ years}} = \$2,000$ per year   2. $\dfrac{\$60,000}{20 \text{ years}} = \$3,000$ per year

3. $\dfrac{\$10,500 - \$1,500}{50,000 \text{ miles}} = \dfrac{\$9,000}{50,000 \text{ miles}} = 18\text{¢ per mile} \times 11,520 \text{ miles} = \$2,074$

4. $\dfrac{\$3,000}{500,000 \text{ gallons}} = \$.006$ per gallon $\times 27,000$ gallons $= \$162$

5.

Year	Cost	Depreciation	Accumulated Depreciation	Book Value
1	$8,450	$1,690	$1,690	$6,760
2	8,450	1,352	3,042	5,408

Declining balance rate = 100%/No. yrs × 2 = 100%/10 = 10% × 2 = 20%
1st year: $8,450 × 20% = $1,690
2nd year: $6,760 × 20% = $1,352

6.

Year	Cost	Depreciation	Accumulated Depreciation	Book Value
1	$87,000	$6,960	$ 6,960	$80,040
2	87,000	6,403	13,363	73,637

Declining balance rate = 100%/No. yrs × 2 = 100%/25 = 4% × 2 = 8%
1st year: $87,000 × 8% = $6,960
2nd year: $80,040 × 8% = $6,403

7.
$$\begin{array}{r} 3 \\ 2 \\ +1 \\ \hline 6 \end{array} = \frac{2}{6} \times (\$2,000 - \$200) = \$600$$

8. $\dfrac{n(n+1)}{2} = \dfrac{6(6+1)}{2} = \dfrac{6(7)}{2} = \dfrac{42}{2} = 21$

$$\begin{array}{r} 6 \\ 5 \\ 4 \\ 3 \\ 2 \\ +1 \\ \hline 21 \end{array} = \frac{3}{21} \times (\$14,750 - \$1,500) = \$1,893$$

9. $\dfrac{\$199 - \$15}{4 \text{ years}} = \dfrac{\$184}{4 \text{ years}} = \$46 \times \dfrac{3}{12} = \$12$

10. $\dfrac{\$450}{10 \text{ years}} = \$45 \times \dfrac{9}{12} = \$34$

11. Declining balance rate = 100%/No. yrs $\times$ 2 = 100%/5 = 20% $\times$ 2 = 40%

$9,420 $\times$ 40% = $3,768 $\times$ $\dfrac{10}{12}$ = $3,140

12. Declining balance rate = 100%/No. yrs $\times$ 2 = 100%/40 = 2.5% $\times$ 2 = 5%

$250,000 $\times$ 5% = $12,500 $\times$ $\dfrac{8}{12}$ = $8,333

13. $\dfrac{n(n+1)}{2} = \dfrac{10(10+1)}{2} = 55$

$\dfrac{10}{55} \times (\$15,700 - \$2,700) \times \dfrac{2}{12} = \$394$

14. $\dfrac{n(n+1)}{2} = \dfrac{25(25+1)}{2} = 325$

$\dfrac{25}{325} \times (\$24,850 - \$1,350) \times \dfrac{6}{12} = \quad\$\ \ 904$

$\dfrac{24}{325} \times (\$24,850 - \$1,350) \times \dfrac{6}{12} = \underline{+\quad 868}$

$$\$1,772$$

15. 22% $\times$ $5,300 = $1,166 Second year depreciation

16. 38% $\times$ $900 = $342 Second year depreciation

Exercise 65A: Straight-Line Depreciation

Name _____

Course/Sect. No. _____   Score _____

Instructions: Find the annual depreciation for the year given using the straight-line method. (Round the answers to whole dollars.)

Self-Check Exercise (Answers on back)

	Cost	Residual Value	Estimated Life	For Year
(1)	$6,000	–0–	12 years	6
(2)	$155,000	$5,000	20 years	10
(3)	$5,985	–0–	15 years	2
(4)	$4,050	$150	13 years	10

(1) _____

(2) _____

(3) _____

(4) _____

(5) _____

(5) Steve bought a dune buggy that cost $5,200. It was only expected to last 4 years and have a scrap value of $200. What would be the third year depreciation?

(6) _____

(6) Kelly bought a stereo system that cost $945. It would have no resale value after its expected life of 7 years. What would be the *book value* after 3 years?

(1) _____ $500 _____

(2) _____ $7,500 _____

(3) _____ $399 _____

(4) _____ $300 _____

(5) _____ $1,250 _____

(6) _____ $540 _____

Exercise 65B: Straight-Line Depreciation

Name_____

Course/Sect. No._____   Score_____

Instructions: Find the annual depreciation for the year given using the straight-line method. (Round all answers to whole dollars.)

	Cost	Residual Value	Estimated Life	For Year
(1) _____	**(1)** $3,200	$300	5 years	3
(2) _____	**(2)** $20,000	$2,000	10 years	4
(3) _____	**(3)** $23,150	$750	14 years	8
(4) _____	**(4)** $17,320	$1,100	20 years	19

(5) _____

(5) Sandra bought a sewing machine for $1,240, expecting it to last for 8 years. It would have no residual value. How much would it have depreciated after 3 years?

(6) _____

(6) Mike purchased a printing calculator for $250. After using it for 5 years he could trade it in for $25. What annual depreciation would he have?

Exercise 66A: Sum-of-the-Years'-Digits Method

Name _____

Course/Sect. No. _____   Score _____

Instructions: Using the sum-of-the-years'-digits method, find the depreciation for the year given. (Round all answers to whole numbers.)

		Cost	Residual Value	Estimated Life	For Year
(1) _____	(1)	$12,000	$1,000	8 years	5
(2) _____	(2)	$8,250	$250	10 years	7
(3) _____	(3)	$10,430	–0–	10 years	3
(4) _____	(4)	$850	–0–	3 years	2
(5) _____	(5)	$1,100	$100	4 years	3
(6) _____	(6)	$9,875	$1,500	7 years	7
(7) _____	(7)	$5,195	$500	3 years	1
(8) _____	(8)	$77,125	$2,750	30 years	20

Self-Check Exercise (Answers on back)

(1) _____ $1,222 _____

(2) _____ $582 _____

(3) _____ $1,517 _____

(4) _____ $283 _____

(5) _____ $200 _____

(6) _____ $299 _____

(7) _____ $2,348 _____

(8) _____ $1,759 _____

Exercise 66B: Sum-of-the-Years'-Digits Method

Name_____

Course/Sect. No._____   Score_____

Instructions: Using the sum-of-the-years'-digits method, find the depreciation for the year given. (Round all answers to whole dollars.)

		Cost	Residual Value	Estimated Life	For Year
(1) _____	(1)	$4,000	$500	5 years	3
(2) _____	(2)	$6,500	–0–	7 years	2
(3) _____	(3)	$3,400	$200	4 years	1
(4) _____	(4)	$5,000	$1,000	6 years	4
(5) _____	(5)	$25,000	$2,000	20 years	11
(6) _____	(6)	$2,750	$150	9 years	6
(7) _____	(7)	$6,000	–0–	4 years	2
(8) _____	(8)	$11,010	$2,500	12 years	10

Exercise 67A: Declining Balance Depreciation

Name _____

Course/Sect. No. _____   Score _____

Instructions: Find the *first year* depreciation using the double-declining balance method. (Round all answers to whole dollars.)

Self-Check Exercise (Answers on back)

	Cost	Residual Value	Estimated Life
(1) _____	(1) $3,400	$500	8 years
(2) _____	(2) $1,550	$250	10 years
(3) _____	(3) $6,180	–0–	25 years
(4) _____	(4) $15,000	$3,000	10 years
(5) _____	(5) $6,400	$1,500	5 years
(6) _____	(6) $12,400	$1,000	8 years
(7) _____	(7) $24,000	$3,000	20 years
(8) _____	(8) $13,300	$500	5 years

(1) _____ $850 _____

(2) _____ $310 _____

(3) _____ $494 _____

(4) _____ $3,000 _____

(5) _____ $2,560 _____

(6) _____ $3,100 _____

(7) _____ $2,400 _____

(8) _____ $5,320 _____

Exercise 67B: Declining Balance Depreciation

Name _____

Course/Sect. No. _____   Score _____

Instructions: Find the *first year* depreciation using the double-declining balance method. (Round all answers to whole dollars.)

		Cost	Residual Value	Estimated Life
(1) _____	(1)	$2,000	–0–	5 years
(2) _____	(2)	$4,000	$1,000	10 years
(3) _____	(3)	$5,000	$1,500	20 years
(4) _____	(4)	$2,195	$200	4 years
(5) _____	(5)	$8,250	$1,250	10 years
(6) _____	(6)	$16,000	$2,000	4 years
(7) _____	(7)	$9,150	–0–	10 years
(8) _____	(8)	$18,100	–0–	16 years

Exercise 68A: Units-of-Production Depreciation

Name_____

Course/Sect. No._____   Score_____

Instructions: Find the depreciation for the units used. (Round all answers to whole dollars.)

Self-Check Exercise (Answers on back)

	Cost	Residual Value	Estimated Life	Units Used
(1) _____	**(1)** $5,200	$800	400 hours	52 hours
(2) _____	**(2)** $980	–0–	1600 units	410 units
(3) _____	**(3)** $4,100	$725	45,000 miles	6,035 miles
(4) _____	**(4)** $5,400	–0–	1,080,000 gallons	124,300 gallons

(5) _____

(5) A purt machine that produces glids will produce 25,000 in its expected lifetime. It costs $16,375,000 and will be worth $2,500 as scrap when it has done its job. How much has it depreciated after making 10,500 glids?

(6) _____

(6) White Saw Mill Co. buys large saw blades for $485 each. When dull, they are sold to a recycling concern for $65. A saw blade will make 35,000 cuts in logs in its useful life. How much is the depreciation for 41 days if it makes an average of 63 cuts per day?

(1) _____ $572 _____

(2) _____ $251 _____

(3) _____ $453 _____

(4) _____ $622 _____

(5) _____ $6,876,450 _____

(6) _____ $31 _____

Exercise 68B: Units-of-Production Depreciation

Name _____

Course/Sect. No. _____   Score _____

Instructions: Find the depreciation for the units used. (Round all answers to whole dollars.)

		Cost	Residual Value	Estimated Life	Units Used
(1)	_____	**(1)** $16,000	–0–	8,000 units	1,800 units
(2)	_____	**(2)** $8,275	$1,000	60,000 miles	12,235 miles
(3)	_____	**(3)** $1,850	$50	3,000 hours	620 hours
(4)	_____	**(4)** $12,000	$1,000	880,000 copies	26,713 copies

(5) _____

(5) A car costing $900 is supposed to last 72,000 miles. It will be worthless after that. What would be the depreciation for a year if it was operated for 18,600 miles?

(6) _____

(6) Acme buys vacuum pumps for $868. When they are worn out, they can be traded in for $85. A pump will run for an estimated 9,000 hours. How much does it depreciate in 42 days if it runs continuously?

Exercise 69A: Accumulated Depreciation and Book Value

Name _____

Course/Sect. No. _____   Score _____

Instructions: Find the accumulated depreciation and the book value after 2 years. (Round to whole dollars.)

Self-Check Exercise (Answers on back)

Method	Cost	Estimated Life	Residual Value
Declining balance	$8,500	5 years	$500

(1) _____   **(1)** Acc. dep.

(2) _____   **(2)** Book value

Sum-of-years'-digits	$2,450	6 years	$300

(3) _____   **(3)** Acc. dep.

(4) _____   **(4)** Book value

Straight line	$3,800	20 years	–0–

(5) _____   **(5)** Acc. dep.

(6) _____   **(6)** Book value

Units-of-production	$10,560	70,000 miles	$900

(1st year = 12,000 miles; 2nd year = 11,820 miles)

(7) _____   **(7)** Acc. dep.

(8) _____   **(8)** Book value

(9) _____   **(9)** ABC, Inc., bought $25,200 worth of office furniture, planning to keep it eight years. The trade-in value will be $1,500. Using the double-declining balance method, what will be the book value of the furniture after 3 years?

(10) _____   **(10)** Joe's car is in its third year, and he hopes to trade it in for $400 at the end of the fifth year. The car cost $8,600 new. Using the double-declining balance method, find the book value after the third year.

(1) _$5,440_

(2) _$3,060_

(3) _$1,126_

(4) _$1,324_

(5) _$380_

(6) _$3,420_

(7) _$3,287_

(8) _$7,273_

(9) _$10,631_

(10) _$1,858_

Exercise 69B: Accumulated Depreciation and Book Value

Name_____

Course/Sect. No._____   Score_____

Instructions: Find the accumulated depreciation and the book value after 2 years. (Round to whole dollars.)

	Method	Cost	Estimated Life	Residual Value
	Straight line	$12,000	4 years	$2,000
(1)_____	**(1)** Acc. dep.			
(2)_____	**(2)** Book value			
	Sum-of-years'-digits	$6,000	8 years	–0–
(3)_____	**(3)** Acc. dep.			
(4)_____	**(4)** Book value			
	Declining balance	$5,750	4 years	$1,250
(5)_____	**(5)** Acc. dep.			
(6)_____	**(6)** Book value			

Units-of-production $28,400   1,080,000 hrs. $1,400
(1st year = 2,550 hours; 2nd year = 4,100 hours)

(7)_____ **(7)** Acc. dep.

(8)_____ **(8)** Book value

(9)_____ **(9)** An $11,000 air conditioning unit is expected to last 12 years, after which it can be scrapped for $800. Using the straight-line method, how much total depreciation has accumulated after 5 years?

(10)_____ **(10)** The XYZ Co. bought a delivery van for $12,600 with plans to trade it in after 5 years. The estimated trade-in value is $1,500. Using the sum-of-the-years'-digits method, find out how much the van will depreciate the first 3 years.

Exercise 70A: Partial-Year Depreciation

Name _____

Course/Sect. No. _____   Score _____

Instructions: Find the depreciation for the year given. (Round to whole dollars.)

Self-Check Exercise (Answers on back)

(1) _____

(1) Using the double-declining balance method, find the depreciation in the year the $2,400 asset was bought if it was purchased on August 3, had a residual value of $500, and had an estimated life of 4 years.

(2) _____

(2) Cost = $4,500; Estimated life = 12 years; Trade-in value = $300. Use the straight-line method to find the depreciation from November 11 to December 31.

(3) _____

(3) Cost = $10,250; Estimated life = 7 years, Sales value = $1,050; Method: Sum-of-the-years'-digits; Purchased: September 2. Find the partial-year depreciation on December 31.

(4) _____

(4) Method: Sum-of-the-years'-digits; Purchase date: March 29; Cost = $12,500; Residual value = $1,200; Estimated life = 4 years. Find the partial-year depreciation on December 31.

(5) _____

(5) On October 22, M. U. Zick bought a piano for $2,425 to use for teaching. Since the students are hard on a piano, it will last him only five years. At that time the trade-in value is expected to be $500. Use the sum-of-the-years'-digits method to find the first calendar year's depreciation.

(6) _____

(6) On April 18 Doc Hurt bought a trumpet, which he estimated would last 8 years. The trumpet cost $1,525 and would not be disposed of. Use the sum-of-the-years'-digits method to find the depreciation for the first calendar year.

(1) _____ $500 _____

(2) _____ $58 _____

(3) _____ $767 _____

(4) _____ $3,390 _____

(5) _____ $107 _____

(6) _____ $226 _____

Exercise 70B: Partial-Year Depreciation

Name_____

Course/Sect. No. _____   Score_____

Instructions: Find the depreciation for the year given. (Round to whole dollars.)

---

(1) _____

(1) Using the straight-line method, find the depreciation in the year the $850 asset was bought if it was purchased on March 18, had no residual value, and had an estimated life of 4 years?

(2) _____

(2) Using the sum-of-the-years'-digits method, find the depreciation in the year the $13,850 asset was bought if it was purchased on May 5, had a residual value of $1,100, and had an expected life of 5 years.

(3) _____

(3) Cost = $62,500; Estimated life = 10 years; Residual value = $300. Use the straight-line method to find the depreciation from November 11 to December 31.

(4) _____

(4) Method: Double-declining balance; Purchase date: January 24; Cost = $16,400; Residual value = 0; Estimated life = 8 years. Find the partial-year depreciation on December 31.

(5) _____

(5) On June 12 Stephanie bought a typewriter for $675 which she expected would last for 10 years. Use the straight-line method to find the partial-year depreciation and assume it will have no scrap value.

(6) _____

(6) The high school bought a pitching machine on February 19 for $1,300. They usually last 4 years, at which time they are sold to the local Little League for $100. Use the double-declining balance method.

Exercise 71A: Accelerated Cost Recovery System (ACRS)

Name_____

Course/Sect. No._____   Score_____

Instructions: Find the depreciation for the year given in each example.

Self-Check Exercise (Answers on back)

	Cost	Classification	For Which Year
(1) _____	(1) $10,000	3-year asset	3
(2) _____	(2) $800	5-year asset	4
(3) _____	(3) $1,200	3-year asset	1
(4) _____	(4) $103,000	5-year asset	5
(5) _____	(5) $60,000	5-year asset	2
(6) _____	(6) $77,500	5-year asset	2
(7) _____	(7) $52,100	5-year asset	4
(8) _____	(8) $34,800	5-year asset	5

(9) _____

(9) A piece of equipment that cost $6,500 was classified as 5-year property. What *book value* would it have after 3 years?

(10) _____

(10) For a 3-year asset that cost $15,000, how much *more* would be depreciated the second year than the first year?

(1) _____$3,700_____

(2) _____$168_____

(3) _____$300_____

(4) _____$21,630_____

(5) _____$13,200_____

(6) _____$17,050_____

(7) _____$10,941_____

(8) _____$7,308_____

(9) _____$2,730_____

(10) _____$1,950_____

Exercise 71B: Accelerated Cost Recovery System (ACRS)

Name _____

Course/Sect. No. _____   Score_____

Instructions: Find the depreciation for the year given in each example.

	Cost	Classification	For Which Year
(1) _____	**(1)** $5,400	3-year asset	2
(2) _____	**(2)** $1,450	5-year asset	2
(3) _____	**(3)** $25,000	5-year asset	1
(4) _____	**(4)** $3,000	3-year asset	3
(5) _____	**(5)** $84,200	5-year asset	3
(6) _____	**(6)** $400	3-year asset	1
(7) _____	**(7)** $600	3-year asset	3
(8) _____	**(8)** $4,050	3-year asset	2

(9) _____

(9) A delivery truck classified as 3-year property was purchased on November 19, 1983. What would be the *accumulated depreciation* on December 31, 1984, if it cost $12,400?

(10) _____

(10) What total percentage of a 5-year asset would have been depreciated after 4 years?

Chapter

13

Inventory

Preview of Terms

Cost of Goods Sold   The difference between the cost of all goods available for sale and the cost of all goods that are still unsold; also referred to as *cost of merchandise sold;* what the merchandise you sold cost you.

FIFO   Abbreviation for *first-in, first-out;* a description of the way merchandise is sold.

Inventory   The total number of items and dollar value for merchandise still in stock.

Inventory Turnover   The number of times inventory has been sold (turned over) during a period.

LIFO   Abbreviation for *last-in, first-out;* a description of the way merchandise is sold.

Inventory is short for *merchandise inventory* and represents all goods that are on the shelves or in the warehouse that have not yet been sold. For a merchandising business, inventory normally represents one of the largest assets (things of value) a store has. This being the case, complete and accurate records concerning this merchandise must be kept.

The sale of this merchandise inventory provides the principal source of income, and the deduction of remaining inventory from revenue in the Income Statement is customarily larger than all other deductions (expenses) combined.

A typical business will *take inventory,* that is, count the number of each item of merchandise, only once a year—at the end of the year. These items will then be multiplied by their unit costs to determine their values. This is called *ending inventory valuation.* An error in the value of the end-of-year inventory valuation can cause a significant overstatement or understatement of the net income and company assets.

INVENTORY TURNOVER

In addition to determining the value of ending inventory, a business may also be interested in establishing its turnover. The *inventory turnover* is the number of times a year the merchandise is sold, repurchased, and put up for sale again. Usually, the larger the turnover, the better.

INVENTORY TURNOVER AT RETAIL

The turnover is computed on retail, or selling price, if the inventory valuation is taken at selling price rather than at cost price.

$$\frac{\text{Net sales}}{\text{Average inventory}} = \text{Inventory turnover at retail}$$

You can determine *net sales* by subtracting sales discounts or sales returns from sales. If both are known, subtract both; otherwise, subtract only the one that is known. You can calculate *average inventory* by adding the number of inventory amounts available and dividing the total by the number of inventories that were added. (Most likely, there will only be a beginning-of-the-year inventory and an end-of-the-year inventory. If that is the case, add them and divide the sum by 2, to get the average inventory.)

Example 1:

The company sales were $420,000, and the sales discounts were $20,000. The January 1 (beginning) inventory was $43,000, and the December 31 (ending) inventory was $37,000. Find the inventory turnover at retail (selling price).

$$
\begin{array}{rl}
\$420,000 & \text{Sales} \\
-\quad 20,000 & \text{Sales discounts} \\
\hline
\$400,000 & \text{Net sales}
\end{array}
$$

$$
\begin{array}{rl}
\$ 43,000 & \text{Beginning inventory} \\
+\quad 37,000 & \text{Ending inventory} \\
\hline
\$ 80,000 \div 2 = {} & \$40,000 \text{ Average inventory}
\end{array}
$$

$$\text{Retail turnover} = \frac{\text{Net sales}}{\text{Average inventory}} = \frac{\$400,000}{\$40,000} = 10 \text{ times*}$$

(*The turnover is always expressed with times, meaning how many times a year it is bought and sold.)

Practice Problems

Compute the inventory turnover at retail (selling price). Round to a tenth.

1. The January 1 inventory was $50,000, and the December 31 inventory was $62,000. The sales for the year were $300,000, the sales discounts $15,000, and the sales returns $5,000.

2. The sales for the year were $38,500, and the inventory was taken at retail (selling price) at the end of each quarter as follows: $12,500; $15,625; $13,075; and $14,000.

INVENTORY TURNOVER AT COST

Most often, the inventory turnover is computed based on cost. To determine this turnover, you must find the *cost of goods sold* (COGS).

$$
\begin{array}{l}
\text{Beginning inventory} \\
\underline{+ \text{ Purchases (net)}} \\
\text{Goods available for sale} \\
\underline{- \text{ Ending inventory}} \\
\text{Cost of goods sold}
\end{array}
$$

$$\text{Inventory turnover at cost} = \frac{\text{Cost of goods sold}}{\text{Average inventory}}$$

Example 2:

The beginning inventory was $50,000, purchases were $100,000, and the ending inventory was $30,000. Find the inventory turnover at cost.

$ 50,000	Beginning inventory
+ 100,000	Purchases
$150,000	Goods available for sale
− 30,000	Ending inventory
$120,000	Cost of goods sold
$ 50,000	Beginning inventory
+ 30,000	Ending inventory
$ 80,000 ÷ 2 = $40,000	Average inventory

$$\text{Inventory turnover at cost} = \frac{\text{Cost of goods sold}}{\text{Average inventory}}$$

$$= \frac{\$120,000}{\$40,000}$$

$$= 3 \text{ times}$$

Example 3:

Find the inventory turnover at cost. Round to a tenth.

Sales		$27,000
Less: Sales returns		− 3,000
Net sales		$24,000
Inventory, January 1	$ 6,000	
Purchases	+ 18,000	
Goods available for sale	$24,000	
Inventory, December 31	− 8,000	
Cost of goods sold		− 16,000
Gross profit		$ 8,000

$ 6,000	Beginning inventory
+ 8,000	Ending inventory
$14,000 ÷ 2 = $7,000	Average inventory

$$\text{Inventory turnover at cost} = \frac{\text{Cost of goods sold}}{\text{Average inventory}}$$

$$= \frac{\$16,000}{\$7,000}$$

$$= 2.3 \text{ times}$$

Find the inventory turnover at cost. Round to a tenth.

3. The beginning and ending inventories were $62,000 and $48,000, respectively. If the cost of goods sold was $152,000, find the inventory turnover at cost.

4. Inventory, January 1 = $18,400; Purchases = $31,500; Inventory, December 31 = $26,400; Sales = $117,200.

INVENTORY VALUATION

The three most commonly used methods of inventory valuation are:

1. FIFO (First-in, First-out). Theoretically this method is used when the first items of merchandise purchased (first in) are the first items sold (first out). Businesses that handle perishable goods probably sell them on a FIFO basis.

2. LIFO (Last-in, First-out). When the last items purchased are the first ones sold, the LIFO method theoretically is the way inventory should be valued.

3. Weighted Average. The weighted average method is supposed to be a compromise between the FIFO and the LIFO methods. Instead of hypothetically assuming goods are sold a certain way, this method assumes all of the items cost the same (the average of the units available for sale).

Note:

A business could sell goods on a FIFO basis and use the LIFO method of inventory valuation, or vice versa. The method assumed when goods are sold does not have to be the same method used to value the inventory.

We will use a single example to explain and compare these three methods of ending inventory valuation.

	Units	Unit Cost	Total Cost
Beginning inventory	4	$100	$ 400
Purchase 1	5	102	510
Purchase 2	7	106	742
Purchase 3	5	108	540
Purchase 4	+ 4	110	+ 440
Goods available for sale	25		$2,632
Ending inventory (units left)	− 7		
Goods sold	18		

FIFO METHOD

You must determine which 7 units of the 25 units are left. If we assume the FIFO method, the most recent units purchased (last ones) would be the ones left. The FIFO refers to the way the units are *sold*. Therefore, the 7 units left would be:

4 units at $110 = $440	From Purchase 4
+3 units at $108 = + 324	From Purchase 3
7 $764	Ending inventory (units left)

Note: *The 18 units sold, starting at the top of the example on the first-in, first-out basis, would be the 4 units in the beginning inventory, plus the 5 from Purchase 1, plus the 7 from Purchase 2, plus 2 from Purchase 3. The total of these units sold—18—would leave the 7 units in the ending inventory.*

LIFO METHOD

Assuming the LIFO method is used, the last units purchased were the first units sold; the next-to-last units purchased were the next units sold; and so on. That would mean that the seven units left would be:

3 units at $102 = $306	From Purchase 1
+4 units at $100 = + 400	From beginning inventory
7 $706	Ending inventory (units left)

Note: *The 18 units sold would come first from the 4 units in Purchase 4, plus the 5 from Purchase 3, plus the 7 from Purchase 2, plus 2 from Purchase 1. That would leave the 7 units that are in the ending inventory (units left).*

WEIGHTED AVERAGE METHOD

In the weighted average method, you average all of the units available for sale by dividing the total inventory cost by the total number of units to give the average cost per unit purchased.

$$\text{Weighted average} = \frac{\text{Total inventory cost}}{\text{Total no. units}}$$

$$= \frac{\$2,632}{25 \text{ units}}$$

$$= \$105.28 \text{ per unit}$$

7 units *left* × $105.28 per unit = $736.96 Ending inventory

As you can see, even though all three methods used the same circumstances, the ending inventories differed.

	FIFO	LIFO	Weighted Average
Ending inventory	$764	$706	$736.96

Thus, a business could alter its profit simply by changing its method of valuing inventory.

RETAIL METHOD

When using the FIFO, LIFO, and weighted average methods to determine the ending inventory, you must take a physical count of all of the remaining merchandise. This is most often very time-consuming. Department stores and many other retail establishments therefore *estimate* the ending inventory so that they can prepare financial statements on a monthly or quarterly basis and not have to take a physical count. The *retail method* is designed for this purpose.

Generally, cost information is not as readily available as retail (sales) information in a retail establishment. In the retail method, you find the ending inventory at its retail valuation (selling price) and convert it to its cost valuation. For this method to be an accurate estimate, the markup on all merchandise items must be similar and consistent.

Example 4:

The inventory on October 1 was valued at $20,000 at cost and $32,500 at retail. During October, merchandise was purchased that cost $40,000, and it was valued at retail at $67,500. The sales for the month of October (at retail) were $55,000. Find the estimated cost of ending inventory.

Step 1: Find the cost and retail value of merchandise available for sale.

	Cost	Retail
Inventory, October 1	$20,000	$ 32,500
Purchases	+ 40,000	+ 67,500
Merchandise available for sale	$60,000	$100,000

Step 2: Find the ratio of cost to retail for merchandise available for sale. This is the *inventory cost ratio.*

$$\frac{\text{Available for sale at cost}}{\text{Available for sale at retail}} = \frac{\$60,000}{\$100,000} = 60\%$$

Note:

This means 60% of every sales dollar is used to buy the merchandise, leaving a 40% markup.

Step 3: Find the ending inventory at retail.

$100,000	Merchandise available for sale at retail
− 55,000	Merchandise sold (sales) at retail
$ 45,000	Merchandise left (ending inventory at retail)

Step 4: Convert the ending inventory at retail to ending inventory at cost.

$45,000	Ending inventory at retail
× 60%	Inventory cost ratio
$27,000	Ending inventory at cost

The complete process follows:

	Cost	Retail
Inventory, October 1	$20,000	$ 32,500
Purchases	+ 40,000	+ 67,500
Merchandise available for sale	$60,000	$100,000
Sales for October (at retail)		− 55,000
Inventory, October 31 (at retail)		$ 45,000
Inventory cost ratio*		× 60%
Inventory, October 31 (at cost)		$ 27,000

$$\left(*\text{Inventory cost ratio: } \frac{\$60,000}{\$100,000} = 60\%\right)$$

Practice Problems

Use the following data for Problems 5, 6, and 7.

	Units	Unit Cost
Beginning inventory	100	$5.00
Purchase 1	80	4.50
Purchase 2	120	4.00
Purchase 3	+ 90	3.75
Goods available for sale	390	
Ending inventory (units left)	−117	
Goods sold	273	

5. Find the value of ending inventory and the cost of goods sold using the FIFO method.

6. Find the value of ending inventory and the cost of goods sold using the LIFO method.

7. Find the value of ending inventory and the cost of goods sold using the weighted average method.

8. Using the following information, find the value of ending inventory at cost using the retail method.

	Cost	Retail
Merchandise inventory, May 1	$145,000	$184,000
Purchases for May	60,000	72,250
Sales for May		120,000

Practice Problem Solutions

1.
$$\begin{array}{r} 50,000 \\ +\ \underline{62,000} \\ \$112,000 \div 2 = \$56,000 \end{array}$$

$$\begin{array}{rl} \$300,000 & \\ -\ \ \ 5,000 & \text{Sales discounts} \\ -\ \underline{15,000} & \text{Sales returns} \\ \$280,000 & \end{array}$$

$$\frac{\$280,000}{\$56,000} = 5 \text{ times}$$

2.
$$\begin{array}{r} \$12,500 \\ 15,625 \\ 13,075 \\ +\ \underline{14,000} \\ \$55,200 \div 4 = \$13,800 \end{array}$$

$$\frac{\$38,500}{\$13,800} = 2.8 \text{ times}$$

3.
$$\begin{array}{r} \$\ 62,000 \\ +\ \underline{48,000} \\ \$110,000 \div 2 = \$55,000 \end{array}$$

$$\frac{\$152,000}{\$55,000} = 2.8 \text{ times}$$

4.
$$\begin{array}{r} \$18,400 \\ +\ \underline{31,500} \\ \$49,900 \\ -\ \underline{26,400} \\ \$23,500 \end{array}$$

$$\begin{array}{r} \$18,400 \\ +\ \underline{26,400} \\ \$44,800 \div 2 = \$22,400 \end{array}$$

$$\frac{\$23,500}{\$22,400} = 1.0 \text{ times}$$

5.
$$\begin{array}{llr} 27 \text{ at } \$4.00 = & & \$108.00 \\ \underline{90} \text{ at } \$3.75 = & + & \underline{337.50} \\ 117 & & \$445.50 \quad \text{Ending inventory} \end{array}$$

$$\begin{array}{rl} \$1,677.50 & \text{Merchandise available for sale} \\ -\ \ \underline{445.50} & \text{Ending inventory} \\ \$1,232.00 & \text{Cost of goods sold} \end{array}$$

6.
$$\begin{array}{llr} 17 \text{ at } \$4.50 = & & \$\ 76.50 \\ +\underline{100} \text{ at } \$5.00 = & + & \underline{500.00} \\ 117 & & \$576.50 \quad \text{Ending inventory} \end{array}$$

$$\begin{array}{rl} \$1,677.50 & \text{Merchandise available for sale} \\ -\ \ \underline{576.50} & \text{Ending inventory} \\ \$1,101.00 & \text{Cost of goods sold} \end{array}$$

7.
$$\frac{\$1,677.50}{390 \text{ units}} = \$4.30 \text{ per unit}$$

117 left $\times$ \$4.30 per unit = \$503.10   Ending inventory

$$\begin{array}{rl} \$1,677.50 & \text{Merchandise available for sale} \\ -\ \ \underline{503.10} & \text{Ending inventory} \\ \$1,174.40 & \text{Cost of goods sold} \end{array}$$

8.

	Cost	Retail
Inventory, May 1	$145,000	$184,000
Purchases	+ 60,000	+ 72,250
Available for sale	$205,000	$256,250
Sales		− 120,000
Inventory, May 31 (at retail)		$136,250
Inventory cost ratio*		× 80%
Inventory, May 31 (at cost)		$109,000

$$*\text{Inventory cost ratio} = \frac{205,000}{256,250} = 80\%$$

Exercise 72A: Inventory Turnover at Retail

Name _____

Course/Sect. No. _____ Score _____

Instructions: Find the inventory turnover at retail. (Round to a tenth.)

Self-Check Exercise (Answers on back)

	Inventories Taken	Sales	Sales Discounts	Sales Returns
(1) _____	(1) #1 = $140,000 #2 = $103,000	$240,000	$10,000	$10,000
(2) _____	(2) #1 = $51,000 #2 = $47,000	$93,200	$1,200	$1,600
(3) _____	(3) #1 = $65,300 #2 = $65,450	$65,900	$2,400	–0–
(4) _____	(4) #1 = $102,400 #2 = $124,600 #3 = $116,200	$667,620	$2,300	$1,800

(5) _____   **(5)** The January 1, 1983, inventory, based on selling price, was $125,638.22. It was $96,685.79 on January 1, 1984. The total sales for 1983 were $678,231.34. What was the turnover for 1983?

(6) _____   **(6)** Leed's Co. gave discounts of $7,652 on gross sales of $193,483 during 1983. The January 1, 1983, inventory was $46,352, and the January 1, 1984, inventory was $38,462. What was the inventory turnover for 1983?

(1) _____1.8 times_____

(2) _____1.8 times_____

(3) _____1.0 times_____

(4) _____5.8 times_____

(5) _____6.1 times_____

(6) _____4.4 times_____

Exercise 72B: Inventory Turnover at Retail

Name_____

Course/Sect. No._____ Score_____

Instructions: Find the inventory turnover at retail. (Round to a tenth.)

	Inventories Taken	Sales	Sales Discounts	Sales Returns
(1) _____	**(1)** #1 = $82,000 #2 = $64,000 #3 = $70,000	$125,400	–0–	$5,400
(2) _____	**(2)** #1 = $17,250 #2 = $20,750	$74,000	–0–	–0–
(3) _____	**(3)** #1 = $110,250 #2 = $121,000 #3 = $102,375	$360,000	$12,000	$7,800
(4) _____	**(4)** #1 = $7,450 #2 = $6,150 #3 = $9,500	$9,270	$500	$300

(5) _____

(5) The store has beginning-of-month inventories (at retail) of $2,120, $1,875, and $2,070 for three consecutive months. Total sales for the third month were $18,600. What was the inventory turnover for that month?

(6) _____

(6) Ron's lemonade stand had consecutive weekly inventories of $9.35, $7.65, and $11.55. During the third week, his sales were $47.80, less discounts of $6.35 for members of his soccer team. What was the turnover that week?

Exercise 73A: Cost of Goods Sold

Name _____

Course/Sect. No. _____   Score_____

Instructions: Find the cost of goods sold.

Self-Check Exercise (Answers on back)

	Beginning Inventory	Ending Inventory	Net Purchases
(1) _____	(1) $72,000	$81,000	$117,000
(2) _____	(2) $13,128	$10,764	$8,415
(3) _____	(3) $38,150	$34,200	$16,510
(4) _____	(4) $64,120	$80,470	$93,190

(5) _____

(5) The frame shop had $4,711 worth of raw materials for making frames at the beginning of the year. During the year, they bought $19,382 worth of materials. The ending inventory was $2,912. What was the cost of goods sold for the year?

(6) _____

(6) The cost of goods sold for the year for Wicker's Store was $204,377.62. The beginning inventory was $73,451.19, and the ending inventory was $89,417.31. How much did they *purchase* during the year?

(1) _____$108,000_____

(2) _____$10,779_____

(3) _____$20,460_____

(4) _____$76,840_____

(5) _____$21,181_____

(6) _____$220,343.74_____

Exercise 73B: Cost of Goods Sold

Name _____

Course/Sect. No. _____   Score _____

Instructions: Find the cost of goods sold.

		Beginning Inventory	Ending Inventory	Net Purchases
(1) _____	**(1)**	$35,000	$55,000	$80,000
(2) _____	**(2)**	$7,500	$9,000	$11,400
(3) _____	**(3)**	$95,500	$103,600	$84,300
(4) _____	**(4)**	$40,000	$50,000	$30,000

(5) _____ **(5)** A neighborhood cosmetics salesperson had inventories of $315.25 at the beginning of June and $112 at the end of June, based on cost. He bought $315 worth of cosmetics during June. What was the cost of goods sold for June?

(6) _____ **(6)** Total purchases for resale of First Bookstore for the last half of the year were $228,348. The midyear inventory was $45,126, and the end-of-year inventory was $29,385. What was the cost of goods sold?

Exercise 74A: Inventory Turnover at Cost

Name _____

Course/Sect. No. _____   Score _____

Instructions: Find the inventory turnover at cost. (Round to a tenth.)

Self-Check Exercise (Answers on back)

	Net Sales	Beginning Inventory	Ending Inventory	Net Purchases
(1) _____	**(1)** $205,300	$84,120	$95,870	$64,500
(2) _____	**(2)** $55,000	$15,000	$4,000	$27,000
(3) _____	**(3)** $412,000	$28,320	$24,519	$117,438
(4) _____	**(4)** $109,375	$60,050	$18,175	$24,115

(5) _____

(5) The company began in May with $87,350 worth of goods on hand. During May, they bought $22,300 worth of goods. On June 1, the inventory was $31,675. What was the inventory turnover during May?

(6) _____

(6) First Plumbing Supply bought $6,752.84 worth of goods in September. If they began the month with $4,181.16 and ended with $5,618.61 in inventory, what was the inventory turnover?

(1) _____.6 times_____

(2) _____4.0 times_____

(3) _____4.6 times_____

(4) _____1.7 times_____

(5) _____1.3 times_____

(6) _____1.1 times_____

Exercise 74B: Inventory Turnover at Cost

Name_____

Course/Sect. No._____   Score_____

Instructions: Find the inventory turnover at cost. (Round to a tenth.)

	Net Sales	Beginning Inventory	Ending Inventory	Net Purchases
(1) _____ **(1)**	$372,000	$155,000	$170,000	$72,000
(2) _____ **(2)**	$317,995	$104,000	$81,000	$115,000
(3) _____ **(3)**	$94,100	$44,300	$16,000	$38,500
(4) _____ **(4)**	$188,000	$75,000	$90,000	$180,000

(5) _____

(5) After beginning the year with $78,300 inventory and purchasing $246,500 worth of goods during the year, Mr. Lewis ended the year with $62,300 in inventory. What was his inventory turnover for the year?

(6) _____

(6) Great Game Co. ended the third quarter of the year with $638,500 in inventory. They had bought $709,475 of goods during the quarter. If they started the quarter with $816,250, what was the inventory turnover for the quarter?

Exercise 75A: Inventory Valuation

Name _____

Course/Sect. No. _____   Score_____

Instructions: Answer the questions as required.

	Units	Unit Cost	Total Cost
Beginning Inventory	35	$2.00	$ 70.00
Purchase 1	15	2.10	31.50
Purchase 2	8	1.90	15.20
Purchase 3	+12	1.60	+ 19.20
Available for sale	70		$135.90
Units left	−23		
Units sold	−47		

(1) _____   **(1)** Find the ending inventory at FIFO.

(2) _____   **(2)** Find the ending inventory at LIFO.

(3) _____   **(3)** Determine the ending inventory at weighted average. (Round the cost per unit to cents.)

On January 1, Wilson's had 2,000 cases of tennis balls, for which they paid $80 each. During the year, they made purchases as follows: March 8—1,000 cases at $85; June 20—750 cases at $74; August 9—2,000 cases at $87.50. They sold 3,400 cases during the year. What is the value of their year-end inventory?

(4) _____   **(4)** At FIFO

(5) _____   **(5)** At LIFO

(6) _____   **(6)** At weighted average

(1) _____ $40.70 _____

(2) _____ $46.00 _____

(3) _____ $44.62 _____

(4) _____ $200,900 _____

(5) _____ $189,750 _____

(6) _____ $194,345 _____

Exercise 75B: Inventory Valuation

Name _____

Course/Sect. No. _____   Score _____

Instructions: Answer the questions as required.

	Units	Unit Cost	Total Cost
Beginning Inventory	200	$75	$15,000
Purchase 1	40	78	3,120
Purchase 2	20	80	1,600
Purchase 3	30	74	2,220
Purchase 4	+ 10	82	820
Available for sale	300		$22,760
Units left	−125		
Units sold	−175		

(1) _____

(1) Find the ending inventory at FIFO.

(2) _____

(2) Find the ending inventory at LIFO.

(3) _____

(3) Find the ending inventory at weighted average. (Round the cost per unit to hundredths.)

On May 1, during a sugar shortage, Ron's Lemonade Stand had on hand 4 pounds of sugar, for which Ron had paid 92¢ per pound. During May, he bought 5 pounds at $1.00, 10 pounds at 98¢, and 8 pounds at $1.05. He used 14 pounds of sugar during May. Find the value of the ending inventory.

(4) _____

(4) At FIFO

(5) _____

(5) At LIFO

(6) _____

(6) At weighted average

Exercise 76A: Ending Inventory, Cost of Goods Sold, and Gross Profit

Name_____

Course/Sect. No._____   Score_____

Instructions: Find each of the answers as indicated. (Round the unit price to cents.)

Self-Check Exercise (Answers on back)

	Quantity	Unit Cost	Total Cost
Beginning inventory	953	$18.75	$17,868.75
Purchase #1	874	19.00	16,606.00
Purchase #2	750	19.10	14,325.00
Purchase #3	+560	18.45	+ 10,332.00
	3147		$59,131.75
	−1131 left		
	2016 sold		

Inventory method: FIFO
Units sold: 2016
Net sales: $62,351

(1) _____ **(1)** Find the ending inventory valuation.

(2) _____ **(2)** Find the cost of goods sold.

(3) _____ **(3)** Find the gross profit.

	Quantity	Unit Cost	Total Cost
Beginning inventory	4,600	$1.38	$6,348
Purchase #1	2,300	1.40	3,220
Purchase #2	5,000	1.35	6,750
Purchase #3	+1,600	1.47	+ 2,352
	13,500		$18,670
	− 3,600 left		
	9,900 sold		

Inventory method: LIFO
Units sold: 9,900
Net sales: $27,335

(4) _____ **(4)** Find the ending inventory valuation.

(5) _____ **(5)** Find the cost of goods sold.

(6) _____ **(6)** Find the gross profit.

(1) $21,047.10

(2) $38,084.65

(3) $24,266.35

(4) $4,968

(5) $13,702

(6) $13,633

Exercise 76B: Ending Inventory, Cost of Goods Sold, and Gross Profit

Name _____

Course/Sect. No. _____   Score _____

Instructions: Find the answers as indicated. (Round the unit price to cents.)

	Quantity	Unit Cost	Total Cost
Beginning inventory	20	$12.00	$240.00
Purchase #1	10	11.90	119.00
Purchase #2	12	12.20	146.40
Purchase #3	+ 9	12.30	+ 110.70
	51		$616.10
	−16 left		
	35 sold		

Inventory method: LIFO
Units sold: 35
Net sales: $1,200

(1) _____ **(1)** Find the ending inventory valuation.

(2) _____ **(2)** Find the cost of goods sold.

(3) _____ **(3)** Find the gross profit.

	Quantity	Unit Cost	Total Cost
Beginning inventory	185	$6.25	$1,156.25
Purchase #1	172	6.70	1,152.40
Purchase #2	163	6.50	1,059.50
Purchase #3	+141	7.00	+ 987.00
	661		$4,355.15
	−211 left		
	450 sold		

Inventory method: Weighted average
Units sold: 450
Net sales: $5,400

(4) _____ **(4)** Find the ending inventory valuation.

(5) _____ **(5)** Find the cost of goods sold.

(6) _____ **(6)** Find the gross profit.

Exercise 77A: Retail Method

Name _____

Course/Sect. No. _____   Score_____

Instructions: Find the estimated cost of the ending inventory using the retail method.

(1) _____

	At Cost	At Retail
(1) Beginning inventory	$143,440	$332,000
Net purchases	$100,408	$222,200
Net sales		$284,300

(2) _____

	At Cost	At Retail
(2) Beginning inventory	$20,000	$40,000
Net purchases	$30,000	$60,000
Net sales		$64,300

(3) _____

	At Cost	At Retail
(3) Beginning inventory	$75,000	$170,000
Net purchases	$25,000	$ 80,000
Net sales		$190,000

(4) _____

	At Cost	At Retail
(4) Beginning inventory	$13,400	$16,500
Net purchases	$11,050	$24,250
Net sales		$21,825

(5) _____

(5) The January 1 inventory at cost was $20,000, and at retail it was $32,200. The January net purchases at cost were $13,150 and at retail they were $18,800. If the retail sales for January were $15,000, what was the percentage of inventory cost of retail?

(6) _____

(6) The merchandise available for sale at cost was $54,560. The sales at retail were $42,000. The merchandise available for sale at retail was $88,000. Find the ending merchandise inventory estimated at cost.

(1)   _$118,576_

(2)   _$17,850_

(3)   _$24,000_

(4)   _$11,355_

(5)   _65%_

(6)   _$28,520_

Exercise 77B: Retail Method

Name _____

Course/Sect. No. _____   Score _____

Instructions: Find the estimated cost of the ending inventory using the retail method.

		At Cost	At Retail
(1) _____	**(1)** Beginning inventory	$420,000	$600,000
	Net purchases	$156,000	$200,000
	Net sales		$356,450
(2) _____	**(2)** Beginning inventory	$102,000	$120,000
	Net purchases	$228,000	$280,000
	Net sales		$184,000
(3) _____	**(3)** Beginning inventory	$18,000	$24,000
	Net purchases	$12,000	$16,000
	Net sales		$27,800
(4) _____	**(4)** Beginning inventory	$80,000	$350,000
	Net purchases	$85,000	$400,000
	Net sales		$500,000

(5) _____ **(5)** The merchandise available for sale at cost was $28,500, and at retail it was $95,000. If the sales for the period were $18,000, what was the estimated cost of the ending inventory?

(6) _____ **(6)** The percentage of inventory cost on retail price is 70%. If the end-of-period inventory at retail was $550,000, what was the estimated end-of-period inventory at cost?

Test Yourself — Unit 5

(1) An asset cost $6,250 and was purchased on February 18. Using the straight-line method, find the depreciation for that year. Assume no residual value and an estimated life of five years. Round to whole dollars.

(2) A delivery truck was bought on June 15 for $9,000. It is expected to be used for 60,000 miles and have a $1,500 trade-in value. Using the units-of-production method, find the depreciation for the year it was purchased if it was driven 5,200 miles that year. Round to whole dollars.

(3) A machine cost $16,220 and was expected to last 15 years and have a residual value of $2,300. Use the declining balance method (at twice the straight-line rate) to determine the book value of the machine after the second year. Round your answer to whole dollars.

(4) Use the sum-of-the-years'-digits method to find the depreciation for the sixth year on a cabinet that cost $1,200, was expected to last eight years, and had no residual value. Round to whole dollars.

(5) An asset classified as five-year property was purchased for $12,500. What would be the *book value* after the third year?

(6) The inventory was taken four times during the year as follows: $31,200; $27,800; $29,520; $33,400. The sales were $95,360, and the sales discounts were $850. Find the inventory turnover at retail. Round to a tenth.

(7) The inventory on January 1 was $55,030, there were net purchases for the year of $175,000, and the inventory on December 31 was $43,960. Find the cost of goods sold.

(8) The inventory on September 1 was $102,525, and on September 30 it was $96,460. If the net purchases were $72,440, find the inventory turnover at cost. Round to a tenth.

Answer the next three questions using the following information:

	Units	Unit Cost
Beginning inventory	12	$8.25
Purchase 1	8	8.32
Purchase 2	6	8.21
Purchase 3	11	8.10

(9) Assuming the FIFO method is used, what is the ending inventory if 16 units were sold?

(10) Assuming the LIFO method is used, what is the ending inventory if 16 units were left?

(11) Assuming the weighted average method is used, what is the ending inventory if 20 units were sold? Round the cost per unit to the nearest cent.

(12) Find the value of the ending inventory at cost using the retail method and the following information:

	Cost	Retail
Beginning inventory	$27,890	$42,000
Purchases	33,600	52,600
Sales		29,500

Unit

6

Checks and Measurements Used in Business

A Preview of What's Next

The bank statement shows a balance that is considerably different from the balance in your checkbook. Can you reconcile the difference?

Do you know the difference between *outstanding checks, NSF checks,* and *canceled checks?*

What is the proper way to express a *turnover?*

Do you know the difference between the *arithmetic mean, median,* and the *mode?*

How do you divide an inch into 10 equal parts?

Objectives

To be effective and efficient in business it is increasingly more important to be able to gather and interpret business information. From this information, the activities of a business can be measured and intelligent decisions can be made for future actions. When you have successfully completed this unit you will be able to:

1. Understand a bank statement.
2. Reconcile the bank balance with the checkbook balance by preparing a reconciliation statement.
3. Distinguish between the *mean, median,* and *mode* in a set of data.
4. Construct pie, bar, and line graphs.
5. Convert within the metric system.
6. Convert from metric or to metric.
7. Compute various analyses for financial statements and express them properly.

Chapter

14

Bank Reconciliation

Preview of Terms

Bank Statement A monthly statement sent to each depositor that itemizes checks written, deposits made, and service charges, and shows the current balance each time a transaction was posted to the account.

Canceled Check A check that has been cashed, returned to your bank for collection, and subsequently returned to you in your bank statement.

Check Stub The portion of a check that is torn off and that contains details concerning the account balance, the amount of the check, and an area for recording amounts of deposits.

Deposit In-transit A deposit that has been mailed in (is on its way to the bank); sometimes referred to as a *late deposit.*

Deposit Slip A form on which the depositor itemizes the currency, coins, and checks that are being deposited.

Electronic Funds Transfer System A system using computers to automatically transfer funds from one account to another. No checks are written.

NSF Check A check that has been rejected because there was not enough money in the account to honor it. Short for *not sufficient funds.*

Outstanding Check A check that has been written but has not yet been returned to your bank for collection.

Reconciliation Statement A form whereby the bank balance of your checking account (the bank statement balance) is reconciled with the checkbook (the check stubs) balance of your checking account.

Signature Card An authorized signature on a bank card that the bank uses to verify signatures on checks.

CHECKING ACCOUNTS

To open a checking account, an individual or a business must sign a signature card that authorizes the bank to recognize those signatures on all checks. A valid check must have an authorized signature on it. At this time an account number is issued and preprinted checks are mailed to the depositor to use. Examples of the signature card, the deposit slip, and a check with its stub are shown below.

Name of Account Address	_WILSON'S DEPT. STORE_ _111 MAIN STREET_ _ANYWHERE, USA_	First State Bank Anywhere, USA

Account Number
 246-121 **Checking Account Signature Card**

The undersigned agree to the terms and conditions of this institution and authorize this institution to recognize the signatures below to transact business in this account.

Depositor's Signature _____ _Ed Wilson_ _____

Depositor's Signature _____

BANK SIGNATURE CARD

Wilson's Department Store
111 Main Street
Anywhere, USA

Date _August 4_ 19 XX

First State Bank

246-121   1620

Currency	200	00
Coins	45	20
C H E C K S A.R. JONES	50	00
Judy SMITH	125	15
MARY TAFLINGER	32	00
Total	452	35
Less: Cash Received		
Total Deposit	452	35

DEPOSIT SLIP

No. 2106 Bal. Fwd. _982.15_ Deposit _____ Total _____ This Check _40.00_ Bal. Fwd. _942.15_	2106 Wilson's Department Store 111 Main Street Anywhere, USA _July 5_ 19 XX Pay to the order of _Seers & Rowbuck_ $ _40 XX/100_ _Forty and no/100_ _____ Dollars For _payment on account_ _Ed Wilson_ 112101713 246-121

SAMPLE CHECK WITH STUB

BANK RECONCILIATION

In business, you will need a checking account to pay your bills and provide you with a record of those payments. Once a month, the bank will send you a statement. This statement will be a record of the activity of your account for the past month. This *bank statement* will show at least the following:

1. A few columns for checks that were written by you, cashed by the payee, and returned to your bank for collection. These are *canceled checks* and have already been subtracted from your balance when you receive the statement. All of the canceled checks will be included with the bank statement. You should put them in order by check number to see how many checks in your checkbook have not cleared the bank. Those checks that you have written but have not yet been received by your bank are *outstanding checks.*

2. At least one column for recording all of the deposits you have made for the month.

3. A column (generally the last column on the right-hand side of the statement) reserved for recording the up-to-date balance of your account each day a check came in or a deposit was made.

4. Usually, your beginning monthly balance and your balance at the end of the month; sometimes, a total of all of the checks and a total of all of the deposits.

5. If applicable, a charge for handling your account, in the checks column, labeled *SC,* for example, for *service charge.* (There should be an explanation—a legend—of these abbreviations at the bottom of the statement.)

A representative bank statement is shown on page 490 for your reference.

The bank's balance for your account will almost always be different from your checkbook balance. This is to be expected and is not necessarily the result of any error on your part or by the bank. The process of bringing the two balances into agreement is called *reconciling.* A formal presentation is called a *reconciliation statement.* For accounting reasons, it is better to prepare the reconciliation statement in the following format:

Bank balance	XXXX	Checkbook Balance	XXXX
Add:	+ X	Add:	+ X
Deduct:	– X	Deduct:	– X
Adjusted bank balance	XXXX	Adjusted checkbook balance	XXXX

If everything is in order and done correctly, the adjusted balances will always be the same. The statement is then said to be reconciled.

Note that there are four main sections—an add and a deduct section on both sides. The most common reasons for entries are discussed in the following paragraph. (Upon investigation, you will probably discover some outstanding checks, some late deposits, a service charge, an insufficient funds charge, and possibly a note collected. In each case, you should determine which set of books—the bank's or your checkbook—already contains the adjustment. Your adjustment then will be to do the same thing on the other side of the statement.)

BANK BALANCE SIDE

ADD SECTION

A *late deposit,* or *deposit in-transit,* is a deposit that has been made after the statement was mailed out. The bank will add this information to your balance on the next month's statement; therefore, your adjustment should be to add it to the bank balance side of the reconciliation statement. (Remember, you already added it to your checkbook balance when you made the deposit; therefore, you need to add it to the bank side now.)

ACCOUNT NUMBER 225 871	STATEMENT DATE 01 13 8X	Mr. and Mrs. Clifford Jones 507 Rose Drive Anywhere, U.S.A.

CHECKS AND OTHER DEBITS			DEPOSITS	DATE	BALANCE
			BAL LAST STATEMENT		550.70
16.31	20.00	160.00		12-11	354.39
15.66	15.87			12-12	322.46
11.00	13.23			12-13	298.63
18.00	50.00	14.20		12-16	216.43
4.85	10.00			12-17	201.58
			361.48		
40.00	5.00SC		1,074.61	12-20	1,592.67
14.25	23.12	50.00		12-23	1,505.30
20.10	14.80	21.00			
23.24				12-24	1,426.16
159.00	167.67			12-26	1,099.49
10.68	14.91	150.50		12-27	923.40
6.31			55.00	12-30	972.09
12.00	12.87	261.28	100.00	1- 3	785.94
6.37	7.20	30.00		1- 6	742.37
30.00	19.92	42.26	125.00		
43.90				1- 7	731.29
13.71				1- 8	717.58
100.00				1-10	617.58
18.75	20.00	10.00			
20.00	91.71		200.00	1-13	657.12

BALANCE LAST STATEMENT	TOTAL AMOUNT OF CHECKS PAID	NUMBER OF CHECKS PAID	No. OF DEPOSITS	TOTAL AMOUNT OF DEPOSITS MADE	ACTIVITY CHARGES	BALANCE THIS STATEMENT
550.70	1,804.67	43	6	1,916.09	5.00	657.12

DM DEBIT MEMO	CM CREDIT MEMO	LS LIST POST	OD OVERDRAWN	SC SERVICE CHARGE

DEDUCT SECTION

Outstanding checks should be listed and added. The total will then be subtracted from the bank balance. (You already subtracted these checks from your checkbook when you wrote the checks.)

CHECKBOOK BALANCE SIDE

ADD SECTION

The most common reason for an adjustment on the checkbook balance side is for a note collection made by the bank on behalf of your business. The note owed to the business is collected for the business by the bank and added to the bank balance at the time of collection. (Since the note already has been added to the bank balance, your adjustment is to add it to your checkbook balance.)

DEDUCT SECTION

The bank service charge has already been deducted from your account when you receive the statement. You should therefore deduct it from your checkbook balance. As a part of earlier deposits of cash and checks collected from customers, some of the checks to you may not have been "good." These checks are returned *NSF (not sufficient funds).* Since the bank could not collect, they in turn reduce your balance by the amount they originally added to your account. A memo will be included in the bank statement notifying you of the subtraction because of an NSF check. You need to subtract the amount from the checkbook balance so that your checkbook will agree with the bank balance.

Errors can occur on your check stubs and on the bank statement. A variety are possible, and most can be reasoned out logically. Reconstruct what should have been done, compare it with what was done, and make the appropriate adjustment.

SAMPLE PROBLEM

From the bank statement on page 490 and the following information, the sample reconciliation statement was prepared:

Bank statement balance, $657.12   Checkbook balance, $564.14
Service charge, $5.00   Note collected by bank, $361.48
NSF check, $335.00   Late deposit, $350.00
Outstanding checks, $421.50

Reconciliation Statement

Bank balance	$657.12	Checkbook balance	$564.14
Add: Late dep.	+ 350.00	Add: Note collection	+ 361.48
Ded.: Out. checks	− 421.50	Ded.: Ser. charge	− 5.00
		NSF check	− 335.00
Adjusted bank bal.	$585.62	Adjusted checkbook bal.	$585.62

Practice Problems

1. The monthly bank statement of the ABC Co. showed a balance of $1,204.50. The company's checkbook balance was $470.11. The bank had collected a note for the ABC Co. for $500. The following checks were determined to be outstanding: No. 804 for $72.50, No. 807 for $33.15, No. 808 for $3.36, No. 811 for $104, and No. 815 for $55.13. The bank service charge was $3.75, and a $30 check was returned to the ABC Co. because of insufficient funds. Prepare a bank reconciliation statement.

2. The XYZ Co. had the following information concerning their monthly bank account. Prepare a bank reconciliation statement.
   Outstanding checks, $173.50
   Checkbook balance, $550.11
   Bank balance, $468.11
   Bank service charge, $5.50
   Deposit in-transit, $250

3. The bank statement for the MNO Co. indicates a balance of $4,182.08 as of May 31. The balance in their checkbook on the same date is $3,177.08. Upon inspection, it was found that a deposit on May 30 for $800 was not recorded on the bank statement; there were outstanding checks for $1,021, $502, $68.25, and $172.50; bank service charges of $3.75 were not recorded in the check stubs; and check No. 398 for $527.36 was recorded incorrectly on the check stub as $572.36. Prepare a reconciliation statement.

PASSBOOK CHECKING ACCOUNTS

Savings and loan institutions are now competing with the commercial banks for checking account deposits. As an incentive to new customers, many are offering to pay interest on the checking account rather than deducting a service charge. To be able to do this, they save money on mailing costs by not including the canceled checks in the monthly statement. Instead, each check number and amount is shown on the statement. Other than accounting for these two differences, the reconciliation procedure would be the same.

Note:

The interest rates and methods used would vary from savings and loan to savings and loan. For all of our examples and problems, we will assume a 6% annual rate (.5% monthly, or 6% ÷ 12) is paid on the account balance as of the 15th of each month.

PASSBOOK CHECKING ACCOUNT STATEMENT

Year to date interest credited 38.20					Balance 7/22 265.65	
Date	Check No./Amount		Deposits	Balance	Check No.	Amount
7/23	#213	7.85		257.80	#211	13.20
7/24	#212	150.00		107.80	#212	150.00
7/24	#217	15.00		92.80	#213	7.85
7/25	#218	35.13		57.67	#215	482.60
8/1	#211	13.20	1,800.00	1,844.47	#217	15.00
8/3	#215	482.60		1,361.87	#218	35.13
8/5	#219	107.34		1,254.53	#219	107.34
8/5	#223	95.00		1,159.53	#221	50.00
8/15	#221	50.00		1,109.53	#223	95.00
8/15	Int. cr.		5.55*	1,115.08	#225	175.20
8/20	#225	175.20		939.88		
Statement Date	Total Amount for Canceled Checks		Total Amount for Deposits Made		New Balance	Interest Earned this Statement
8/22	1,131.32		1,800.00		939.88	5.55

*The account balance of $1,109.53 on the 15th of the month multiplied by .5% (.005) is $5.55 interest earned (int. cr.). This amount is added to the account balance.

Note:

Notice in the last two columns on the right how the checks are listed for you in numerical order. This is done to help you reconcile since you have no canceled checks. In this statement example, checks numbered 214, 216, 220, 222, and 224 would be outstanding checks as of the statement date.

Practice Problem

4. From the passbook checking account statement below, determine the interest earned during the month and the new balance as of the statement date. Assume interest of 6% (.5% monthly) is paid on the account balance as of the 15th of the month.

Year to date interest credited 1.08				Balance 7/1 435.24	
Date	**Check No./Amount**	**Deposits**	**Balance**	**Check No.**	**Amount**
7/3	#1014 12.01			#1012	7.85
7/5	#1012 7.85			#1013	50.00
7/7	#1019 21.50			#1014	12.01
7/9	#1015 116.26			#1015	116.26
7/10	#1017 35.00			#1016	300.00
7/12	#1013 50.00			#1017	35.00
7/12	#1021 4.29			#1018	100.00
7/14	#1018 100.00			#1019	21.50
7/15	Int. cr.			#1020	550.00
7/18	#1024 65.00	650.00		#1021	4.29
7/21	#1016 300.00			#1022	3.89
7/26	#1022 3.89			#1024	65.00
8/1		650.00			
8/2	#1020 550.00				

Statement Date	Total Amount for Canceled Checks	Total Amount for Deposits Made	New Balance	Interest Earned this Statement
8/2	1,265.80	1,300.00		

ELECTRONIC FUNDS TRANSFERS

The more traditional way of transferring money by the use of checks is gradually being replaced by transferring funds instantly using computers. In this system a plastic bank card is usually used to get access to a computer so the amount can be immediately deducted from one account and added to another account. This obviously would eliminate "bad checks" since a transaction would not be processed unless the funds were available. One of the problems that has not been satisfactorily worked out is the unauthorized use of the plastic cards. Until some of these problems have been solved, it appears that checks will continue to be used for transferring funds.

Practice Problem Solutions

1.

Bank balance	$1,204.50	Checkbook balance	$470.11	
Add:		Add: Note	+500.00	
Deduct:		Deduct:		
Outstanding checks	− 268.14	Service charge	− 3.75	
		NSF check	− 30.00	
Adjusted bank balance	$ 936.36	Adjusted checkbook balance	$936.36	

2.

Bank balance	$468.11	Checkbook balance	$550.11
Add: Deposit	+250.00	Add:	
Deduct:		Deduct:	
Outstanding checks	−173.50	Service charge	− 5.50
Adjusted bank balance	$544.61	Adjusted checkbook balance	$544.61

3.

Bank balance	$4,182.08	Checkbook balance	$3,177.08
Add:		Add: Check No. 398 error	+ 45.00*
Late deposit	+ 800.00	Deduct: Service charge	− 3.75
Deduct:		Adjusted checkbook balance	$3,218.33
Outstanding checks	−1,763.75		
Adjusted bank balance	$3,218.33		

(*Too much was subtracted in the checkbook when check No. 398 was recorded [$572.36 was deducted rather than the proper amount of $527.36—$45 too much]. To correct for the error, the $45 will have to be added back to the checkbook balance.)

4.

Year to date interest credited	1.08				Balance 7/1	435.24
Date	**Check No./Amount**		**Deposits**	**Balance**	**Check No.**	**Amount**
7/3	#1014	12.01		423.23	#1012	7.85
7/5	#1012	7.85		415.38	#1013	50.00
7/7	#1019	21.50		393.88	#1014	12.01
7/9	#1015	116.26		277.62	#1015	116.26
7/10	#1017	35.00		242.62	#1016	300.00
7/12	#1013	50.00		192.62	#1017	35.00
7/12	#1021	4.29		188.33	#1018	100.00
7/14	#1018	100.00		88.33	#1019	21.50
7/15	Int. cr.		.44	88.77	#1020	550.00
7/18	#1024	65.00	650.00	673.77	#1021	4.29
7/21	#1016	300.00		373.77	#1022	3.89
7/26	#1022	3.89		369.88	#1024	65.00
8/1			650.00	1019.88		
8/2	#1020	550.00		469.88		

Statement Date	Total Amount for Canceled Checks	Total Amount for Deposits Made	New Balance	Interest Earned this Statement
8/2	1,265.80	1,300.00	469.88	.44

Exercise 78A: Bank Reconciliation

Name_____

Course/Sect. No._____   Score_____

Instructions: Find the reconciled amount.

(1) _____

Self-Check Exercise (Answers on back)

(1) The bank statement and the transactions for October in the checkbook are shown below. Prepare a bank reconciliation statement and find the reconciled amount.

Bank Statement

Checks		Deposits	Date	Balance
			Oct. 1	$3,500.00
		400.00	Oct. 4	3,900.00
100.00			Oct. 6	3,800.00
175.00	12.50	225.00	Oct. 11	3,837.50
354.28			Oct. 15	3,483.22
60.00 NSF			Oct. 20	3,423.22
		750.00	Oct. 21	4,173.22
13.35	110.00		Oct. 24	4,049.87
87.14	3.50 SC		Oct. 27	3,959.23

Checkbook Records

October	1 balance	$3,500.00	October 21 deposit	$750.00
October	2 check 402	100.00	October 21 check 408	13.35
October	4 deposit	400.00	October 22 check 409	87.14
October	4 check 403	12.50	October 23 check 410	110.00
October	7 check 404	175.00	October 28 deposit	385.00
October	11 deposit	225.00	October 29 check 411	12.00
October	12 check 405	354.28	October 30 deposit	135.50
October	14 check 406	72.25	October 31 check 413	54.95
October	19 check 407	38.50	October 31 balance	4,365.53

(2) _____

(2) Bank balance, $2,150; Outstanding checks, $857; Service charge, $7; Checkbook balance, $2,355; Late deposit, $1,020; NSF check, $35. Find the reconciled amount.

(3) _____

(3) Bank balance, $6,110.32; Service charge, $2; Checkbook balance, $6,052.72; Outstanding checks, $214.60; Late deposit, $300; Note collection charge, $5; Returned check (NSF), $100; Note collected, $250. Find the reconciled amount.

(1) _$4,302.03_

(2) _$2,313_

(3) _$6,195.72_

Exercise 78B: Bank Reconciliation Statement

Name _____

Course/Sect. No. _____   Score _____

(1) From the information given, prepare a bank reconciliation statement in the space that follows (page 501).

Bank Statement

Checks and Debits			Deposits	Date	Balance
				9/1	Last Statement 401.50
28.00	12.00			9/5	61.50
7.35				9/7	54.15
82.20	16.31	8.04	500.00	9/8	747.60
20.00				9/10	727.60
13.33	6.15			9/12	708.12
15.00	8.90			9/15	684.22
40.00	7.35 NSF			9/20	636.87
100.50			750.00	9/23	1286.37
9.00	325.00	34.60		9/26	917.77
45.00	60.00	1.25		9/28	811.52
15.25	17.30			9/29	778.97
12.87	3.00 SC			9/30	763.10

Balance Last Statement	Total Amount of Checks Paid	Number of Checks Paid	Total Amount of Deposits	Balance This Statement
$101.50	$878.05	22	$1,250.00	$763.10

NSF = Not Sufficient Funds CM = Credit Memo SC = Service Charge

Check Stubs

No. 430		No. 433		No. 436	
Bal Fwd	401.50	Bal Fwd	_____	Bal Fwd	_____
Deposit	_____	Deposit	_____	Deposit	_____
Total	_____	Total	_____	Total	_____
This Check	28.00	This Check	16.31	This Check	20.00
Bal Fwd	_____	Bal Fwd	_____	Bal Fwd	_____

No. 431		No. 434		No. 437	
Bal Fwd	_____	Bal Fwd	_____	Bal Fwd	_____
Deposit	_____	Deposit	_____	Deposit	_____
Total	_____	Total	_____	Total	_____
This Check	6.15	This Check	8.04	This Check	13.33
Bal Fwd	_____	Bal Fwd	_____	Bal Fwd	_____

No. 432		No. 435		No. 438	
Bal Fwd	_____	Bal Fwd	_____	Bal Fwd	_____
Deposit	_____	Deposit	500.00	Deposit	_____
Total	_____	Total	_____	Total	_____
This Check	12.00	This Check	82.20	This Check	15.00
Bal Fwd	_____	Bal Fwd	_____	Bal Fwd	_____

Exercise 78B: Bank Reconciliation Statement (Continued)

Name _____

Course/Sect. No. _____   Score _____

(1) Continued

Check Stubs

No. 439		No. 445		No. 451	
Bal Fwd	_____	Bal Fwd	_____	Bal Fwd	_____
Deposit	_____	Deposit	_____	Deposit	_____
Total	_____	Total	_____	Total	_____
This Check	40.00	This Check	1.25	This Check	17.30
Bal Fwd	_____	Bal Fwd	_____	Bal Fwd	_____

No. 440		No. 446		No. 452	
Bal Fwd	_____	Bal Fwd	_____	Bal Fwd	_____
Deposit	_____	Deposit	_____	Deposit	_____
Total	_____	Total	_____	Total	_____
This Check	8.90	This Check	7.35	This Check	67.89
Bal Fwd	_____	Bal Fwd	_____	Bal Fwd	_____

No. 441		No. 447		No. 453	
Bal Fwd	_____	Bal Fwd	_____	Bal Fwd	_____
Deposit	_____	Deposit	_____	Deposit	_____
Total	_____	Total	_____	Total	_____
This Check	9.00	This Check	150.00	This Check	13.00
Bal Fwd	_____	Bal Fwd	_____	Bal Fwd	_____

No. 442		No. 448		No. 454	
Bal Fwd	_____	Bal Fwd	_____	Bal Fwd	_____
Deposit	750.00	Deposit	_____	Deposit	_____
Total	_____	Total	_____	Total	_____
This Check	45.00	This Check	100.50	This Check	15.25
Bal Fwd	_____	Bal Fwd	_____	Bal Fwd	_____

No. 443		No. 449		No. 455	
Bal Fwd	_____	Bal Fwd	_____	Bal Fwd	_____
Deposit	_____	Deposit	_____	Deposit	_____
Total	_____	Total	_____	Total	_____
This Check	34.60	This Check	12.87	This Check	3.30
Bal Fwd	_____	Bal Fwd	_____	Bal Fwd	_____

No. 444		No. 450		No. 456	
Bal Fwd	_____	Bal Fwd	_____	Bal Fwd	_____
Deposit	_____	Deposit	_____	Deposit	200.00
Total	_____	Total	_____	Total	_____
This Check	60.00	This Check	325.00	This Check	75.00
Bal Fwd	_____	Bal Fwd	_____	Bal Fwd	_____

Exercise 78B: Bank Reconciliation (Continued)

Name _____

Course/Sect. No. _____   Score _____

Find the amount necessary to reconcile.

(1) _____   **(1)** Prepare the reconciliation statement below.

(2) _____   **(2)** Service charge, $5; Bank balance, $7,085; Checkbook balance, $5,800; Note collection charge, $20; Outstanding checks, $410; Deposit in-transit, $150; Note collected by bank, $970 plus interest income of $80. Find the reconciled amount.

(3) _____   **(3)** Bank balance, $7,503.82; Checkbook balance, $10,250.56; Late deposits, $2,416.48, $784; Service charge, $4.50; Outstanding checks, $17.50, $302.20, $3.59, $84.00, $19.95, $31.00. Find the reconciled amount.

Exercise 79A: Passbook Checking Accounts

Name _____

Course/Sect. No. _____   Score _____

Instructions: Assume a 6% annual rate for interest (.5% monthly) is paid on the account balance as of the 15th of the month.

Self-Check Exercise (Answers on back)

(1) _____   **(1)** What interest should have been credited to the account on January 15?

(2) _____   **(2)** What should be the bank balance (new balance) on the January 31 statement date?

(3) _____   **(3)** At what amount would the statement be reconciled with the checkbook?

Year to date interest credited					Balance Dec. 31	431.17
Date	**Check No./Amount**		**Deposits**	**Balance**	**Check No.**	**Amount**
Jan. 1			1,200.00		#4010	338.00
Jan. 2	#4012	20.50			#4012	20.50
Jan. 5	#4013	16.25			#4013	16.25
Jan. 6	#4010	338.00			#4015	5.37
Jan. 10	#4015	5.37			#4016	15.71
Jan. 14	#4021	169.42			#4018	532.60
Jan. 15			1,200.00		#4019	41.00
Jan. 15	Int. cr.				#4021	169.42
Jan. 17	#4019	41.00			#4022	71.10
Jan. 21	#4016	15.71				
Jan. 24	#4018	532.60				
Jan. 27	#4022	71.10				
Statement Date	Total Amount for Canceled Checks		Total Amount for Deposits Made	New Balance	Interest Earned this Statement	
Jan. 31						

From the depositor's checkbook records, the following are determined to be outstanding checks:

#4011	$ 25.00	#4020	$ 9.09
#4014	205.49	#4023	145.55
#4017	13.67		

(1) _____$11.41_____

(2) _____$1,632.63_____

(3) _____$1,233.83_____

Exercise 79B: Passbook Checking Accounts

Name_____

Course/Sect. No._____   Score_____

Instructions: From the information provided in the statement below, complete the passbook checking account statement by filling in the correct amounts for each blank.

Year to date interest credited 24.92				Balance May 1 842.20	
Date	**Check No./Amount**	**Deposits**	**Balance**	**Check No.**	**Amount**
May 2	#723	13.50	150.00	_____	#722	_____
May 5	#730	117.39		_____	#723	13.50
May 6	#725	8.45		_____	#725	8.45
May 8			225.00	_____	#727	25.00
May 12	#722	_____		1,028.66	#729	43.17
May 12	#731	84.49		_____	#730	117.39
May 13	#727	25.00		_____	#731	84.49
May 15	Int. cr.		_____	_____	#733	_____
May 20	#735	13.50		_____	#734	39.75
May 24	#733	_____		_____	#735	13.50
May 25	#729	43.17	75.00	1,724.02		
May 27			_____	_____		
May 29	#734	39.75		_____		

Statement Date	Total Amount for Canceled Checks	Total Amount for Deposits Made	New Balance	Interest Earned this Statement
May 30	612.53	565.00	_____	4.60

What would be the new balance on the May 30 statement date?

_____

Chapter

15

Statistics and Graphs

Preview of Terms

Mean   The average for a set of data.

Median   The middle value in a distribution, above and below which are an equal number of values.

Mode   The value in a set of data that occurs most often.

MEASURES OF CENTRAL TENDENCY

Numerical information can sometimes be misleading and many times is misinterpreted. Consider the following examples.

Assume Homer is interviewing for a new job. Each of two companies pay the same starting salary of $10,000. Company A claims the average work week for the position Homer is seeking is 35 hours per week, whereas Company B claims theirs is 45 hours per week. The first impression would be to select Company A over Company B. However, before making a decision, it is important to know more about the *distribution* of the numerical information. The hours for Company A may range from 20 to 65 hours per week, while Company B hours may range from 38 to 48.

Assume also that Karen is choosing between two jobs. Company X states her job classification would place her within a $5,000 salary range, while Company Y boasts a salary range of $30,000. Karen's first impression would be that there is more opportunity for salary advancement with Company Y. Upon closer inspection it is found that Company X has salaries that average $12,000 (with a range between $9,000 and $14,000), while Company Y has salaries that average $8,000 (with a range between $4,000 and $34,000). This is another example that illustrates that a better decision can be made when you have more information about the *distribution* of the numerical information.

The single most important measure describing numerical information concerns the location of the center of the data. The three most common statistical measures used concerning central location are the *mean*, the *median*, and the *mode.*

THE MEAN

The *arithmetic mean* is more commonly called the *average.* It is found by adding the items of numerical information and dividing that sum by the total number of items added.

Example 1: Gary made an 85, 80, 94, and 78 on his four unit tests. Calculate his average.

$$\text{Sum of items} = 85 + 80 + 94 + 78 = 337$$
$$\text{Total items} = 4$$

$$\text{Mean (average)} = \frac{\text{Sum of items}}{\text{Total items}} = \frac{337}{4} = 80.25$$

Example 2: The inventory of merchandise was taken quarterly as follows: March 31 = $45,230; June 30 = $62,780; September 30 = $54,030; December 31 = $50,140. Find the arithmetic mean, or the average inventory.

$$
\begin{array}{l}
\$45,230 \\
62,780 \\
54,030 \\
+\ 50,140 \\
\hline
\$212,180
\end{array}
\qquad
\text{Mean (average)} = \frac{\$212,180}{4}
$$

$$= \$53,045$$

THE MEDIAN

The median is also called the *middle value* in a set of numbers arranged from high to low or from low to high (descending or ascending order). If there are an odd number of items, the median is the number exactly at the halfway point. When there are an even number of items, the median falls halfway between the two middle values. No matter how many items there are, the median is always located in the numerical order so that the number of items on each side of the median are equal.

LOCATING THE MEDIAN WHEN THERE IS AN ODD NUMBER OF ITEMS

Use the formula $\frac{(n+1)}{2}$, where n is the number of items in the group.

Example 3: Find the median of the following numbers: 1, 9, 7, 2, 8, 5, 3.

Step 1: Order them from high to low (or low to high)

9, 8, 7, 5, 3, 2, 1

Step 2: Count the number of items in the group. In this example there are 7 (7 then becomes n in the formula).

Step 3: Substitute the known value of n and solve.

$$\frac{(n + 1)}{2} = \frac{(7 + 1)}{2} = \frac{8}{2} = 4$$

The result of this step tells you how far the median is located from either end of the grouped data. In this example it is fourth from either end.

$$9, 8, 7, \underline{5}, 3, 2, 1$$

LOCATING THE MEDIAN WHEN THERE IS AN EVEN NUMBER OF ITEMS

When ordered from high to low there will be *two* middle values if there is an even number of items. The median is located halfway between these two middle values. If you have 8 values, the median is located between the 4th and 5th values. If you have 30 values, the median is located between the 15th and 16th values.

Example 4: Find the median of the following salaries: $27,000; $18,000; $42,500; $12,800; $8,200; $102,000; $39,000; $92,000; $13,000; $75,000.

Step 1: Order them from high to low (or low to high)

$8,200
$12,800
$13,000
$18,000
$27,000
$39,000
$42,500
$75,000
$92,000
$102,000

Step 2: Determine the two middle values by using $n/2$, where n is the number of items.

$$\frac{n}{2} = \frac{10}{2} = 5$$

The middle values will be the nth item and the one following it, in this instance the 5th and the 6th items. In Step 1 the 5th and the 6th items are $27,000 and $39,000, respectively.

Step 3: The median is halfway between the two middle values.

$$\$27,000 + \$39,000 = \$66,000 \div 2 = \$33,000$$

$33,000 is halfway between the two middle values. Therefore, it is the median in this example.

THE MODE

The mode is the value that occurs most often in a group of items. The mode may be a poor measure of central location since it may not appear near the center of the data.

Example 5:   Find the mode from the following data: 20, 24, 25, 26, 27, 28, 29, 29, 29, 30. The number "29" appears most often in these values and is therefore the *mode.*

GRAPHS

Graphs are used in business to supplement statistics and financial information. They allow the reader to grasp relationships of parts to a whole more easily and quickly. The three most commonly used graphs are pie graph, bar graph, and line graph.

PIE GRAPH

This is called a *pie graph* because it is shaped like a pie with sections in it resembling slices. It is constructed by determining the ratio of each segment to the total and multiplying by 360 degrees (the total number of degrees in a complete circle). The computed degrees of each segment can then be drawn into the circle (or pie) by using a protractor. The primary purpose of the pie graph is to present the component parts of a whole.

Example 6:   The XYZ Co. last year had sales of $6,000,000; cost of goods sold of $5,300,000; taxes of $350,000; dividends paid out of $150,000; and retained earnings of $200,000. Develop a pie graph to illustrate this.

Step 1: Divide each of the components by sales of $6,000,000 to determine their proportion to the total.

Cost of Goods Sold =	$5,300,000 ÷ $6,000,000* =	.883 =	88.3%
Taxes =	350,000 ÷ 6,000,000 =	.058 =	5.8%
Dividends =	150,000 ÷ 6,000,000 =	.025 =	2.5%
Retained Earnings =	200,000 ÷ 6,000,000 =	+ .033 =	+ 3.3%
	$6,000,000	1.000	99.9%

*Remember, you can set your calculator for a constant divisor or a constant multiplier, which will save you time in problems such as these.

Step 2: Multiply each component proportional amount by 360 degrees. (Round to a whole degree.)

$$\text{Cost of Goods Sold} = .883 \times 360°* = 318°$$
$$\text{Taxes} = .058 \times 360° = 21°$$
$$\text{Dividends} = .025 \times 360° = 9°$$
$$\text{Retained Earnings} = .033 \times 360° = \underline{+\ 12°}$$
$$360°$$

Step 3: Use your protractor to measure off the proper degrees for each component and label the parts. (The segments can be illustrated in this case either by dollar amounts or by percentages.)

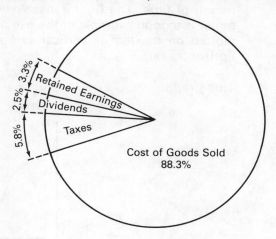

BAR GRAPH

The bar graph is usually used to show changes in the same items over selected time periods. A vertical or horizontal scale can be used. The smallest value on the scale should always be located at the bottom left corner. (This value is usually zero.) When time comparisons are being made on the same items, a legend must be given to identify the bars.

Example 7:   A vertical bar graph.

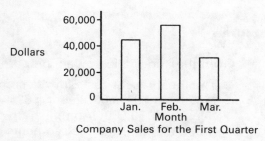

Example 8:   A horizontal bar graph.

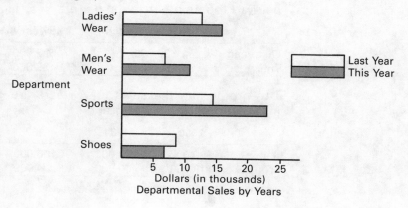

To be accurate, the construction of bar graphs should be on graph or scaled paper. The bars should be the same width and the spacing between bars should be uniform.

LINE GRAPH

This graph is used most often to indicate the fluctuations of items over a period of time. The Dow Jones Average of the stock market in your daily paper is a good example of the use of a line graph. By custom, amounts are plotted on the left or vertical axis and time is plotted on the bottom or horizontal axis.

Example 9:   Line graph.

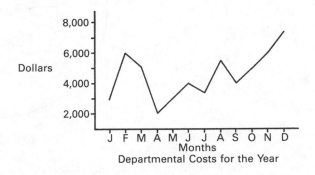

Departmental Costs for the Year

The construction of a line graph is similar to that of a bar graph. For the line graph, locate and plot each value from the left axis as it pertains to the proper time element on the bottom axis. After all points are plotted, connect them with lines.

Practice Problems

From the information given, prepare the necessary graph.

4. The most recent costs for the company were broken down as follows: Materials, $375,000; Labor, $625,000; Overhead, $125,000; Selling expenses, $150,000; Administrative expenses, $50,000. Prepare a pie graph and indicate the segments using percents.

5. The company advertising by branches for the past two years is given below. Prepare a horizontal bar graph to illustrate the information.

	Eastern Branch	Midwest Branch	Western Branch
Last year	$35,000	$27,000	$55,000
This year	52,000	33,000	80,000

6. The profits and the sales for 5 years are given below. Illustrate the information using a line graph.

	1980	1981	1982	1983	1984
Sales	$80,000	$70,000	$106,000	$121,000	$140,000
Profits	13,000	10,000	24,000	37,000	42,000

Practice Problem Solutions

1. Mean = 2 + 4 + 8 + 8 + 10 + 11 + 13 = 56 ÷ 7 numbers = 8.
Median = Halfway between the 7 numbers would be the fourth $[\frac{n+1}{2} = \frac{7+1}{2}]$
number, which is 8.
Mode = The value that occurs most often is 8.

2. Mean = $8 + $8 + $10 + $12 + $15 + $15 + $15 + $21 + $32 + $42
+ $50 + $50 + $60 = $338 ÷ 13 numbers = $26.
Median = Halfway between 13 numbers would be the seventh $[\frac{n+1}{2} = \frac{13+1}{2}]$
number, which is $15.
Mode = The value that occurs most often is $15.

3. Mean = 6,000 + 6,000 + 8,000 + 15,000 + 20,000 + 50,000 = 105,000
÷ 6 numbers = 17,500.
Median = Halfway between 6 numbers would be the third $(n/2 = 6/2)$
number and the one following it, the fourth. The median would be halfway
between these two middle values. $\frac{8,000+15,000}{2} = 11,500$
Mode = The value that occurs most often is 6,000.

4.

Materials	$375,000 ÷ $1,325,000 =	28.3% × 360° = 102°
Labor	625,000 ÷ 1,325,000 =	47.2% × 360° = 170°
Overhead	125,000 ÷ 1,325,000 =	9.4% × 360° = 34°
Selling expenses	150,000 ÷ 1,325,000 =	11.3% × 360° = 41°
Administrative expenses	+ 50,000 ÷ 1,325,000 =	+ 3.8% × 360° = 14°
	$1,325,000	100.0%

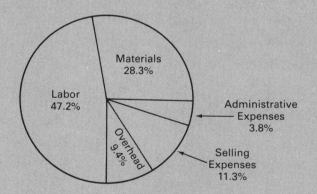

5.

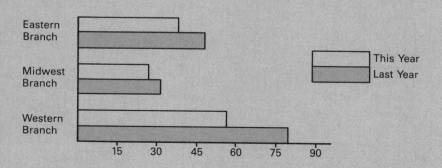

6.

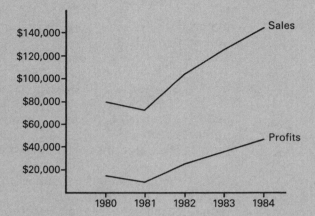

Exercise 80A: Measures of Central Tendency

Name _____

Course/Sect. No. _____   Score _____

Instructions: Find the mean, median, or mode.

Self-Check Exercise (Answers on back)

Group 1 Scores: 10, 20, 80, 40, 20, 30, 60, 50, 20, 50
Group 2 Scores: 104, 102, 110, 106, 104, 120, 117

(1) _____   **(1)** Use group 1 scores and find the mean.

(2) _____   **(2)** Use group 2 scores and find the mean.

(3) _____   **(3)** Use group 1 scores and find the median.

(4) _____   **(4)** Use group 2 scores and find the median.

(5) _____   **(5)** Use group 1 scores and find the mode.

(6) _____   **(6)** Use group 2 scores and find the mode.

(7) _____   **(7)** The mail route covered 125 houses in a 5-day week. How many did it average each day?

(8) _____   **(8)** The odometer on the car showed you had traveled 820 miles in 2.5 days. How many miles did you average each day?

(9) _____   **(9)** The board lengths ordered were: 8 ft., 10 ft., 12 ft., 6 ft., 8 ft., 10 ft., 4 ft., 12 ft., 8 ft., 4 ft. What was the mode in that group of data?

(10) _____   **(10)** Seven households were randomly selected to find out their income earned. The results were: $20,000; $100,000; $60,000; $30,000; $42,000; $55,000; $72,000. What was the median income of these data?

(1) _____38_____

(2) _____109_____

(3) _____35_____

(4) _____106_____

(5) _____20_____

(6) _____104_____

(7) _____25_____

(8) _____328_____

(9) _____8 ft._____

(10) _____$55,000_____

Exercise 80B: Measures of Central Tendency

Name_____

Course/Sect. No._____ Score_____

Instructions: Find the mean, median, or mode.

	Observation I				Observation II				
25	8	37	19	43	31¢	27¢	33¢	29¢	22¢
7	80	51	60	28	47¢	22¢	38¢	39¢	
25	73								

(1) _____ **(1)** Use observation I and find the mean.

(2) _____ **(2)** Use observation II and find the mean.

(3) _____ **(3)** Use observation I and find the median.

(4) _____ **(4)** Use observation II and find the median.

(5) _____ **(5)** Use observation I and find the mode.

(6) _____ **(6)** Use observation II and find the mode.

(7) _____ **(7)** The delivery truck made 23 trips during the 5-day work week. How many deliveries did it average a day?

(8) _____ **(8)** The shoe clerk in the ladies department brought out the following priced shoes: $33, $18.95, $28, $59.95, $44. What was the median priced shoe?

(9) _____ **(9)** The auto "counter" that was laid across the road registered the following number of cars this week: 115, 83, 97, 72, 83, 150, 181. What would be the mode for these data?

(10) _____ **(10)** From March to April the company stock sold for $50 a share, from April to May it sold for $52.50 a share, and from May to June it sold for $49.75 a share. What was the average price per share during this 3-month period?

Exercise 81A: Graphs

Name _____

Course/Sect. No. _____   Score _____

Instructions: Work each as instructed.

Self-Check Exercise (Answers on back)

The total sales for the four departments are given below. Find the number of degrees (rounded to a whole degree) that each department would represent in a pie graph.

(1) _____   **(1)** Department A: $180,000
(2) _____   **(2)** Department B: $ 45,000
(3) _____   **(3)** Department C: $212,000
(4) _____   **(4)** Department D: $106,000

(5) Construct a vertical bar graph to illustrate the company net profits for the following years:
1984: $550,000
1983: $475,000
1982: $325,000
1981: $400,000
1980: $300,000

(6) Construct a line graph to indicate the number of units sold for each year.

	1983	1982	1981	1980
ABC Company	25,000	17,000	19,000	15,000
XYZ Company	13,000	12,800	12,500	12,000

(1) _____ *119°*
(2) _____ *30°*
(3) _____ *141°*
(4) _____ *70°*

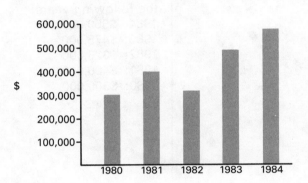

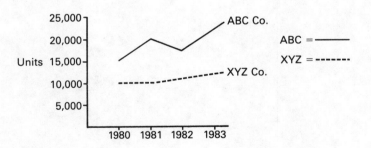

Name _____

Course/Sect. No. _____   Score _____

Instructions: Work each as instructed.

_____

(1) The salaries of the manager and three employees are given below. Prepare a pie graph to illustrate the parts of the total payroll they make up. Indicate how many degrees (to a whole degree) each would represent in the graph.

Manager	= $20,160	Manager	_____
Employee A =	$13,440	Employee A	_____
Employee B =	$ 9,600	Employee B	_____
Employee C =	$ 4,800	Employee C	_____

(2) The cost of gasoline increased over a 3-year period as follows: 1982 = $1.05/gal., 1983 = $1.12/gal., 1984 = $1.19/gal. Construct a *horizontal bar graph* to illustrate this.

(3) The Dowd Chemical Co. and the Fillips Petroleum Co. stock prices for the first three months (January, February, March) are shown below. Construct a *line graph* making the Fillips Petroleum Co. line a dotted line.

	Stock Prices/Share		
	Jan.	**Feb.**	**Mar.**
Dowd Chemical Co.	$31	$25	$20
Fillips Petroleum Co.	$72	$74	$79

Chapter

16

Metric Measurement

Preview of Terms

Metric Systems
A system of weights and measures based on the decimal system where there is only one basic unit for measuring a quantity and all other units are related to that basic unit by factors of 10. For example, our base unit for money is dollars and all other money measurements are measured in factors of 10.

Basic Units
Each category of measurement has a base unit that can be divided into multiples or submultiples. Examples include *gram* (for weight), *meter* (for length), and *liter* (for volume).

Prefixes
The prefixes (e.g., *milli-*, *centi-*, *kilo-*) used in front of the basic unit names are ways to indicate which power of 10 the multiple or submultiple represents. For example, *kilo* means 10^3 or 1,000; *deka* means 10^1 or 10.

The United States has committed itself to a gradual and voluntary conversion to the metric system. The most rapid advances will most likely occur in the field of business. The standards of measurement that will be reviewed in this text are length, weight, temperature, and volume.

Quantity	Metric Base Unit	Symbol
Length	Meter	m
Weight	Gram	g
Temperature	*Celsius	°C
Volume (liquid)	Liter	l

*The official metric temperature scale is Kelvin, but most countries still use the Celsius scale. The United States still uses the Fahrenheit scale.

The American system of weights and measures has evolved into a jumbled and confusing collection of inches, feet, ounces, pounds, gallons, and so on. In some cases even the same unit can refer to different things.

- How do you divide an inch into 10 equal parts?
- What is the price per pound of 7 ounces of food?
- What is the difference between "ounces" of milk and "ounces" of meat?

In the metric system, however, measurements are all based on 10. You can change from one unit to another by multiplying or dividing by 10. Each system of measurement only has one basic unit of measure and the size of the base unit is determined by reference to prefixes using the powers of 10.

Base Unit

Prefix =	Kilo	Hecto	Deka	Meter	Gram	Liter	Deci	Centi	Milli
Abbrev. =	k	h	dk	m	g	l	d	c	mm

The prefixes are used in front of the basic metric unit to indicate the size of the base unit.

Example 1:

A **centi**meter means .01 meters or 1/100 of a meter.
A **deka**liter means 10 liters.
5 **kilo**grams means 5 × 1000 grams or 5000 grams.

CONVERTING WITHIN THE METRIC SYSTEM

Determine which direction on the metric scale you are moving and either multiply or divide using the methods of conversion that follow:

Methods of Conversion

Kilo	Hecto	Deka	Base Unit	Deci	Centi	Milli
1000	100	10	1	.1	.01	.001

When moving $\xrightarrow{\text{left to right}}$ multiply by 10 for each position.

When moving $\xleftarrow{\text{right to left}}$ divide by 10 for each position.

Example 2:

How many centimeters (cm) are there in 3.14 meters (m)?

Since the direction of conversion is from left to right, you would multiply by 10 for each position you moved through.

Meter   Decimeter   Centimeter

$$| \xrightarrow{\times 10} | \xrightarrow{\times 10} |$$

$$3.14 \times 10 = 31.4 \times 10 = 314$$

Note:

Remember, since 10 × 10 = 100, you can move the decimal two places to the right (314.) when multiplying.

Example 3:

How many hectometers (hm) are there in 7560 decimeters (dm)?

Since the direction of conversion is from right to left, you would divide by 10 for each position you moved through.

Hectometer   Dekameter   Meter   Decimeter

$$| \xleftarrow{\div 10} | \xleftarrow{\div 10} | \xleftarrow{\div 10} |$$

$$7560 \div 10 = 756 \div 10 = 75.6 \div 10 = 7.56$$

Note:

Remember, since 10 × 10 × 10 = 1000, you can move the decimal 3 places to the left (7.560) when dividing.

Practice Problems

1. How many meters are there in 1.011 kilometers?
2. How many centigrams are there in .456 grams?
3. How many liters are there in 2511 milliliters?
4. How many kilometers are there in 35,284 decimeters?

CONVERTING *FROM* METRIC AND *TO* METRIC

Until the public can instantly recognize the metric equivalent of our traditional U.S. measures (or vice-versa), it will be necessary to refer to conversion tables. The following are approximate conversion tables.

Length

From U.S. to Metric		
When You Know	**Multiply By**	**To Find**
inches	2.5	centimeters
feet	30	centimeters
yards	.9	meters
miles	1.6	kilometers

From Metric to U.S.		
When You Know	**Multiply By**	**To Find**
centimeters	.4	inches
meters	3.3	feet
meters	1.1	yards
kilometers	.6	miles

Example 4: How many meters are there in 100 yards?

100 yards × .9 = 90 meters

Example 5: How many miles are there in 325 kilometers?

325 kilometers × .6 = 195 miles

Weight

From U.S. to Metric		
When You Know	**Multiply By**	**To Find**
ounces	28	grams
pounds	.45	kilograms
short tons (2000 lbs)	.9	metric tons

From Metric to U.S.		
When You Know	**Multiply By**	**To Find**
grams	.035	ounces
kilograms	2.2	pounds
metric tons	1.1	short tons

Example 6:   How many grams are there in 16 ounces?

$$16 \text{ ounces} \times 28 = 448 \text{ grams}$$

Example 7:   How many pounds are there in 150 kilograms?

$$150 \text{ kilograms} \times 2.2 = 330 \text{ pounds}$$

Volume (Liquid)

| | From U.S. to Metric | |
When You Know	Multiply By	To Find
ounces	29.6	milliliters
quarts	.9	liters
gallons	3.8	liters

| | From Metric to U.S. | |
When You Know	Multiply By	To Find
milliliters	.034	ounces
liters	1.06	quarts
liters	.264	gallons

Example 8:   How many liters are there in 12 quarts?

$$12 \text{ quarts} \times .9 = 10.8 \text{ liters}$$

Example 9:   How many gallons are there in 20 liters?

$$20 \text{ liters} \times .264 = 5.28 \text{ gallons}$$

TEMPERATURE

Absolute zero (0) on the scales for Fahrenheit (F), Celsius (C), and Kelvin (K), respectively, are −460°F, −273°C, and 0°K. To convert from Fahrenheit to Celsius or from Celsius to Fahrenheit use the following conversion factors.

Fahrenheit to Celsius
(degrees Fahrenheit − 32) × 5/9 = degrees Celsius

Celsius to Fahrenheit
(degrees Celsius × 9/5) + 32 = degrees Fahrenheit

Example 10:   How many degrees Celsius is 84°F?

$$(84°F - 32) \times 5/9 = °C$$
$$52 \times 5/9 = 28.9°C \text{ (Rounded to a tenth)}$$

Example 11:   How many degrees Fahrenheit is 17°C?

$$(17°C \times 9/5) + 32 = °F$$
$$30.6 + 32 = 62.6°F$$

Practice Problems

5. How many kilometers are there in 2,500 miles?
6. How many ounces are there in 120 grams?
7. How many milliliters are there in 15 ounces?
8. How many degrees Celsius is 95°F?

Practice Problem Solutions

1. Moving from kilometers to meters → requires multiplication.

Kilo   Hecto   Deka   Meter

$\rightarrow \times 10 \rightarrow \times 10 \rightarrow \times 10$

1.011 Kilometers $\times 10 = 10.11 \times 10 = 101.1 \times 10 = 1011$

or

$10 \times 10 \times 10 = 1000 \times 1.011 = 1011$ meters

2. Moving from grams to centigrams → requires multiplication.

Grams   Deci   Centi

$\rightarrow \times 10 \rightarrow \times 10$

.456 grams $\times 10 \times 10 = 45.6$ centigrams

or

$10 \times 10 = 100 \times .456 = 45.6$ centigrams

3. Moving from milliliters to liters ← requires division.

Liter   Deci   Centi   Milli

$\div 10 \leftarrow \div 10 \leftarrow \div 10\leftarrow$

2511 milliliters $\div 10 = 251.1 \div 10 = 25.11 \div 10 = 2.511$ liters

or

$10 \times 10 \times 10 = 1000 \qquad 2511 \div 1000 = 2.511$ liters

4. Moving from decimeters to kilometers ← requires division.

Kilo   Hecto   Deka   Meter   Deci

$\div 10 \leftarrow \div 10 \leftarrow \div 10 \leftarrow \div 10 \leftarrow$

$35{,}284 \div 10 = 3528.4 \div 10 = 352.84 \div 10 = 35.284 \div 10 = 3.5284$

or

$10 \times 10 \times 10 \times 10 = 10{,}000 \qquad 35{,}284 \div 10{,}000 = 3.5284$ kilometers

5. 2,500 miles $\times 1.6 = 4{,}000$ kilometers

6. 120 grams $\times .035 = 4.2$ ounces

7. 15 ounces $\times 29.6 = 444$ milliliters

8. $(95°F - 32°) \times 5/9 = C$
$63° \times 5/9 = 35°C$

Exercise 82A: Converting within the Metric System

Name _____

Course/Sect. No. _____   Score _____

Instructions: Convert the following as instructed.

---

(1) _____

(2) _____

(3) _____

(4) _____

(5) _____

(6) _____

(7) _____

(8) _____

(9) _____

(10) _____

Self-Check Exercise (Answers on back)

(1) .0348 g = ___ mg   **(2)** 5044 mm = ___ m

(3) 23 l = ___ kl   **(4)** .187 kg = ___ g

(5) 350.46 dg = ___ hg   **(6)** 78 km = ___ dam

(7) 14.83 m = ___ cm   **(8)** .001002 kg = ___ mg

(9) The distance between two cities was 2,500 km. How many *meters* would that be?

(10) The hungry teenager ate 32,000 centigrams of steak. If the price of the steak was 1.5¢ per gram, how much was spent on the teenager's steak?

(1) _____34.8_____

(2) _____5.044_____

(3) _____.023_____

(4) _____187_____

(5) _____.35046_____

(6) _____7,800_____

(7) _____1,483_____

(8) _____1,002_____

(9) ___2,500,000___

(10) _____$4.80_____

Exercise 82B: Converting within the Metric System

Name _____

Course/Sect. No. _____   Score _____

Instructions: Convert the following as instructed.

(1) _____	**(1)** 1.504 l = ____ ml **(2)** 7.115 m = ____ mm
(2) _____	
(3) _____	**(3)** 8828 g = ____ kg **(4)** 68 m = ____ km
(4) _____	
(5) _____	**(5)** .1234567 kg = ____ mg **(6)** .4062 km = ____ cm
(6) _____	
(7) _____	**(7)** 124.47 ml = ____ dl **(8)** .005 m = ____ dm
(8) _____	
(9) _____	**(9)** How many 500 milliliter cups are in 1 liter?
(10) _____	**(10)** A tiger could leap 3.5 meters. How many dekameters could he jump?

Exercise 83A: Converting from Metric to U.S.

Name_____

Course/Sect. No._____   Score_____

Instructions: Convert the following as instructed, using the approximate conversion tables.

---

(1) _____

(2) _____

(3) _____

(4) _____

(5) _____

(6) _____

(7) _____

(8) _____

(9) _____

(10) _____

Self-Check Exercise (Answers on back)

(1) 65 meters = ____ feet   **(2)** 651 kilometers = ____ miles

(3) 25 metric tons = ____ short tons   **(4)** 40.5 liters = ____ gallons

(5) 40.5 milliliters = ____ ounces   **(6)** 1.5 hectograms = ____ ounces

(7) 7020 centigrams = ____ ounces **(8)** 800 centimeters = ____ yards

(9) The bed of the foreign pickup truck was guaranteed to hold 27,000 hectograms. How many pounds would that be?

(10) The company received a shipment of three boxes that weighed 18.5 kilograms, 990 dekagrams, and 35 pounds. How many total *pounds* did the three boxes weigh?

(1) _____214.5_____

(2) _____390.6_____

(3) _____27.5_____

(4) _____10.692_____

(5) _____1.377_____

(6) _____5.25_____

(7) _____2.457_____

(8) _____8.8_____

(9) _____5,940_____

(10) _____97.48_____

Exercise 83B: Converting from Metric to U.S.

Name _____

Course/Sect. No. _____   Score _____

Instructions: Convert the following as instructed, using the approximate conversion tables.

(1) _____   **(1)** 204 centimeters = ____ inches   **(2)** 65 meters = ____ yards

(2) _____

(3) _____   **(3)** 10.5 kilograms = ____ pounds   **(4)** 580 grams = ____ ounces

(4) _____

(5) _____   **(5)** 40.5 liters = ____ quarts   **(6)** 35 decimeters = ____ inches

(6) _____

(7) _____   **(7)** .85 kiloliters = ____ ounces   **(8)** 13 dekameters = ____ yards

(8) _____

(9) _____   **(9)** The cable from England was priced at $1.25 per foot but was quoted as 100 meters on the package. How much did the package of cable cost?

(10) _____   **(10)** The gasoline pump listed the price to be $.32 per liter. The tank was filled at a cost of $21.60. How many gallons were purchased?

Exercise 84A: Converting from U.S. to Metric

Name _____

Course/Sect. No. _____   Score _____

Instructions: Convert the following as instructed, using the approximate conversion tables.

(1) _____	Self-Check Exercise (Answers on back)

(1) _____

(2) _____

(3) _____

(4) _____

(5) _____

(6) _____

(7) _____

(8) _____

(9) _____

(10) _____

Self-Check Exercise (Answers on back)

(1) 15.5 feet = ____ centimeters   **(2)** 530 miles = ____ kilometers

(3) 38 ounces = ____ grams   **(4)** 12.25 quarts = ____ liters

(5) 15.7 gallons = ____ liters   **(6)** 25 ounces = ____ kilograms

(7) 1000 yards = ____ dekameters   **(8)** 3 quarts = ____ deciliters

(9) The box to be mailed weighed 1.75 pounds. The overseas postage rate was 5¢ per dekagram. What did it cost to mail it?

(10) The new car was advertised in terms of kilometers per gallon. It was driven 190 miles on 10 gallons. How many kilometers per gallon would that be?

(1) _465_

(2) _848_

(3) _1064_

(4) _11.025_

(5) _59.66_

(6) _.7_

(7) _90_

(8) _27_

(9) _$3.94_

(10) _30.4_

Exercise 84B: Converting from U.S. to Metric

Name _____

Course/Sect. No. _____   Score _____

Instructions: Convert the following as instructed, using the approximate conversion tables.

(1) _____	**(1)** 38 inches = ____ centimeters **(2)** 75 yards = ____ meters
(2) _____	
(3) _____	**(3)** 18.5 short tons = ____ metric tons **(4)** 185 pounds = ____ kilograms
(4) _____	
(5) _____	**(5)** .55 ounces = ____ milliliters **(6)** 10 feet = ____ meters
(6) _____	
(7) _____	**(7)** 3472 ounces = ____ liters **(8)** 200 pounds = ____ hectograms
(8) _____	
(9) _____	**(9)** Kendi is $5\frac{1}{2}$ feet tall. How many meters is that?
(10) _____	**(10)** Lauri was instructed to be sure to buy at least 35 quarts of bottled soft drinks for the party. She could only find liter containers. What minimum number of liter bottles should she buy?

Exercise 85A: Temperature Conversions

Name _____

Course/Sect. No. _____   Score_____

Instructions: Convert the following as instructed. (Round to a tenth of a degree if necessary.)

(1) _____	
(2) _____	
(3) _____	
(4) _____	
(5) _____	
(6) _____	
(7) _____	
(8) _____	

Self-Check Exercise (Answers on back)

(1) 50° F = ____° C   **(2)** 92° F = ____° C

(3) 75.5° F = ____° C   **(4)** 17° F = ____° C

(5) 31° C = ____° F   **(6)** −10° C = ____° F

(7) 25.5° C = ____° F   **(8)** 20° C = ____° F

(1) _____10°_____

(2) _____33.3°_____

(3) _____24.2°_____

(4) _____−8.3°_____

(5) _____87.8°_____

(6) _____14°_____

(7) _____77.9°_____

(8) _____68°_____

Exercise 85B: Temperature Conversions

Name _____

Course/Sect. No. _____   Score _____

Instructions: Convert the following as instructed. (Round to a tenth of a degree if necessary.)

(1) _____	**(1)** 86° F = ____° C
	(2) 113° F = ____° C
(2) _____	
	(3) 40° F = ____° C
(3) _____	**(4)** 68° F = ____° C
(4) _____	
(5) _____	**(5)** 7.2° C = ____° F
	(6) 16.2° C = ____° F
(6) _____	
(7) _____	**(7)** 40° C = ____° F
	(8) −30° C = ____° F
(8) _____	

Chapter

17

Statement Analysis

Preview of Terms

Balance Sheet　A statement of financial position that shows that total assets are equal to total liabilities plus owner's equity.

Capital　The equity the owner has in a business; the residual value of the business after all debts are paid.

Current Assets　All assets whose future benefit is expected to occur within a year.

Gross Profit　The difference between the selling price and the cost price; markup. Sales − Cost of goods sold = Gross profit.

Horizontal Analysis　An analysis of statements where percentages are compared over time.

Income Statement　The statement that summarizes revenues and expenses for a period and ends with the net income for the period.

Liabilities　The debts of a business; what the business owes.

Liquidity　The capability of converting assets into cash with relative ease.

Partnership　The contractual arrangement between two or more persons to run a business and share resources jointly.

Plant Assets　Those assets that can be expected to benefit the business for more than a year.

Quick Assets　Assets that can quickly be converted to cash. Specifically, they are *cash, all receivables,* and *marketable securities.*

Sole Proprietorship　A business where all of the owner's equity belongs to one person; a single-owner business.

Solvency　Ability to meet the debts as they become due.

Stockholders' Equity	For a corporate form of business, the paid-in capital plus retained earnings.
Working Capital	The current assets left after all current liabilities have been taken care of.

The two basic financial statements are the Balance Sheet and the Income Statement. The contents of these statements provide important information and are many times the basis for crucial decisions. They can be analyzed in great detail, or the information of interest to a specific group can be selected for analysis. Basically, the analysis is important for the following broad reasons:

1. To reveal the solvency of the business, that is, the company's ability to meet its debts as they come due
2. To reveal the company's profitability
   a. The amount and trend of a company's earnings
   b. The relationship of earnings to assets needed
3. To reveal the company's stability
   a. A continued demand for the company's goods or services
   b. A stable relationship between revenues and expenses
   c. Enough net income to pay continuing dividends

For this course, it will not be necessary to analyze in great detail or to interpret the results. These activities are left to courses in accounting.

BALANCE SHEET

The solvency of a business is presented in the Balance Sheet. This statement is a reflection of the basic accounting equation:

$$\text{Assets} = \text{Liabilities} + \text{Capital}.$$

- *Assets* are the items of worth of a business; what the business *owns.*
- *Liabilities* are the debts of the business; what the business *owes.*
- *Capital* is the ownership rights in the business; what is left of assets after all debts are paid.

These three major sections of the Balance Sheet are subdivided into specific types. Assets are usually either current or plant assets. They are *current* assets if they can be expected to be used up or converted to cash within one year (the period of time accountants use for *current*). Within the classification of current, the assets are customarily listed in the order of their liquidity. *Liquidity* refers to how quickly an asset can be converted to cash. *Plant* assets are those assets that will last for more than one year and will be depreciated (except for land, which is not depreciated).

Likewise, liabilities are subdivided into current and long-term liabilities. *Current* liabilities are due to be paid within one year, and they, too, are customarily listed according to how quickly they are due to be paid—the earliest due date first, and so on. *Long-term* liabilities are debts not due within the current year and generally are limited to long-term notes such as mortgages.

The capital section reveals the owner's (or owners') investment in the business. A sole proprietorship or partnership will have *capital accounts* for each owner. A corporation will usually have a stockholders' equity section that replaces the capital section in the other forms of business organization. In the stockholders' equity section, the types of stock sold, the number of shares outstanding, and the profits that have been kept in the business (retained earnings) are listed.

A Balance Sheet for the A-1 Co. is provided below. Notice that the total assets must always be equal to the total of the liabilities and stockholders' equity. Familiarize yourself with the items within the Balance Sheet so that you can find them more easily when later analyses are performed.

A-1 Company
Comparative Balance Sheet*

Assets	1984		1983	
Current assets				
Cash	$ 110,400		$ 102,600	
Marketable securities	52,000		57,000	
Accounts receivable (net)	387,200		303,000	
Merchandise inventory	620,000		476,000	
Prepaid expenses	12,800		11,400	
Total current assets		$1,182,400		$ 950,000
Long-term investments				
Investment in B-1 Co.		217,000		190,000
Plant assets				
Equipment (net)	832,300		617,500	
Buildings (net)	917,250		1,163,750	
Land	118,750		118,750	
Total plant assets		1,868,300		1,900,000
Total assets		3,267,700		3,040,000
Liabilities				
Current liabilities				
Accounts payable	432,500		400,000	
Salaries payable	35,000		50,000	
Taxes payable	18,500		19,000	
Total current liabilities		486,000		469,000
Long-term liabilities				
Mortgage note payable, due 1985	170,000		170,000	
Bonds payable, 5%, due 1990	570,000		570,000	
Total long-term liabilities		740,000		740,000
Total liabilities		1,226,000		1,209,000
Stockholders' Equity				
Preferred stock, 6% cumulative, $100 par	475,000		475,000	
Common stock, $25 par	475,000		475,000	
Retained earnings	1,091,700		881,000	
Total stockholders' equity		2,041,700		1,831,000
Total liabilities and stockholders' equity		3,267,700		3,040,000

*The following data are from the Balance Sheet for 1982: Accounts receivable (net)—$211,000; Long-term investments—$166,250; Total assets—$2,940,000; Total stockholders' equity—$1,775,000.

INCOME STATEMENT

The profitability of a business is presented on the Income Statement. Basically, the Income Statement includes all of the revenues and expenses of a business as of a certain date. There are several major classifications.

- *Net sales* are the gross revenue from the sale of merchandise adjusted downward by discounts and returned goods.

- *Cost of merchandise sold* is a breakdown of the flow of inventory that determines what the merchandise you sold cost you.

- *Gross profit on sales* is the net sales less the cost of merchandise sold. *Gross profit* and *markup* mean the same thing.

- *Operating expenses* are the expenses of operating a business. They are usually subdivided into *selling expenses* (those expenses that pertain to sales) and *general expenses* (those expenses that pertain to administration).

- *Net income* is the income found by subtracting the total operating expenses (and taxes if applicable) from gross profit. It is also called *net profit.*

If the business is a corporation, the operating income will be reduced by income taxes levied on the business. (A sole proprietorship or partnership pays no income tax. The owners must claim any business income on their personal income tax returns.)

An example of an Income Statement for the A-1 Co. is given on page 549. Notice the major section totals of net sales, cost of merchandise sold, gross profit on sales, and net income.

HORIZONTAL ANALYSIS

Year-to-year changes in the amounts and the percentages of the items on the Balance Sheet and Income Statement can be computed and compared by performing horizontal analysis. *Horizontal analysis* is a comparison of the same items on statements of two or more years (or periods), and is generally expressed both in dollar amounts and as percentages. The earlier or earliest item being compared is used as the base. This analysis is helpful in revealing relationships and trends. It requires a *comparative Balance Sheet* or *comparative Income Statement.*

Whether the comparison is made on the Balance Sheet or on the Income Statement, the rate of change (rate of increase or decrease) is found as follows:

$$\text{Rate of increase (decrease)} = \frac{\text{Amount of change}}{\text{Earliest amount}}$$

This formula was presented in the chapter on "Percentages, Proportions, and Ratios."

A comparative Balance Sheet and a comparative Income Statement for the B-1 Co. are provided on page 550. The following four examples use these statements.

Example 1:   *Current asset analysis* — The amount of change from 1982 to 1983 for the B-1 Co. for current assets was a $10,000 increase. Since 1982 is earlier than 1983, the 1982 amount is used as the base.

A-1 Company
Comparative Income Statement

	Dec. 31, 1984		Dec. 31, 1983	
Sales		$3,567,200		$2,935,500
Less: Sales returns and allowances	$ 83,100		$ 80,000	
Sales discounts	6,600	89,700	5,500	85,500
Net sales		3,477,500		2,850,000
Cost of merchandise sold				
Merchandise inventory, Jan. 1	476,000		360,000	
Purchases (net)	2,817,000		2,111,000	
Merchandise available for sale	3,293,000		2,471,000	
Merchandise inventory, Dec. 31	503,000		476,000	
Cost of merchandise sold		2,790,000		1,995,000
Gross profit on sales		687,500		855,000
Operating expenses				
Selling expenses	375,100		427,500	
General expenses	120,000		133,000	
Total operating expenses		495,100		560,500
Net operating income		192,400		294,500
Other income				
Gain on sale of plant assets	11,580		8,550	
Other expenses				
Interest expense	7,200	4,380	38,000	29,450
Income before income tax		196,780		265,050
Income tax		94,454		126,350
Net income		102,326		138,700

$$\text{Rate of increase} = \frac{\$10,000}{\$120,000} = .0833333 = 8.3\%$$

Note: *It is customary to round off all statement analysis rates to one-tenth of 1%.*

Example 2: *Current liability analysis* — The amount of change from 1982 to 1983 for the B-1 Co. for current liabilities was a $10,000 decrease. The 1982 amounts are used as bases since 1982 is the earlier year.

$$\text{Rate of decrease} = \frac{(\$10,000)}{\$60,000} = (.1666666) = (16.7\%)$$

Example 3: *Cost of goods sold analysis* — The amount of change of $47,000 for the B-1 Co. for the cost of goods sold (the same as the cost of merchandise sold) divided by the earlier amount of $49,000 gives the percentage of increase.

$$\text{Rate of increase} = \frac{\$47,000}{\$49,000} = .9591836 = 95.9\%$$

Example 4: *Net income analysis* — The amount of change of $6,000 for the B-1 Co. for net income divided by the earlier amount of $23,000 gives the percentage of increase.

$$\text{Rate of increase} = \frac{\$6,000}{\$23,000} = .2608695 = 26.1\%$$

Horizontal Analysis
B-1 Company
Comparative Balance Sheet (condensed)

	1983	1982	Amount of Increase or Decrease	% of Increase or Decrease
Assets				
Current assets	$130,000	$120,000	$10,000	8.3
Plant assets	69,000	60,000	9,000	15.0
Total assets	199,000	180,000	19,000	10.6
Liabilities				
Current liabilities	50,000	60,000	(10,000)	(16.7)
Long-term liabilities	85,000	70,000	15,000	21.4
Total liabilities	135,000	130,000	5,000	3.8
Capital				
Ken Hardcastle, capital	64,000	50,000	14,000	28.0
Total liabilities and capital	199,000	180,000	19,000	10.6

Horizontal Analysis
B-1 Company
Comparative Income Statement (condensed)

	Dec. 31, 1983	Dec. 31, 1982	Amount of Increase or Decrease	% of Increase or Decrease
Net sales	$150,000	$ 90,000	$60,000	66.7
Less: Cost of goods sold	96,000	49,000	47,000	95.9
Gross profit	54,000	41,000	13,000	31.7
Less: Operating expenses	25,000	18,000	7,000	38.9
Net income	29,000	23,000	6,000	26.1

Practice Problems

From the statements for the A-1 Co. (pages 547 and 549), perform the horizontal analysis on the following, indicating the amount and percentage of change:

1. Total plant assets.   **2.** Total current liabilities.   **3.** Net sales.

VERTICAL ANALYSIS

In *vertical analysis,* each item is expressed as a percentage of a significant total *in that separate statement.* Vertical analysis is useful in comparing one company with another since it brings them to a "common size."

On the Income Statement, it is customary to compare all items to net sales, which is always 100%. This means that each item for that year will be divided by the total of net sales.

On the Balance Sheet, each item is compared to total assets. Since total assets and total liabilities and capital are always equal, they will both always equal 100% in the analysis.

Note: *For vertical analysis, it is not necessary to have comparative statements because all of the comparisons are done within the same statement for that year.*

The following examples use the information found in the two statements provided for the C-1 Co. for 1983, found below and on the next page.

Example 5: *Cash vertical analysis* — The total for cash of $2,000 is divided by the significant total for the Balance Sheet, which is the figure for total assets. In this example, total assets are $40,000.

$$\frac{\text{Cash}}{\text{Total assets}} = \frac{\$2,000}{\$40,000} = .05 = 5.0\%$$

Example 6: *Total liabilities vertical analysis* — The amount of $22,000 for total liabilities is divided by total assets of $40,000.

$$\frac{\text{Total liabilities}}{\text{Total assets}} = \frac{\$22,000}{\$40,000} = .55 = 55.0\%$$

Vertical Analysis
C-1 Company
Comparative Balance Sheet

Assets	1984	1983	
Current Assets			
Cash	$ 3,600	$ 2,000	5.0%
Notes receivable	1,200	800	2.0
Accounts receivable	2,000	1,200	3.0
Merchandise inventory	5,800	4,000	10.0
Total current assets	12,000	8,000	20.0
Plant assets			
Buildings (net)	18,500	20,000	50.0
Equipment (net)	13,700	12,000	30.0
Total plant assets	32,200	32,000	80.0
Total assets	44,800	40,000	100.0
Liabilities			
Current liabilities			
Notes payable	750	500	1.3
Accounts payable	1,700	1,500	3.8
Total current liabilities	2,450	2,000	5.0
Long-term liabilities			
Mortgage note payable	20,000	20,000	50.0
Total liabilities	22,450	22,000	55.0
Capital			
Ken Hardcastle, capital	22,350	18,000	45.0
Total liabilities and capital	44,800	40,000	100.0

Vertical Analysis
C-1 Company
Income Statement (condensed)

	Dec. 31, 1984	Dec. 31, 1983	
Net sales	$63,500	$60,000	100.0%
Less: Cost of goods sold	42,000	40,000	66.7
Gross profit	21,500	20,000	33.3
Operating expenses			
Selling expenses	3,200	3,000	5.0
General expenses	4,750	5,000	8.3
Total operating expenses	7,950	8,000	13.3
Net income	13,550	12,000	20.0

Example 7:   *Gross profit vertical analysis* — The gross profit amount of $20,000 is divided by the significant total in the Income Statement, which is the figure for net sales. In this example, net sales are $60,000.

$$\frac{\text{Gross profit}}{\text{Net sales}} = \frac{\$20,000}{\$60,000} = .3333333 = 33.3\%$$

Example 8:   *General expenses vertical analysis* — The total of $5,000 for general expenses is divided by the net sales of $60,000.

$$\frac{\text{General expenses}}{\text{Net sales}} = \frac{\$5,000}{\$60,000} = .0833333 = 8.3\%$$

Practice Problems

From the statements for the C-1 Co. shown above, perform vertical analysis on the following data for 1984.

4. Merchandise inventory.   **5.** Cost of goods sold.

6. Accounts payable.   **7.** Net income.

SPECIAL ANALYSES

Some selected ratios and analyses follow. You should understand that many others are commonly used and that the result of each ratio or amount is more meaningful as a trend or when compared to a previous ratio or to an industry-wide average.

WORKING CAPITAL

This represents the amount that would be left from current assets if current liabilities were all paid. Generally, the larger the better, but too much probably means funds that are idle instead of producing income.

Working capital = Current assets − Current liabilities

CURRENT RATIO

This figure provides a more dependable indication of solvency than working capital. A gradual increase in the current ratio is usually a sign of improved financial strength. In the past, a ratio of 2 to 1 has been accepted as standard.

$$\text{Current ratio} = \frac{\text{Current assets}}{\text{Current liabilities}}$$

ACID-TEST RATIO

This ratio is sometimes called *Quick Ratio.* It measures the instant debt-paying ability of a company. It is stronger than the current ratio as a test of a company's current position. The minimum acceptable ratio has traditionally been 1 to 1.

$$\text{Acid-test ratio} = \frac{\text{Quick assets}}{\text{Current liabilities}}$$

MERCHANDISE INVENTORY TURNOVER

This figure indicates the number of times merchandise is purchased, sold, and repurchased during the year. A larger turnover is more favorable.

$$\text{Merchandise inventory turnover} = \frac{\text{Cost of merchandise sold}}{\text{Average merchandise inventory}}$$

RATIO OF PLANT ASSETS TO LONG-TERM LIABILITIES

This provides an indication of the potential ability of the business to borrow additional funds on a long-term basis.

$$\text{Ratio} = \frac{\text{Plant assets}}{\text{Long-term liabilities}}$$

The following examples use the Balance Sheet and the Income Statement for the A-1 Co. for 1983, found on pages 547 and 549, respectively.

Example 9: Find the working capital.

$$\text{Working capital} = \$950,000 - \$469,000$$
$$= \$481,000$$

Example 10: Find the current ratio.

$$\text{Current ratio} = \frac{\$950,000}{\$469,000}$$
$$= 2.0 \text{ to } 1$$

Note: *For all ratios, when you divide and round to one-tenth, you express your answer "to 1."*

Example 11: Find the acid-test ratio.

$$\text{Acid-test ratio} = \frac{(\$102,600 + \$57,000 + \$303,000)}{\$469,000}$$

$$= \frac{\$462,600}{\$469,000}$$

$$= 1.0 \text{ to } 1$$

Example 12: Find the merchandise inventory turnover.

$$\text{Merchandise inventory turnover} = \frac{\$1,995,000}{\$418,000*}$$

$$= 4.8 \text{ times}$$

$$\frac{*\text{Beginning merchandise inventory} + \text{Ending merchandise inventory}}{2}$$

Example 13: Find the ratio of plant assets to long-term liabilities.

$$\text{Ratio} = \frac{\$1,900,000}{\$740,000}$$

$$= 2.6 \text{ to } 1$$

Practice Problems

Perform the special analysis for each of the following, using the statements for the A-1 Co. for 1984, found on pages 547 and 549.

8. Working capital.   **9.** Current ratio.   **10.** Acid-test ratio.

11. Merchandise inventory turnover.

12. Ratio of plant assets to long-term liabilities.

Practice Problem Solutions

1. $\dfrac{1,868,300 - 1,900,000}{1,900,000} = \dfrac{(31,700)}{1,900,000} = 1.7\%$ decrease

Amount of change = $31,700 decrease

2. $\dfrac{486,000 - 469,000}{469,000} = \dfrac{17,000}{469,000} = 3.6\%$ increase

Amount of change = $17,000 increase

3. $\dfrac{3,477,500 - 2,850,000}{2,850,000} = \dfrac{627,500}{2,850,000} = 22.0\%$ increase

Amount of change = $627,500 increase

4. $\dfrac{5,800}{44,800} = 12.9\%$ **5.** $\dfrac{42,000}{63,500} = 66.1\%$

6. $\dfrac{1,700}{44,800} = 3.8\%$ **7.** $\dfrac{13,550}{63,500} = 21.3\%$

8. $1,182,400 - $486,000 = $696,400   **9.** $\dfrac{1,182,400}{486,000} = 2.4$ to 1

10. $\dfrac{(110,400 + 52,000 + 387,200)}{486,000} = \dfrac{549,600}{486,000} = 1.1$ to 1

11. $\dfrac{2,790,000}{\dfrac{(476,000 + 503,000)}{2}} = \dfrac{2,790,000}{489,500} = 5.7$ times   **12.** $\dfrac{1,868,300}{740,000} = 2.5$ to 1

Exercise 86A: Horizontal Analysis

Name _____

Course/Sect. No. _____   Score _____

Instructions: From the condensed Balance Sheet, complete the horizontal analysis on each item—the amount and percentage (to one-tenth of 1%) of change.

_____

Self-Check Exercise (Answers on the back)

Comparative Balance Sheet

	Assets	1984	1983	Amount of Change	% of Change
(1)	Cash	$ 12,525	$ 8,962	_____	_____
(2)	Accounts receivable	6,540	4,231	_____	_____
(3)	Merchandise inventory	38,929	30,417	_____	_____
(4)	Buildings (net)	65,000	67,250	_____	_____
(5)	Land	100,000	100,000	_____	_____
(6)	Total assets	222,994	210,860	_____	_____
	Liabilities				
(7)	Accounts payable	4,350	3,875	_____	_____
(8)	Notes payable	8,900	2,420	_____	_____
(9)	Mortgage note payable	75,500	78,000	_____	_____
(10)	Total liabilities	88,750	84,295	_____	_____
	Stockholders' Equity				
(11)	Common stock	80,000	65,000	_____	_____
(12)	Retained earnings	54,244	61,565	_____	_____
(13)	Total stockholders' equity	134,244	126,565	_____	_____
(14)	Total liabilities and stockholders' equity	222,994	210,860	_____	_____

	Amount of Change	% of Change
(1)	3563	39.8
(2)	2309	54.6
(3)	8512	28.0
(4)	(2250)	(3.3)
(5)	—	—
(6)	12,134	5.8
(7)	475	12.3
(8)	6480	267.8
(9)	(2500)	(3.2)
(10)	4455	5.3
(11)	15,000	23.1
(12)	(7321)	(11.9)
(13)	7679	6.1
(14)	12,134	5.8

Exercise 86B: Horizontal Analysis

Name_____

Course/Sect. No._____   Score_____

Instructions: From the condensed Income Statement, complete the horizontal analysis on each item—the amount and percentage (to one-tenth of 1%) of change.

Comparative Income Statement

		1984	1983	Amount of Change	% of Change
(1)	Sales	$162,800	$150,420	_____	____
(2)	Less: Sales discounts	2,415	1,830	_____	____
(3)	Net sales	160,385	148,590	_____	____
(4)	Less: Cost of goods sold	84,720	75,730	_____	____
(5)	Gross profit	75,665	72,860	_____	____
(6)	Operating expenses				
(7)	Selling expenses	9,450	8,315	_____	____
(8)	General expenses	7,820	6,425	_____	____
(9)	Total expenses	17,270	14,740	_____	____
(10)	Net income before taxes	58,395	58,120	_____	____
(11)	Income taxes	28,030	27,898	_____	____
(12)	Net income	30,365	30,222	_____	____

Exercise 87A: Vertical Analysis

Name _____

Course/Sect. No. _____   Score _____

Instructions: From the condensed Balance Sheet, complete the vertical analysis on each item. Round to one-tenth of 1%.

Self-Check Exercise (Answers on the back)

Balance Sheet

				%
(1) _____	**(1)**	**Assets** Cash	$ 52,500	____
(2) _____	**(2)**	Accounts receivable	38,250	____
(3) _____	**(3)**	Merchandise inventory	120,400	____
(4) _____	**(4)**	Buildings (net)	350,000	____
(5) _____	**(5)**	Equipment (net)	100,000	____
(6) _____	**(6)**	Land	250,000	____
(7) _____	**(7)**	Total assets	911,150	____
		Liabilities		
(8) _____	**(8)**	Accounts payable	23,350	____
(9) _____	**(9)**	Notes payable	88,525	____
(10) _____	**(10)**	Mortgage payable	400,000	____
(11) _____	**(11)**	Total liabilities	511,875	____
		Capital		
(12) _____	**(12)**	H. M. Smith, capital	399,275	____
(13) _____	**(13)**	Total liabilities and capital	911,150	____

(1) _____5.8_____

(2) _____4.2_____

(3) _____13.2_____

(4) _____38.4_____

(5) _____11.0_____

(6) _____27.4_____

(7) _____100.0_____

(8) _____2.6_____

(9) _____9.7_____

(10) _____43.9_____

(11) _____56.2_____

(12) _____43.8_____

(13) _____100.0_____

Exercise 87B: Vertical Analysis

Name _____

Course/Sect. No. _____   Score _____

Instructions: From the condensed Income Statement, complete the vertical analysis on each item. Round to one-tenth of 1%.

Income Statement

				%
(1) _____	**(1)**	Sales	$425,000		____
(2) _____	**(2)**	Less: Sales return	12,000		____
(3) _____	**(3)**	Net sales		$413,000	____
(4) _____	**(4)**	Merchandise inventory, Jan. 1	135,000		____
(5) _____	**(5)**	Purchases (net)	82,000		____
(6) _____	**(6)**	Merchandise available for sale	217,000		____
(7) _____	**(7)**	Merchandise inventory, Dec. 31	116,000		____
(8) _____	**(8)**	Cost of merchandise sold		101,000	____
(9) _____	**(9)**	Gross profit		312,000	____
		Operating expenses			
(10) _____	**(10)**	Selling expenses	102,000		____
(11) _____	**(11)**	General expenses	80,000		____
(12) _____	**(12)**	Total expenses		182,000	____
(13) _____	**(13)**	Net income		130,000	____

Exercise 88A: Selected Ratios and Analyses

Name _____

Course/Sect. No. _____   Score _____

Instructions: From the information provided, find the answers for each of the following. Round to one-tenth in each case.

Self-Check Exercise (Answers on back)

	1984	1983		1984	1983
Cash	$ 12,500	$ 16,000	Salaries payable	$ 1,250	$ 1,000
Marketable securities	28,250	35,000	Taxes payable	3,300	3,500
Notes receivable	8,000	15,000	Notes payable	18,500	12,000
Prepaid insurance	2,000	2,000	Accounts payable	42,250	20,000
Merchandise inventory	65,200	75,500			
Supplies	275	350			
Total current assets	$139,575	$173,350	Total current liabilities	$65,300	$36,500

(1) _____   **(1)** Find the working capital for 1983.

(2) _____   **(2)** Find the current ratio for 1983.

(3) _____   **(3)** Find the acid-test ratio for 1983.

	1984	1983	1982
Accounts receivable	$ 44,570	$38,725	$35,500
Merchandise inventory	104,315	82,898	75,330
Notes receivable	15,735	10,880	13,200
Cost of goods sold	67,384	52,808	40,315
Gross profit	69,913	48,917	42,006
Net income	47,738	35,520	24,000
Plant assets	63,821	55,230	50,100
Long-term liabilities	37,500	37,500	17,500

(4) _____   **(4)** Find the merchandise inventory turnover for 1983.

(5) _____   **(5)** Find the ratio of plant assets to long-term liabilities for 1983.

(1) _____ $136,850 _____

(2) _____ 4.7 to 1 _____

(3) _____ 1.8 to 1 _____

(4) _____ .7 times _____

(5) _____ 1.5 to 1 _____

Exercise 88B: Selected Ratios and Analyses

Name _____

Course/Sect. No. _____   Score _____

Instructions: From the information provided, find the answers for each of the following. Round to one-tenth in each case.

	1984	1983		1984	1983
Cash	$ 12,500	$ 16,000	Salaries payable	$ 1,250	$ 1,000
Marketable securities	28,250	35,000	Taxes payable	3,300	3,500
Notes receivable	8,000	15,000	Notes payable	18,500	12,000
Prepaid insurance	2,000	2,000	Accounts payable	42,250	20,000
Merchandise inventory	65,200	75,500			
Supplies	275	350			
Total current assets	$139,575	$173,350	Total current liabilities	$65,300	$36,500

(1) _____ **(1)** Find the working capital for 1984.

(2) _____ **(2)** Find the current ratio for 1984.

(3) _____ **(3)** Find the acid-test ratio for 1984.

	1984	1983	1982
Accounts receivable	$ 44,570	$38,725	$35,500
Merchandise inventory	104,315	82,898	75,330
Notes receivable	15,735	10,880	13,200
Cost of goods sold	67,384	52,808	40,315
Gross profit	69,913	48,917	42,006
Net income	47,738	35,520	24,000
Plant assets	63,821	55,230	50,100
Long-term liabilities	37,500	37,500	17,500

(4) _____ **(4)** Find the merchandise inventory turnover for 1984.

(5) _____ **(5)** Find the ratio of plant assets to long-term liabilities for 1984.

TEST YOURSELF — UNIT 6

(1) Prepare a bank reconciliation statement and find the reconciled amount.
- Bank service charge — $2.25
- Bank balance — $637.95
- Late deposit — $350
- Checkbook balance — $730.20
- Outstanding checks — No. 40 for $80, No. 43 for $62.50, No. 49 for $100, No. 51 for $17.50

(2) Prepare a bank reconciliation statement and find the reconciled amount.
- Bank balance — $15,200.30
- Checkbook balance — $15,562.22
- Returned check — $82.30
- Deposit in-transit — $2,300
- Outstanding checks — $520.38
- Note collection — $1,500

(3) From the following grade scores, find:
- **(A)** The mean.
- **(B)** The median.
- **(C)** The mode.

80, 70, 90, 70, 85, 95, 55, 92, 65

(4) Convert the following within the metric system.
- **(A)** 1.675 meters to _____ millimeters.
- **(B)** 56 meters to _____ kilometers.

(5) How many degrees Celsius in 70 degrees F?

(6) How many degrees Fahrenheit in 34 degrees C?

(7) From the information below, find:
- **(A)** The current ratio.
- **(B)** Merchandise inventory turnover at cost.

(Round your answers to a tenth.)

Current assets	160,000	Beginning inventory	400,000
Plant assets	800,000	Ending inventory	600,000
Current liabilities	40,000	Net purchases	800,000
Long-term liabilities	300,000	Net sales	1,000,000

Unit
7

Insurance, Taxes, Stocks, and Bonds

A Preview of What's Next

What percentage of your house replacement cost does it take to be fully insured for a fire insurance loss?

The tax rate is determined by the amount of taxes needed and the total assessed valuations. Would $12.50 per thousand be the same as .0125 mill?

If you knew the assessed valuation and the tax rate, could you find the property tax?

How much would it cost to purchase a $300,000, 20-year term life insurance policy?

Objectives

A business or individual has expenses for property owned other than the cost of purchasing that property. Fire insurance is needed to minimize the loss in case of fire damage. The insurer and the insured share in fire losses in some ratio, depending on the contract purchased.

Additional expenses are levied annually in the form of property taxes. These property taxes come generally from several different taxing units, such as school districts, city governments, and county governments.

When you have successfully completed this unit, you will be able to:

1. Determine the maximum amount paid by an insurance company in case of fire loss.
2. Compute the proportion of a fire loss that is paid by the insurance company when there is a coinsurance clause.
3. Compute the assessed valuation when the assessment rate and market value are given.
4. Compute the tax rate on property.
5. Express the tax rate in dollars per hundred or in mills.
6. Calculate the property tax when the tax rate, the market value, and the assessment rate are known.
7. Compute the annual cost of insurance based on a rate per $100.
8. Compute the monthly premium on a life insurance policy.
9. Compute the purchase price on a stock quotation.
10. Calculate dividends per share for each type of stock.
11. Compute the purchase price of a bond issue.
12. Calculate the annual and semiannual bond interest.
13. Determine the approximate bond yield.

Chapter

18

Property Insurance and Life Insurance

Preview of Terms

Beneficiary The individual to whom payments are made by the insurance company.

Coinsurance Clause A clause in a fire insurance policy that stipulates that the insured carry insurance equal to a certain percentage of the replacement value. (The most common percentage is 80.)

Deductible The specified amount that the insurance company will not pay in the event of a loss. It is, for instance, common for automobile insurance to have the insured pay the first $100 of each loss before the insurance company pays anything.

Replacement Cost The cost, under current market prices, to replace the insured asset.

Premium The amount paid for insurance coverage.

Property Insurance Insurance on property for damage from wind, fire, water, vandalism, and (possibly) other sources.

Liability Insurance Insurance to protect the insured against injury or damage that may occur to someone else or their property.

PROPERTY INSURANCE

Property insurance provides protection against loss or damage to property caused by fire, windstorm, riots, aircraft, motor vehicles, vandalism, smoke damage, and many other perils. *Liability insurance* provides protection for injury to persons or to property of others for which you may be held liable. With recent liberalized court interpretations, it is probably wise for a business owner to carry both property and liability insurance.

Liability coverage is usually set for a fixed sum. Those sums are categorized as follows:

Liability Coverages	Amount of Coverage
Personal liability	$2500 for each occurrence
Medical payments to others	$ 500 for each person
Damage to property of others	$ 250 for each occurrence

A typical cost for a policy like this covering your legal liability would be somewhere between $5 and $10 per year.

Fire insurance protects against loss by fire or fire-related causes such as water damage. The cost of fire insurance depends on such factors as the buildings's proximity to a fire hydrant or fire station, the type of structure (brick, wood, etc.), and the contents, if it is a commercial building.

Fire insurance rates are usually quoted in dollars per $100 of insurance. Therefore, if the rate was $.55/$100, insurance coverage of $60,000 would cost $330.

$$\frac{\text{Insurance carried}}{\$100} \times \text{rate} = \text{annual cost}$$

$$\$60,000 \div \$100 = 600 \times \$.55 = \$330$$

Example 1:   The store was insured for $85,500 and the rate was $.42 per $100. What was the cost?

$$\frac{\$85,500}{\$100} \times \$.42 = \text{cost}$$
$$\$855 \times \$.42 = \text{cost}$$
$$\$359.10 = \text{cost}$$

Practice Problems

Find the annual cost for each.

1. The rate per hundred is $.35 and the insurance coverage selected was $120,000.
2. The building was to be insured for $90,000 at $.40/$100.

COINSURANCE

Few fires result in a total loss, so property owners carry as little insurance as possible, but enough to cover a major loss. However, to collect a fire loss in full, the insurance company will require the insured to maintain a minimum percentage of the current replacement cost of the property insured. This percentage is called *coinsurance.* The typical coinsurance percent used is 80%. This means the insured must purchase insurance of at least 80% of the current replacement cost to be fully covered. Any loss that occurs when at least 80% coverage is *not* kept will be partially shared by the insured.

Note: *The insurance company will pay the lesser of (1) the face value of the policy, (2) the actual amount of the loss, or (3) the amount obtained from the co-insurance formula. Therefore, a replacement cost of $50,000 insured up to*

80% ($40,000) would mean coverage of only $40,000 (the face value of the policy) in the event of a total fire loss.

$$\text{Insurance payment} = \frac{\text{Amount of insurance carried}}{(\text{Coinsurance \%})(\text{Replacement cost})} \times \text{Fire loss}$$

Example 2:   Your store has a replacement cost estimated at $220,000, and you have insured it for $176,000. There was a fire loss of $43,000. What would the insurance company pay, assuming there was an 80% coinsurance clause?

$$\text{Insurance payment} = \frac{\$176,000}{(80\%)(\$220,000)} \times \$43,000$$
$$= \$43,000$$

Example 3:   A warehouse valued at $85,000 was insured for 65% of its value. The loss due to fire was $12,000. What amount did the insurance company pay if there was an 80% coinsurance clause?

$$\text{Insurance payment} = \frac{(65\% \times \$85,000)}{(80\%)(\$85,000)} \times \$12,000$$
$$= \$9,750^*$$

(*The insured would have to absorb the remaining loss of $2,250 [$12,000 − $9,750].)

Practice Problems

Find the amount paid by the insurance company in each case.

3. The XYZ Co. has a replacement cost of $850,000, and it was insured for $680,000 with an 80% coinsurance clause. The fire loss was $50,000.

4. The ABC Co. carried 50% insurance coverage on a building valued at $275,000. The fire loss was $22,000, and there was an 80% coinsurance clause.

LIFE INSURANCE

There are three basic kinds of life insurance policies, with many other combinations available.

1. *Whole life* — With this type of coverage, the insured pays premiums until death or until the policy is canceled.

2. *Term* — With this type of coverage, the insured pays premiums for a specified period of time (the *term*). The policy expires at the end of the term period.

3. *Endowment* — With this type of coverage the insured pays premiums for a specified number of years. At the end of that time, the insurance company pays whether the insured lives or dies.

When a policy is taken out, the insured must declare to whom the benefits will be paid in the event of the insured's death. The person designated to receive the benefits is called the *beneficiary*.

Life insurance rates are based on a mortality table that predicts average age of death very accurately. As age increases, the likelihood of death increases and, therefore, the cost (insurance premiums) for insurance coverage will be higher. The life insurance contract then is based on a person's present age, health, and in some extreme situations occupation or extra-hazardous activities. Since a woman's life expectancy is longer than a man's, their rates are not the same at the same age. These variables are built into policy tables and the premiums are generally computed based on how many $1,000 units of coverage are purchased.

Example 4:   If the insured buys a $65,000 policy, how many $1,000 units of insurance were purchased?

$$\$65,000 \div \$1,000 = 65 \text{ units}$$

After the units of coverage have been determined, you multiply the premium per thousand by the units and then add in the policy fee. This will give you the annual premium. To convert that to monthly premiums, you would divide by 12.

Example 5:   Assume the premium for a 20-year-old male is $1.43 per $1,000 and the policy fee is $20. What would the monthly premium for an $80,000 policy be?

$$\$80,000 \div \$1,000 = 80 \text{ Units of insurance}$$

$$80 \text{ Units} \times \$1.43/\text{Unit} = \$114.40/\text{Year} + \$20 \text{ Policy fee} = \$134.40$$

$$\$134.40 \div 12 = \$11.20/\text{Month}$$

To find the premium per thousand, you need to locate the proper age with the appropriate term period. Refer to Table 18.1 and notice that the premium for a 20-year term policy at age 20 is $1.48.

TABLE 18.1
Term policy rate table — male
Policy fee $20
cost per $1000 for
(selected years and amounts)

Age	For 10 Years	For 20 Years	For 30 Years
20	$1.43	$1.48	$1.85
25	1.52	1.70	2.32
30	1.69	2.17	3.18
35	2.17	3.08	4.70
40	3.20	4.69	7.22
45	4.90	7.30	9.05

TABLE 18.2
Whole life policy rate table — female
Policy fee $10
cost per $1000 for
(selected years and amounts)

Age	Premium	Age	Premium
10	$4.10	35	$12.15
15	5.06	40	15.46
20	6.22	45	19.64
25	7.71	50	25.27
30	9.65	55	32.88

Practice Problems

Compute the monthly premium on each.

5. A male aged 30 and a $65,000, 30-year term policy.

6. A male aged 40 and a $125,500, 10-year term policy.

7. A female aged 25 and a $25,000 whole life policy.

8. A female aged 40 and a $100,000 whole life policy.

Practice Problem Solutions

1. $\dfrac{\$120,000}{\$100} \times \$.35 = $ Cost

$1200 \times \$.35 = $ Cost

$\$420 = $ Cost

2. $\dfrac{\$90,000}{\$100} \times \$.40 = $ Cost

$900 \times \$.40 = $ Cost

$\$360 = $ Cost

3. Insurance payment $= \dfrac{\$680,000}{(80\%)(\$850,000)} \times \$50,000$

$= \$50,000$

4. Insurance payment $= \dfrac{\$137,500*}{(80\%)(\$275,000)} \times \$22,000$

$= \$13,750$

*$(50\% \times \$275,000)$

5. $\quad \$65,000 \div \$1,000 = 65$ Units of insurance

65 Units $\times \$3.18/$Unit $= \$206.70 + \$20$ Policy fee $= \$226.70$

$\$226.70 \div 12 = \$18.89/$Month

6. $\$125,500 \div \$1,000 = 125.5$ Units of insurance

125.5 Units $\times \$3.20 = \$401.60 + \$20$ Policy fee $= \$421.60$

$\$421.60 \div 12 = \$35.13/$Month

7. $\$25,000 \div \$1,000 = 25$ Units of insurance

25 Units $\times \$7.71 = \$192.75 + \$10$ Policy fee $= \$202.75$

$\$202.75 \div 12 = \$16.90/$Month

8. $\$100,000 \div \$1,000 = 100$ Units of insurance

100 Units $\times \$15.46 = \$1,546 + \$10$ Policy fee $= \$1,556$

$\$1,556 \div 12 = \$129.67/$Month

Exercise 89A: Property Insurance

Name_____

Course/Sect. No._____   Score_____

Instructions: Find the annual cost for each.

(1) _____	
(2) _____	
(3) _____	
(4) _____	
(5) _____	
(6) _____	
(7) _____	
(8) _____	
(9) _____	
(10) _____	

Self-Check Exercise (Answers on back)

Insurance Coverage	Rate per Hundred	Insurance Coverage	Rate per Hundred
(1) $80,000	$.42	(2) $62,000	$.32
(3) $74,200	$.28	(4) $66,000	$.43
(5) $55,500	$.45	(6) $150,000	$.38
(7) $40,000	$.40	(8) $225,000	$.285

(9) The company changed the rate from hundreds to thousands. It is now $3.10 per thousand dollars. What would be the cost of a policy with replacement value of $80,400?

(10) If the cost was $242.50 based on 50¢ per $100, what was the insured amount?

(1) _$336_

(2) _$198.40_

(3) _$207.76_

(4) _$283.80_

(5) _$249.75_

(6) _$570_

(7) _$160_

(8) _$641.25_

(9) _$249.24_

(10) _$48,500_

Exercise 89B: Property Insurance

Name _____

Course/Sect. No. _____   Score_____

Instructions: Find the annual cost for each.

	Insurance Coverage	Rate per Hundred	Insurance Coverage	Rate per Hundred
(1)	**(1)** $25,000	$.25	**(2)** $45,500	$.60
(2)				
(3)	**(3)** $125,000	$.50	**(4)** $35,000	$.30
(4)				
(5)	**(5)** $38,000	$.55	**(6)** $135,000	$.35
(6)				
(7)	**(7)** $93,000	$.70	**(8)** $250,000	$.25
(8)				

(1) _____

(2) _____

(3) _____

(4) _____

(5) _____

(6) _____

(7) _____

(8) _____

(9) _____

(9) The insurance coverage is based on the replacement cost of $75,500. The rate is sixty-one cents for each hundred dollars. Find the cost.

(10) _____

(10) What is the rate per *hundred* on a cost of $3.10 per thousand?

Exercise 90A: Coinsurance

Name _____

Course/Sect. No. _____   Score _____

Instructions: Find the amount paid by the insurance company in each case. An 80% coinsurance clause means that 80% of the replacement cost is required for "full coverage."

	Self-Check Exercise (Answers on back)			
	Replacement Cost	Fire Loss	Coinsurance	Insurance Carried
(1) _____	(1) $75,000	$60,000	80%	$50,000
(2) _____	(2) $85,000	$10,000	90%	$80,000
(3) _____	(3) $95,000	$70,000	70%	$70,000
(4) _____	(4) $105,000	$50,000	90%	$80,000
(5) _____	(5) $30,000	$15,000	70%	$25,000
(6) _____	(6) $38,500	$12,000	90%	$30,000
(7) _____	(7) $46,250	$5,000	80%	$40,000
(8) _____	(8) $54,750	$50,000	80%	$35,000

(9) _____

(9) ABC's building and equipment were estimated to be worth $2,158,000. There was an 80% coinsurance clause in their fire insurance policy. A fire destroyed all but $70,000 worth of the building and equipment. If ABC was insured for $1,700,000, how much did their insurance company pay?

(10) _____

(10) XYZ Co. carries fire insurance for 60% of the replacement cost of its building, which is $110,000. If they carried insurance of $75,000, what amount would the insurance company pay on a $24,000 fire loss?

(1) _$50,000 Face_

(2) _$10,000 Fire loss_

(3) _$70,000 Face_

(4) _$42,328 Formula_

(5) _$15,000 Fire loss_

(6) _$10,390 Formula_

(7) _$5,000 Fire loss_

(8) _$35,000 Face_

(9) _$1,700,000 Face_

(10) _$24,000 Fire loss_

Exercise 90B: Coinsurance

Name _____

Course/Sect. No. _____   Score_____

Instructions: Find the amount paid by the insurance company in each case. An 80% coinsurance clause means that 80% of the replacement cost is required for "full coverage."

	Replacement Cost	Fire Loss	Coinsurance	Insurance Carried
(1) _____	**(1)** $60,000	$30,000	70%	$50,000
(2) _____	**(2)** $70,000	$40,000	75%	$40,000
(3) _____	**(3)** $80,000	$70,000	80%	$60,000
(4) _____	**(4)** $90,000	$25,000	80%	$75,000
(5) _____	**(5)** $100,000	$50,000	75%	$65,000
(6) _____	**(6)** $125,000	$75,000	85%	$60,000
(7) _____	**(7)** $130,000	$80,000	75%	$100,000
(8) _____	**(8)** $135,000	$90,000	70%	$100,000

(9) _____

(9) Peace Cannery Co. carries fire insurance for $300,000 on their plant, which has a replacement cost of $300,000. Their policy has an 80% coinsurance clause. If the plant is totally destroyed by fire, what expense will Peace Cannery bear in replacing it?

(10) _____

(10) Harry's Hot Dog Stand could be replaced for $3,000. Tom's Insurance Co. wrote a policy for $2,000, insisting on 80% coinsurance. A fire did damage amounting to $2,500. How much did Tom's Insurance Co. pay?

Exercise 91A: Life Insurance

Name _____

Course/Sect. No. _____   Score _____

Instructions: Using the policy rate tables in your book, find the monthly premium. (Include the policy fee charge.)

Self-Check Exercise (Answers on back)

		Male Age	Insurance Coverage	Length of Term
(1) _____	**(1)**	25	$30,000	20 years
(2) _____	**(2)**	45	$60,000	30 years
(3) _____	**(3)**	40	$84,000	10 years
(4) _____	**(4)**	30	$75,000	30 years

		Female Age	Insurance Coverage
(5) _____	**(5)**	40	$25,000
(6) _____	**(6)**	35	$80,000
(7) _____	**(7)**	60	$75,000
(8) _____	**(8)**	55	$50,000

(9) _____   **(9)** The 40-year-old woman wanted a $300,000 whole life policy. What would this coverage cost per year?

(10) _____   **(10)** How much *more* per month would it cost a 45-year-old woman to buy a $100,000, whole life policy than a 45-year-old man to buy a 30-year term policy for $200,000?

(1) _$5.92_

(2) _$46.92_

(3) _$24.07_

(4) _$21.54_

(5) _$33.04_

(6) _$81.83_

(7) _$158.77_

(8) _$137.83_

(9) _$4,648_

(10) _$12.00_

Exercise 91B: Life Insurance

Name_____

Course/Sect. No. _____   Score_____

Instructions: Using the policy rate tables in your book, find the monthly premium. (Include the policy fee charge.)

	Male Age	Insurance Coverage	Length of Term
(1) _____	**(1)** 20	$20,000	10 years
(2) _____	**(2)** 35	$45,500	10 years
(3) _____	**(3)** 30	$100,000	20 years
(4) _____	**(4)** 20	$200,000	30 years

	Female Age	Insurance Coverage
(5) _____	**(5)** 20	$10,000
(6) _____	**(6)** 25	$17,000
(7) _____	**(7)** 15	$2,000
(8) _____	**(8)** 30	$20,000

(9) _____   **(9)** The 30-year-old man wanted as much coverage as possible for under $20 per month. Which length of term coverage would provide the largest policy?

(10) _____   **(10)** Would it be less expensive for a 25-year-old man to buy a 20-year policy for $80,000 or a $65,000 policy for 30 years? (Enter the policy amount that is the least expensive.)

Chapter

19

Property and Sales Taxes

Preview of Terms

Assessed Valuation The value placed upon a piece of property for the purpose of determining the amount of tax the owner will have to pay.

Market Value The sale value of the property in the current market.

Mill One-thousandth of a dollar (one-tenth of a cent); used to express the tax rate.

Sales Tax A tax that the final consumer pays based on the amount of the retail sale.

PROPERTY TAXES

A trained individual hired by governing authorities to determine periodically the value of real estate is a *tax assessor.* The value placed on the property is called the *assessed valuation.* The assessed valuation is based on some percentage of the market value of the property.

Example 1: A business structure with a market value of $50,000 is assessed for tax purposes at 40% of its market value. What is the assessed valuation?

$$\$50,000 \times 40\% = \$20,000 \quad \text{Assessed valuation}$$

TAX RATE

Each governmental unit reviews its budget and determines how much money it needs from tax collections. It then determines the tax rate by dividing the tax collections needed by the total assessed valuation of the property under its jurisdiction.

$$\text{Tax rate} = \frac{\text{Taxes needed}}{\text{Total assessed valuation}}$$

Once the general tax rate is found, it is usually expressed in hundredths or thousandths (mills). A *mill* is a tenth of a cent ($.001).

Example 2:   The city budget reveals a need to raise $65,000 in taxes, and the total assessed valuation for the city is $4,800,000. What will be the tax rate?

Expressed "per $100," the tax rate will be $1.35 per $100 of assessed valuation (simply multiply the decimal rate by 100 and round off). Expressed "per $1,000" (in mills), the tax rate will be $13.54 per $1,000 of assessed valuation (simply multiply the decimal rate by 1,000 and round off).

To find the actual tax to be paid, multiply the tax rate by the assessed valuation.

Example 3:   If your business is assessed at $25,000 and the tax rate is $1.35 per $100, what will your property tax be?

$$\text{Tax} = \frac{\$25,000 \times \$1.35}{100}$$
$$= \$337.50$$

Example 4:   If your business is assessed at $25,000 and the tax rate is 13.54 mills, what will be your property tax?

$$\text{Tax} = \frac{\$25,000 \times 13.54}{1,000}$$
$$= \$338.50$$

Practice Problems

Find the assessed valuation and the tax for each of the following.

1. The market value is $80,000, the assessment rate is 55%, and the tax rate is $2.32 per hundred.

2. The market value is $38,250, the assessment rate is 35%, and the tax rate is 13.07 mills.

SALES TAXES

Most retailers are required to collect sales taxes from the consumer when merchandise is sold. The taxes are then forwarded to the proper governmental unit. The sales tax usually varies between 4% and 7%. To determine the proper amount of sales tax, multiply the amount of the merchandise purchased by the tax rate. The sales tax, of course, will increase the total amount owed from the consumer.

Example 5:   The cash register total for merchandise purchased was $27.84. If the tax rate is 5%, find the amount of the sales tax and the total owed by the buyer.

$$\$27.84 \times 5\% = \$1.39 \text{ Sales tax}$$
$$\$27.84 + \$1.39 = \$29.23 \text{ Total owed}$$

A sales clerk will usually have to complete a sales slip when a customer buys merchandise. In the sales slip the "amount" column is determined by multiplying the number in the "quantity" column by the number in the "unit price" column. (See the Example 6 sales slip.)

Example 6:   The customer bought the following merchandise:

1 blouse, $19.95   1 sweater, $34.95
3 pairs of hose, $3.50 each   2 pairs of slacks, $29.95 each
Assume a 5% sales tax rate.

Grable's Department Store

Customer Name *Ms. S. Hopkins* Date *xx/xx/xx*
Address *204 Pine Street*
         Anywhere, USA
___ Cash Purchase   ✓ Charge Purchase

Quantity	Description	Unit Price	Amount
1	Blouse	19.95	19.95
3	Pairs of hose	3.50	10.50
1	Sweater	34.95	34.95
2	Pairs of slacks	29.95	59.90
		SUBTOTAL	125.30
		SALES TAX	6.27
		TOTAL	131.57

Sales Slip

Practice Problems

Find the sales tax and the total owed for each.

3. Cost of goods purchased, $85.91; Sales tax rate = 4%.

4. The taxable items at the grocery store totaled $103.41 and the sales tax rate is 6%.

Practice Problem Solutions

1. Assessed valuation = $80,000 × $55% = $44,000

    $$\text{Tax} = \frac{\$44,000 \times \$2.32}{100}$$

    = $1,020.80

2. Assessed valuation = $38,250 × 35% = $13,387.50

    $$\text{Tax} = \frac{\$13,387.50 \times \$13.07}{1,000}$$

    = $174.97

3. $85.91 × 4% = $3.44 Sales tax
    $85.91 + $3.44 = $89.35 Total owed

4. $103.41 × 6% = $6.20 Sales tax
    $103.41 + $6.20 = $109.61 Total owed

Exercise 92A: Property Taxes

Name_____

Course/Sect. No._____   Score_____

Instructions: For the first four problems, determine the tax rate that would be levied per $100. For the next four, compute the property tax.

Self-Check Exercise (Answers on back)

	Total Assessed Valuation	Taxes Needed
(1) _____	**(1)** $65,000	$877.50
(2) _____	**(2)** $130,000	$1,495
(3) _____	**(3)** $120,500	$1,506.25
(4) _____	**(4)** $115,000	$1,207.50

	Assessed Valuation	Tax Rate	Assessment Ratio
(5) _____	**(5)** $60,000	$11	Per $1,000
(6) _____	**(6)** $125,000	$.95	Per $100
(7) _____	**(7)** $67,500	$14.50	Per $1,000
(8) _____	**(8)** $200,000	$13.25	Per $1,000

(9) _____ **(9)** In Nottingham, the assessed value is 50% of the market value. The Hood residence has a market value of $93,000. The tax rate is $1.85 per $100 of assessed value. What is the tax bill for the Hoods?

(10) _____ **(10)** Mr. Mizer's sewer tax bill for this year on his $56,000 (assessed value) house was $610. What is the tax rate per $100?

(1) _____ $1.35 _____

(2) _____ $1.15 _____

(3) _____ $1.25 _____

(4) _____ $1.05 _____

(5) _____ $660 _____

(6) _____ $1,187.50 _____

(7) _____ $978.75 _____

(8) _____ $2,650 _____

(9) _____ $860.25 _____

(10) _____ $1.09 _____

Exercise 92B: Property Taxes

Name_____

Course/Sect. No. _____   Score_____

Instructions: For the first four problems, determine the tax rate that would be levied per $100. For the next four, compute the property tax.

(1) _____

Total Assessed Valuation   **Taxes Needed**

(1) $80,000 $1,200

(2) _____

(2) $95,500 $859.50

(3) _____

(3) $77,000 $654.50

(4) _____

(4) $50,000 $500.00

	Assessed Valuation	Tax Rate	Assessment Ratio
(5) _____	(5) $75,000	$1.25	Per $100
(6) _____	(6) $95,500	$12.50	Per $1,000
(7) _____	(7) $100,000	$1.55	Per $100
(8) _____	(8) $88,000	$1.05	Per $100

(9) _____

(9) The Morrow residence is assessed at $81,000. The tax rate for the school district is $1.08 per $100 of assessed valuation. How much school tax will the Morrows pay this year?

(10) _____

(10) In Camptown, the county tax rate is $9.11 per $1,000, the city rate is $8.05 per $1,000, and the school rate is $3.75 per $1,000. The assessed value is 45% of the market value. The Newtons own a rental house with a market value of $55,000 and their own residence is worth $104,000. How much total tax will they pay on the houses they own?

Exercise 93A: Sales Taxes

Name _____

Course/Sect. No. _____   Score _____

Instructions: Find the total owned, *including* the sales tax.

(1) _____	
(2) _____	
(3) _____	
(4) _____	
(5) _____	
(6) _____	
(7) _____	
(8) _____	
(9) _____	
(10) _____	

Self-Check Exercise (Answers on back)

	Taxable Merchandise	Tax Rate		Taxable Merchandise	Tax Rate
(1)	$113.82	$4\frac{1}{2}\%$	(2)	$109	$5\frac{1}{2}\%$
(3)	$238.50	$6\frac{1}{2}\%$	(4)	$137.51	5%
(5)	$67.60	4%	(6)	$174.39	6%
(7)	$4,183.19	5%	(8)	$28.95	6%

(9) The store owner knows that he can multiply the cost of the merchandise by 1.05 and have the total owed by the customer, including the 5% tax. How much was the tax portion if the total owed was $31.45?

(10) What would the customer owe if the tax rate was 5.5% and the items purchased cost $5.00, $3.95, $1.89, $15.50, and $5.25?

(1) $118.94

(2) $115

(3) $254

(4) $144.39

(5) $70.30

(6) $184.85

(7) $4,392.35

(8) $30.69

(9) $1.50

(10) $33.33

Exercise 93B: Sales Taxes

Name_____

Course/Sect. No._____   Score_____

Instructions: Find the total owed, *including* the sales tax.

		Taxable Merchandise	Tax Rate	Taxable Merchandise	Tax Rate
(1) _____	(1) $8.20		4%	(2) $77.81	5%
(2) _____					
(3) _____	(3) $65.99		6%	(4) $17.04	7%
(4) _____					
(5) _____	(5) $2,499.81		6%	(6) $83.44	5%
(6) _____					
(7) _____	(7) $8,950		4%	(8) $.95	4%
(8) _____					

(1) _____

(2) _____

(3) _____

(4) _____

(5) _____

(6) _____

(7) _____

(8) _____

(9) _____

(9) The three items purchased cost $18.50, $3.99, and $11.95. If the tax rate is 4.5%, what would be the total sales tax?

(10) _____

(10) The total taxes collected (at a 6% rate) were $38.90. How much merchandise was sold?

Exercise 94A: Sales Taxes

Name_____

Course/Sect. No._____   Score_____

Instructions: From the information given, complete the bottom portion of the sales slip.

_____

Self-Check Exercise (Answers on back)

Items purchased: 3 towels at $1.99 each
                  4 sheets at $4.95 each
                  1 bedspread at $29.50
                  2 pillow cases at $1.75 each

Sales tax rate: 6%

	Quantity	Description	Unit Price	Amount
(1)	3	Towels	1.99	
(2)	4	Sheets	4.95	
(3)	1	Bedspread	29.50	
(4)	2	Pillow cases	1.75	
(5)			Subtotal =	
(6)			Sales tax =	
(7)			Total =	

Items purchased: 4 washers at $.05 each   4 bolts at $.12 each
                  4 nuts at $.05 each   3 clamps at $.35 each
                  1 screwdriver at $2.95   1 lb. nails at $.40/lb.

Sales tax rate: 5%

	Quantity	Description	Unit Price	Amount
(8)	4	Washers	.05	
(9)	4	Nuts	.05	
(10)	1	Screwdriver	2.95	
(11)	4	Bolts	.12	
(12)	3	Clamps	.35	
(13)	1 lb	Nails	.40	
(14)			Subtotal =	
(15)			Sales tax =	
(16)			Total =	

(1) _____5.97_____
(2) _____19.80_____
(3) _____29.50_____
(4) _____3.50_____
(5) _____58.77_____
(6) _____+3.53_____
(7) _____62.30_____

(8) _____.20_____
(9) _____.20_____
(10) _____2.95_____
(11) _____.48_____
(12) _____1.05_____
(13) _____.40_____
(14) _____5.28_____
(15) _____+.26_____
(16) _____5.54_____

Exercise 94B: Sales Taxes

Name_____

Course/Sect. No._____   Score_____

Instructions: From the information given, complete the bottom portion of the sales slip.

_____

Items purchased: 5 cans of paint at $5.99 each
                   2 paint brushes at $4.50 each
                   1 can of cleaner at $3.99

Sales tax rate: 7%

	Quantity	Description	Unit Price	Amount
(1)	5	Cans of paint	5.99	
(2)	2	Paint brushes	4.50	
(3)	1	Can of cleaner	3.99	
(4)			Subtotal =	
(5)			Sales tax =	
(6)			Total =	

Items purchased: 1 baseball glove at $39.95
                   2 baseballs at $3.50 each
                   2 baseball bats at $19.95 each
                   1 pair of shoes at $24.95

Sales tax rate: 5%

	Quantity	Description	Unit Price	Amount
(7)	1	Baseball glove	39.95	
(8)	2	Baseballs	3.50	
(9)	2	Baseball bats	19.95	
(10)	1	Pair shoes	24.95	
(11)			Subtotal =	
(12)			Sales tax =	
(13)			Total =	

Chapter

20

Stocks and Bonds

Preview of Terms

Bond	A certificate to show evidence of a debt; generally, an I.O.U. payable in five or more years.
Bond Contract Rate	The interest rate specified in the bond issue.
Cumulative Preferred Stock	Stock for which dividends missed in prior years must be paid before any current year dividends can be paid.
Discount	Situation when the selling price of a bond is less than the face amount. If a $1,000 bond sells for $990, the discount is $10.
Dividends	The distribution of earnings to owners of a corporation.
Dividends in Arrears	When owners of cumulative preferred stock are not paid their annual dividends, those dividends are then "owed" and are said to be "in arrears."
Par Value	The face amount of a stock.
Participating Preferred Stock	Preferred stock that can receive more dividends than the amount usually fixed by the stated rate.
Premium	Situation when the selling price of a bond is more than the face amount. If a $1,000 bond sells for $1,015, the premium is $15.
Yield	The effective return on a bond investment.

STOCKS

The ownership of a corporation is divided into transferable units called *stock.* Generally, shares of stock are assigned an arbitrary amount called the *par value.* This stock is purchased through a registered representative called a *stockbroker.* Stock quotations are given in dollar amounts or in some fractional amount of a dollar. The fractional amounts commonly used are some multiple of 12.5 cents ($1/8 = .125$).

Multiples of 12.5 Cents and their Fractional Equivalents

1/8 of $1 = $.125	5/8 of $1 = $.625
2/8 = 1/4 of $1 = $.25	6/8 = 3/4 of $1 = $.75
3/8 of $1 = $.375	7/8 of $1 = $.875
4/8 = 1/2 of $1 = $.50	

STOCK PURCHASES

Therefore, a stock quote of $93\frac{5}{8}$ would be $93.625 \times \$1 = \$93.625$ per share. Twenty shares at that price would cost $1,872.50 (20 × $93.625), plus the broker's commission.

Note: *The total cost of a stock purchase will include a commission for a stockbroker. These commissions vary considerably and will not be included in any situations used in this chapter.*

Example 1: The XYZ Co. stock was selling at $12\frac{3}{8}$. Larry bought 25 shares. How much did he pay?

$$12\frac{3}{8} = \$12.375 \text{ per share}$$
$$\$12.375 \times 25 \text{ shares} = \$309.38$$

Practice Problems

Find the total cost of each stock purchase.

1. 103 shares were purchased at $35\frac{1}{2}$.

2. Aubrey asked his broker to buy 52 shares of TT and A when it was selling at $110\frac{7}{8}$.

TYPES OF STOCK

The types of stock are *common stock* and *preferred stock.* If a corporation has only one type of stock, it is common stock. Common stock is by far the most prevalent. It generally is the only type that allows the right to vote, and therefore the common stockholders are the ones that actually control the corporation.

Preferred stock is usually paid a fixed dividend amount, but it can have several "preferences." These preferences pertain to a prior claim to assets upon liquidation, or to payment of dividends before common shareholders receive any dividends, or to participating or cumulative features.

CUMULATIVE PREFERRED

Owners of this stock have the right to accumulate any dividends that were not paid and receive these dividends in future years. All back dividends are said to be *in arrears.* Dividends in arrears on preferred stock must be paid before the current year preferred dividends are paid and all preferred dividends must be paid before common stockholders receive anything.

NONCUMULATIVE PREFERRED

Preferred stock that does *not* carry the cumulative right is called *noncumulative.* Once a dividend is not paid, it is lost.

PARTICIPATING PREFERRED

Owners of this stock have the right to participate in dividends in excess of the specified amount. The extent of their participation will vary depending on the contract.

NONPARTICIPATING PREFERRED

This stock limits dividend payments to a specified amount.

Note: *For the situations in this chapter, only cumulative preferred stock will be illustrated.*

CASH DIVIDENDS ON STOCK

The return of earnings of a corporation to the owners is called *paying dividends.* The amount each shareholder receives depends on the type of stock and the number of shares owned. The amount the owners of the common shares receive in cash dividends is determined by the board of directors of the corporation. The amount common shareholders receive is paid after any preferred dividends are paid and will usually fluctuate depending on the level of earnings of the corporation that year. Although common stock does not provide the security of preferred stock, it has considerably more potential for larger dividends.

Example 2:   The board of directors declares a regular quarterly cash dividend of $.75 per share on 10,000 shares of common stock. John owns 50 shares. What dividend will John receive?

50 shares × $.75 per share = $37.50 cash dividend

Dividends on preferred stock are expressed either in monetary terms or as a percentage of par. A 5% dividend on $100 par preferred stock can be stated as $5, or as 5% of the $100 par amount.

Example 3:   The XYZ Co. declares a cash dividend of $100,000. The 1,000 shares of preferred stock are to receive 7% of the $100 par value, and the 5,000 shares of common stock are to receive the remainder. What total amount and annual dividend per share does each type of stock generate?

$100,000 Cash dividend
−   7,000 Preferred dividend*
$ 93,000 Common dividend ($93,000 ÷ 5,000 shares = $18.60 per share)

*7% × $100 = $7 per share × 1,000 shares = $7,000

Example 4:

Assume the situation in Example 3 with the exceptions that the 7% preferred stock is cumulative and dividends are in arrears for three years before this year. Determine the total dividends per share. (Remember, the dividends in arrears are paid first, then the regular preferred dividend, and then the common dividend.)

Preferred = 7% × $100 par = $7 per share × 1,000 shares = $7,000

$100,000 Cash dividend
−   28,000 Preferred dividend ($7,000 × 4 years)*
$ 72,000 Common dividend ($72,000 ÷ 5,000 = $14.40 per share)

*Current year + 3 years in arrears

Practice Problems

Find the dividend per share for each problem.

3. The MNO Co. declared a cash dividend of $250,000. The 5,000 shares of 6%, $100 par value noncumulative preferred stock were to be paid first and the remainder was to be paid to the 20,000 shares of common stock.

4. The Bell Company paid the 500 shares of 8%, $50 par value cumulative stock its current dividend plus the dividends owed for two years in arrears. The remaining $150,000 was paid to the 100,000 shares of $20 par common shares.

BONDS

Corporations borrow money on a long-term basis by issuing bonds. A *bond* is a long-term liability to the corporation that has a fixed amount of interest that must be paid each year. The corporation also must pay the face value of the bond at some specified future date (usually at least five years from the issue date).

Ordinarily, the principal, or face value, of each bond is $1,000 or some multiple of $1,000. The interest rate specified is called the *contract rate.* This contract rate will probably be different from the market rate at the time the bonds are issued. If the market rate is higher than the bond contract rate, the bonds will sell for less than $1,000 each (at a *discount*). If the market rate is lower than the bond contract rate, the bonds will sell for more than $1,000 each (at a *premium*).

BOND PURCHASE PRICE

The price of a bond is quoted as a percentage of its face value. A quotation of 95 means 95% of the face value of $1,000, which would be $950 (a *dis-*

count). A quotation of $101\frac{1}{2}$ means 101.5% of the face value of $1,000, which would be $1,015 (a *premium*). When you multiply the bond quotation by the face value of the bond, you have the purchase price of the bond.

Note: *From now on, it will be assumed that the face value of all individual bonds is $1,000.*

Example 5: The MNO Co. sells $10,000 worth of bonds at 98. What amount of cash do they receive?

$$\begin{array}{rl} \$10,000 & \text{Face amount} \\ \times \quad 98\% & \text{Bond quotation} \\ \hline \$\ 9,800 & \text{Purchase price (Cash received by MNO Co.)} \end{array}$$

Example 6: The PQR Co. sold $50,000 worth of bonds at 102. What was the purchase price and how much was the premium?

$$\begin{array}{rl} \$50,000 & \text{Face amount} \\ \times \quad 102\% & \text{Bond quotation} \\ \hline \$51,000 & \text{Purchase price} \end{array} \qquad \begin{array}{rl} \$51,000 & \text{Purchase price} \\ - \ 50,000 & \text{Face amount} \\ \hline \$\ 1,000 & \text{Premium} \end{array}$$

Practice Problems

Find the purchase price and the amount of the premium or discount.

5. The $100,000 bond issue sold at 93.5.

6. $500,000 worth of bonds were purchased at 103.

BOND INTEREST

The bond contract rate is the fixed rate of interest that is paid each year during the life of the bond issue. It is usually quoted as a percentage of the face value (8% would mean 8% × $1,000, or $80 interest). This interest is usually paid twice a year (semiannually). Therefore, in the example just given, the $80 annual interest would be paid as $40 at the end of each six-month period.

Example 7: Find the annual interest on $8,000 worth of bonds with a 6.5% contract rate.

$$\begin{array}{rl} \$8,000 & \text{Face value} \\ \times \quad 6.5\% & \text{Contract rate} \\ \hline \$\ 520 & \text{Annual interest due} \end{array}$$

Example 8: Find the semiannual interest on $25,000 worth of bonds with a 9.0% contract rate.

$$\begin{array}{rl} \$25,000 & \text{Face value} \\ \times \quad 9\% & \text{Contract rate} \\ \hline \$\ 2,250 & \text{Annual interest} \end{array} \qquad \$2,250/2 = \$1,125 \text{ semiannual interest}$$

BOND YIELD

When a bond issue sells for more (premium) or less (discount) than its face value, it will have a different return on the amount invested than the contract rate is paying. This return on the investment is called the *bond yield.* To find the approximate bond yield, divide the annual interest by the purchase price. The yield is expressed as a percentage.

$$\text{Bond yield} = \frac{\text{Bond interest}}{\text{Purchase price}}$$

Example 9:

The annual interest on a $200,000, 7.5% bond issue is $15,000 and it was purchased for $190,000. What is the approximate bond yield rounded to one-tenth of 1%?

$$\begin{aligned}\text{Bond yield} &= \$15,000/\$190,000 \\ &= .0789473 \\ &= 7.9\%\end{aligned}$$

Example 10:

The $100,000, 6% bonds sold for 104. What is the approximate bond yield to one-tenth of 1%?

$$\begin{aligned}\$100,000 \times \quad 6\% &= \$6,000 \text{ Annual interest} \\ \$100,000 \times 104\% &= \$104,000 \text{ Purchase price}\end{aligned}$$

$$\begin{aligned}\text{Bond yield} &= \$6,000/\$104,000 \\ &= .0576923 \\ &= 5.8\%\end{aligned}$$

Practice Problem Solutions

1. 35½ = $35.50 Per share

$35.50 × 103 shares = $3,656.50 Total cost

2. 110⅞ = $110.875 Per share

$110.875 × 52 shares = $5,765.50 Total cost

3.
$250,000 Cash dividend

− 30,000 Preferred dividend (6% × $100 = $6/share × 5000 shares)

$220,000 Common dividend ($220,000 ÷ 20,000 shares = $11 per share)

Preferred = $6/share

Common = $11/share

4. 8% × $50 par = $4/share for preferred stock

$4/share × 500 shares = $2,000 regular preferred annual dividend

$2,000 × 3 years (current year + 2 in arrears) = $6,000

$6,000 total preferred dividend ÷ 500 shares = $12/share

$150,000 ÷ $100,000 common shares = $1.50/share

Preferred = $12/share

Common = $1.50/share

5.

$100,000	Face		$100,000	Face
× 93.5%	Quotation		− 93,500	Purchase price
$ 93,500	Purchase price		$ 6,500	Discount

6.

$500,000	Face		$515,000	Purchase price
× 103%	Quotation		− 500,000	Face
$515,000	Purchase price		$ 15,000	Premium

7.

$200,000	Face
× 7.5%	Contract rate
$ 15,000	Annual interest

8.

$300,000	Face
× 8.5%	Contract rate
$ 25,500	Annual interest

$$\frac{\$25,500}{2} = \$12,750 \text{ Semiannual interest}$$

9. $250,000 × 7% = $17,500 Annual interest

$250,000 × 98% = $245,000 Purchase price

$$\text{Bond yield} = \frac{\$17,500}{\$245,000}$$

$$= .0714285$$

$$= 7.1\%$$

10. $2,000,000 × 9.5% = $190,000 Annual interest

$2,000,000 × 102.5% = $2,050,000 Purchase price

$$\text{Bond yield} = \frac{\$190,000}{\$2,050,000}$$

$$= .0926829$$

$$= 9.3\%$$

Exercise 95A: Stocks

Name _____

Course/Sect. No. _____   Score _____

Instructions: Find the total cost and dividend per share as indicated.

(1) _____

(2) _____

(3) _____

(4) _____

(5) _____

(6) _____

(7) _____

(8) _____

(9) _____

(10) _____

Self-Check Exercise (Answers on back)

For problems 1 through 4, find the total cost of each stock purchase.
(1) 220 shares purchased at $17\frac{1}{4}$.

(2) 15 shares purchased at $483\frac{1}{8}$.

(3) 10 shares purchased at $117\frac{5}{8}$.

(4) 28.5 shares purchased at $28\frac{1}{2}$.

For problems 5 through 8, find the dividends per share for each type of stock.
(5) Preferred = 5%, $100 par, noncumulative, 1,000 shares; common = $100 par, 2,000 shares; cash dividend = $100,000.

(6) Preferred = 7%, $50 par, noncumulative, 3,000 shares; common = $100 par, 10,000 shares; cash dividend = $175,000.

(7) Preferred = 8%, $20 par, cumulative, 2,000 shares; common = $5 par, 20,000 shares; cash dividend = $400,000.

(8) Preferred = 9%, $100 par, cumulative, 5,000 shares; common = $100, 10,000 shares; cash dividend = $200,000.

(9) The company paid the 200, 7%, $100 par cumulative preferred shareholders and then the 500 common shareholders. A total cash dividend of $80,000 was declared and the preferred was one year in arrears before this year. What is the dividend per share for each type of stock?

(10) The 5,000 shares of 5%, $25 par, cumulative preferred stock were 3 years in arrears before this year. There were also 50,000 shares of common stock. The cash dividend was $125,000. What is the dividend per share for each type of stock?

(1) _$3,795_

(2) _$7,246.88_

(3) _$1,176.25_

(4) _$812.25_

(5) _Pref. = $5.00_
Com. = $47.50

(6) _Pref. = $3.50_
Com. = $16.45

(7) _Pref. = $1.60_
Com. = $19.84

(8) _Pref. = $9.00_
Com. = $15.50

(9) _Pref. = $14.00_
Com. = $154.40

(10) _Pref. = $5.00_
Com. = $2.00

Exercise 95B: Stocks

Name _____

Course/Sect. No. _____   Score _____

Instructions: Find the total cost and the dividend per share as indicated.

---

(1) _____

(2) _____

(3) _____

(4) _____

(5) _____

(6) _____

(7) _____

(8) _____

(9) _____

(10) _____

For problems 1 through 4, find the total cost of each stock purchase.

(1) 193 shares purchased at $34\frac{3}{4}$.

(2) 30 shares purchased at $225\frac{7}{8}$.

(3) 64 shares purchased at $8\frac{3}{8}$.

(4) 6 shares purchased at 107.

For problems 5 through 8, find the dividends per share for each type of stock.

(5) Preferred = 4%, $50 par, noncumulative, 6,000 shares; common = $100 par, 6,000 shares; cash dividend = $15,000.

(6) Preferred = 8%, $50 par, cumulative, 400 shares; common = $25 par, 3,000 shares; cash dividend = $8,350.

(7) Preferred = 10%, $100 par, noncumulative, 1,000 shares; common = $50 par, 5,000 shares; cash dividend = $40,000.

(8) Preferred = 6%, $10 par, cumulative, 10,000 shares; common = $100 par, 50,000 shares; cash dividends = $43,500.

(9) ABC Co. paid cash dividends to the 2,000, 6%, $100 par cumulative preferred shareholders including enough to pay off the 2 years in arrears before this year. The 8,000 common shareholders received the rest. The total dividend paid was $76,000. What were the dividends per share for preferred and common?

(10) The EFG Co. paid a cash dividend of $70,000 to the 20,000 common shares and the rest of the dividend was needed to satisfy the 1,000 cumulative, 8%, $50 par preferred stockholders. There were no dividends in arrears. The total dividend was $74,000. Find preferred and common dividends per share.

Exercise 96A: Bonds

Name _____

Course/Sect. No. _____   Score _____

Instructions: Find the purchase price, interest, and approximate yield as required.

Self-Check Exercise (Answers on back)

(1) _____

(1) From a $250,000 bond issue at 96, what was the purchase price of the issue?

(2) _____

(2) $300,000 worth of bonds were purchased at 108. How much was the *premium amount?*

(3) _____

(3) What is the purchase price on a $100,000 bond issue quoted at 97.5?

(4) _____

(4) How much is the *discount* for $4,000 worth of bonds purchased at $98\frac{1}{4}$?

(5) _____

(5) Find the annual bond interest on a $200,000 issue of 5% bonds.

(6) _____

(6) Find the semiannual bond interest on a $50,000 issue of 8% bonds.

(7) _____

(7) What semiannual interest would be paid on a single $1,000 bond with a 7.5% contract rate?

(8) _____

(8) Find the *total 5-year cost* of interest of a $300,000 bond issue with a 9% contract rate.

(9) _____

(9) Find the approximate bond yield accurate to one-tenth of 1% on a $400,000 bond issue sold at 99 if the contract rate was 5%.

(10) _____

(10) Find the approximate bond yield accurate to one-tenth of 1% on an $80,000 bond issue sold at $103\frac{1}{4}$ if the contract rate was $8\frac{1}{2}$%.

(1) _____ $240,000 _____

(2) _____ $24,000 _____

(3) _____ $97,500 _____

(4) _____ $70 _____

(5) _____ $10,000 _____

(6) _____ $2,000 _____

(7) _____ $37.50 _____

(8) _____ $135,000 _____

(9) _____ 5.1% _____

(10) _____ 8.2% _____

Exercise 96B: Bonds

Name _____

Course/Sect. No. _____   Score _____

Instructions: Find the purchase price, interest, and approximate yield as required.

(1) _____   **(1)** What is the purchase price paid for $150,000 worth of bonds at $93\frac{1}{2}$?

(2) _____   **(2)** Find the purchase price on a $2,000,000 bond issue quoted at 104.

(3) _____   **(3)** What is the *premium* on a $25,000 bond issue sold at $105\frac{1}{2}$?

(4) _____   **(4)** Find the purchase price of a $250,000 bond issue bought at $94\frac{3}{4}$.

(5) _____   **(5)** Find the annual bond interest on $125,000 worth of bonds with a 6% contract rate.

(6) _____   **(6)** Find the semiannual bond interest for $4\frac{1}{2}$%, $500,000 bonds.

(7) _____   **(7)** How much *more interest per year* would a 7% bond receive than a $6\frac{1}{2}$% bond (a $1,000 bond)?

(8) _____   **(8)** Find the semiannual interest on $50,000 worth of bonds with a 9% contract rate.

(9) _____   **(9)** Find the approximate bond yield accurate to one-tenth of 1% on a $750,000 bond issue purchased at $95\frac{1}{2}$ if the contract rate is 8%.

(10) _____   **(10)** Find the approximate bond yield accurate to one-tenth of 1% on $350,000 worth of bonds purchased at 93 if the contract rate is 10%.

Test Yourself — Unit 7

(1) The business was insured for $115,000 and the rate was $.35 per $100. What did the insurance cost?

(2) The company warehouse cost $475,000 to replace, and it was insured for $350,000. There was a fire loss of $137,000. If there was an 80% coinsurance clause, what part of the fire loss did the insurance company pay? Round to an even dollar.

(3) If the rate per $100 is 28.5¢, what would the cost be for a $75,000 insurance policy?

(4) Find the monthly cost of a 20-year term policy for a 45-year-old man if the face amount of the policy is $120,000. (Use the "term policy rate table" in your book and include the policy fee.)

(5) A business has a market value of $165,000 and is assessed for tax purposes at 55% of its market value. What is the assessed valuation?

(6) For the business in Question 5, find the property tax if the tax rate is $1.10 per $100.

(7) What would the tax in Question 6 have been if the tax rate were 14.25 mills?

(8) The market value is $87,700, the assessment rate is 40%, and the tax rate is $1.65 per hundred. Find the assessed valuation and the tax.

(9) If the sales tax rate is 6% and the taxable items purchased cost $84.23, what is the amount of the sales tax?

(10) Henry bought 6 shares of 8%, $100 par cumulative preferred stock selling at $25\frac{1}{8}$.

(A) What was his total purchase price?

(B) What total annual dividend should he receive in a year when dividends were paid and his stock was in arrears one year?

(11) Helen bought four $1,000, 7.5% bonds when the quotation was 95.

(A) What was her total purchase price?

(B) What total semiannual interest does she receive?

(C) What is her approximate bond yield? (Round to one-tenth of 1%.)

Practice Set

This practice set features Ed Carson. He begins in Phase I of the practice set by starting a small-scale business on a part-time basis out of his home.

In Phase II the business has grown large enough and Ed is encouraged enough to quit his regular full-time job and expand by buying a building in town and offering a wider line of merchandise for sale.

In Phase III Ed incorporates and expands his line of merchandise even further.

You will be asked to solve the many problems that Ed encounters along the way. Some are relatively simple and others, more challenging. Enter your solutions on the appropriate blank lines that follow each problem.

PHASE I: CARSON'S CUSTOM C.B.'S

Ed Carson decided to take some of his own money, borrow some more, and sell C.B. radios in his spare time from his home. The following situations took place.

1. The State Bank would loan Ed the $3,000 he needed if he would pay 8.5% interest for 2 years, to be paid in monthly installments of $146.25.

At his local credit union he could borrow the $3,000 for $\frac{3}{4}$% per month on the unpaid balance. If he chose to borrow from the credit union, he would pay $100 *more* each month than the interest charged for that month (except that the last installment would be for the amount left). Use the worksheet on the next page.

Which loan would require the least amount of interest? _____

Total interest on State Bank Loan = _____

Total interest on Credit Union Loan = _____

2. Two wholesale supply houses handled the type of C.B.'s Ed wanted to sell, but they had different discount terms.

Wholesaler A would sell him the $2,000 in C.B. merchandise with trade discounts of 5%, 2%, and the balance at 2/10, n/30, F.O.B. destination. Freight on the shipment would be $50.

Wholesaler B would sell him the same amount of merchandise with trade discounts of 3.5%, 3.5%, and the balance at 3/15, n/60, F.O.B. shipping point. Freight on their shipment would be $25.

Credit Union Worksheet

Payment No.	Interest	New Balance	Payment Amount	Unpaid Balance
				3,000
1	22.50*	3,022.50	122.50	2,900
2	21.75	2,921.75	121.75	2,800
3	_____	_____	_____	_____
4	_____	_____	_____	_____
5	_____	_____	_____	_____
6	_____	_____	_____	_____
7	_____	_____	_____	_____
8	_____	_____	_____	_____
9	_____	_____	_____	_____
10	_____	_____	_____	_____
11	_____	_____	_____	_____
12	_____	_____	_____	_____
13	_____	_____	_____	_____
14	_____	_____	_____	_____
15	_____	_____	_____	_____
16	_____	_____	_____	_____
17	_____	_____	_____	_____
18	_____	_____	_____	_____
19	_____	_____	_____	_____
20	_____	_____	_____	_____
21	_____	_____	_____	_____
22	_____	_____	_____	_____
23	_____	_____	_____	_____
24	_____	_____	_____	_____
25	_____	_____	_____	_____
26	_____	_____	_____	_____
27	_____	_____	_____	_____
28	_____	_____	_____	_____
29	_____	_____	_____	_____
30	_____	_____	_____	_____

Total = ⎓⎓⎓

*3,000 × 3/4% = $22.50

Ed would pay within the discount period and choose the wholesaler with the lowest net cost to him. Which wholesaler did he choose and how much did he owe within the discount period? _____

Invoice

Quantity		Unit Price	Total Price
10	Model X C.B. radio & speaker	39.50	395.00
20	Model Y C.B. radio & speaker	59.25	1,185.00
4	Model Z C.B. radio & speaker	95.75	383.00
37 ft	Speaker wire	1.00/ft	37.00
			2,000.00

Wholesaler A: Invoice Price   =   2,000.00

                 Less: Trade Discounts = − _____

                                             _____

                 Less: Discount   = − _____

                                               _____

                 Plus: Freight Charges = _____

                 Net Cost   = _____

Wholesaler B: Invoice Price   =   2,000.00

                    Less: Trade Discounts = − _____

                                               _____

                 Less: Discount   = − _____

                                               _____

                 Plus: Freight Charges = + _____

                 Net Cost   = _____

3. Ed received three models of C.B.'s (Models X, Y, and Z) that cost him $39.50, $59.25, and $95.75, respectively. He wants to sell each model with a 30% markup. The least expensive models will be marked up on *cost* while the top of the line will be marked up on *retail.* What will be their respective selling prices?

                              C.B. Model X cost/unit = $ 39.50

                              Plus 30% *cost* markup   = + _____

                              Regular selling price   = _____

                              C.B. Model Y cost/unit = $ 59.25

                              Plus 30% *cost* markup   = + _____

                              Regular selling price   = _____

                              C.B. Model Z cost/unit = $ 95.75

                              Plus 30% *retail* markup = + _____

                              Regular selling price   = _____

4. For his "Grand Opening," Ed advertised a 20% markdown on all of his merchandise. Based on his previously computed selling prices, what would each model sell for during the grand opening sale?

                           C.B. Model X sales price/unit = _____

                           Less: 20% markdown   = − _____

                           Grand opening selling price   = _____

                           C.B. Model Y sales price/unit = _____

                           Less: 20% markdown   = − _____

                           Grand opening selling price   = _____

                           C.B. Model Z sales price/unit = _____

                           Less: 20% markdown   = − _____

                           Grand opening selling price   = _____

5. During Ed's Grand Opening sale, he sold 7 units of Model X, 1 unit of Model Y, and 3 of Model Z. All others were sold during the year at the 30% markup he established in situation (3).

What was Ed's total gross profit and what percentage of total sales did *each* model provide?

		Model X	Model Y	Model Z	Total
Sales	=	___ +	___ +	___ =	___
Less: Cost of goods sold	=	___ +	___ +	___ =	___
Gross profit	=	___ +	___ +	___ =	___

		% of Total Sales
Model X Sales	= ___	___
Model Y Sales	= ___	___
Model Z Sales	= ___	___

6. After 12 months, Ed needed to determine the value of his inventory and to compute his cost of goods sold. The following is a record of his inventory.

	Model X	Model Y	Model Z
Beginning inventory	10 @ $39.50	20 @ $59.25	4 @ $95.75
Purchase #1	15 @ 39.50	12 @ 61.50	6 @ 95.75
Purchase #2	8 @ 39.50	15 @ 61.50	8 @ 95.75
Purchase #3	12 @ 41.50	9 @ 61.50	12 @ 97.50

Taking an inventory count, there were 14 units left of Model X, 15 of Model Y, and 6 of Model Z. Ed used the LIFO method of inventory valuation.

What was his *total* inventory valuation and his total cost of goods sold?

		Model X	Model Y	Model Z	Total
Cost of goods available for sale	=	___ +	___ +	___ =	___
Less: Ending inventory*	=	___ +	___ +	___ =	___
Cost of goods sold	=	___ +	___ +	___ =	___

*Inventory valuation at LIFO.

7. Ed's December bank statement for the business arrived. He collected the following information.

	Shown in Checkbook	Shown in Bank Statement
November balance	$1,015.69	$1,045.69
Check #402		30.00
Check #403	25.50	25.50
Check #404	175.00	175.00
Check #405	8.86	8.86
Check #406	3.50	
Check #407	15.75	15.75
Deposit	500.00	500.00
Check #408	17.25	17.25
Service charge		2.50
NSF check		55.00
Check #409	38.30	
Deposit	275.00	
December balance	1,506.53	1,215.83

Reconcile his checkbook balance with the bank statement by preparing a reconciliation statement.

December Reconciliation Statement

Bank Balance: _____ Checkbook Balance _____

Add: _____ Add: _____

Deduct: _____ Deduct: _____

_____

Adjusted Balance _____ Adjusted Balance _____

PHASE II: CARSON'S CUSTOM C.B.'S & HI FI'S

Ed's home C.B. shop did better than he had hoped and Ed decided to quit his job, buy a building, and expand his merchandise line to include stereos, tape decks, and video cassette recorders in addition to his C.B.'s.

8. The small building that Ed bought cost him $70,000 on a 25-year, 11% mortgage. He paid $5,000 of that as a down payment with the balance to be amortized monthly.

What was his monthly payment and how much was his loan balance after the first *year?*

Payment No.	Monthly Payment	Payment On Interest	Payment On Principal	Loan Balance
1	____	____	____	____
2	____	____	____	____
3	____	____	____	____
4	____	____	____	____
5	____	____	____	____
6	____	____	____	____
7	____	____	____	____
8	____	____	____	____
9	____	____	____	____
10	____	____	____	____
11	____	____	____	____
12	____	____	____	____

9. To secure the loan, Ed had to insure his property for fire damage. He chose to insure for $50,000 with an 80% coinsurance clause based on the replacement cost of $70,000.

If a fire resulted in damages of $40,000, what amount would the insurance company pay?

$$\frac{\text{Insurance}}{\text{(Coinsurance \%)(Replacement cost)}} \times \text{Fire loss} = \text{Insurance payment}$$

Insurance payment = _____

10. A shipment of Japanese Matochi systems and component parts arrived with metric length and weight descriptions. Ed wanted to know approximately what these metric measurements would convert to in U.S. measures.

U.S. Conversions

Stereo system walnut cabinet, 9.09 kilograms = How many pounds?

10" 4-way speaker, 3,830 grams = How many pounds?

4-hour video cassette recorder, 2,688 grams = How many ounces?

Stereo speaker wire, 7,500 millimeters = How many feet?

C.B. antenna, 90 centimeters = How many inches?

Stereo system cabinet	9.09 kilograms	≅ _____ pounds
4-way speaker	3,830 grams	≅ _____ pounds
Video cassette recorder	2,688 grams	≅ _____ ounces
Stereo speaker wire	7,500 millimeters	≅ _____ feet
C.B. antenna	90 centimeters	≅ _____ inches

11. To operate the store it was necessary to hire two employees. Bob Hinkel was hired on a full-time basis and Mary Evans worked part-time. Both were hourly wage earners. Their payroll information for the first month follows:

	Week	Hours	Regular Rate		Week	Hours	Regular Rate
Bob Hinkel*	1	42	$5.00	Mary Evans*	1	17	$3.25
	2	44			2	21	
	3	43.5			3	13	
	4	41			4	15.5	

*Each is single and claims 1 dependent.

Based on the information given, complete an Employee's Earnings Record for each for the first month the business was open. (Assume time-and-one-half for any overtime and FICA taxes at 6.7% on the first $31,800 earned.)

Bob Hinkel

Hrs	Reg. Pay	O/T Pay	Gross Pay	FICA Tax	Inc. Tax	Total Ded.	Net Pay
42	_____	_____	_____	_____	_____	_____	_____
44	_____	_____	_____	_____	_____	_____	_____
43.5	_____	_____	_____	_____	_____	_____	_____
41	_____	_____	_____	_____	_____	_____	_____

Mary Evans

Hrs	Reg. Pay	O/T Pay	Gross Pay	FICA Tax	Inc. Tax	Total Ded.	Net Pay
17	_____	_____	_____	_____	_____	_____	_____
21	_____	_____	_____	_____	_____	_____	_____
13	_____	_____	_____	_____	_____	_____	_____
15.5	_____	_____	_____	_____	_____	_____	_____

12. Based on the computations in (11), determine the payroll taxes Ed had to pay for the *third* week. The FICA was 6.7% on the first $31,800, the SUT was 3.1% on the first $6,000, and the FUT was .7% on the first $6,000.

3rd Week Gross Pay	FICA = _____
Bob Hinkel _____	SUT = _____
Mary Evans _____	
_____	FUT = _____

13. After consulting with Mary Evans, his part-time employee, Ed agreed she would be put on a straight commission pay basis. She was to be paid using the following sliding monthly scale:

5% on all sales, plus
10% on sales over $1,000, plus
15% on sales over $2,000

Her sales the first month the new scale was used were $2,250. What gross pay did she receive?

$$5\% \text{ commission} = \underline{\hspace{1.5cm}}$$
$$10\% \text{ commission} = \underline{\hspace{1.5cm}}$$
$$15\% \text{ commission} = \underline{\hspace{1.5cm}}$$
$$\text{Gross pay} \quad = \underline{\hspace{1.5cm}}$$

14. In the first 18 months Ed found that the $38,500 gross income was earned from his stereo, C.B., and video recorder departments in a 6:3:5 ratio, respectively. The same departments had expenses of $17,490 in the ratio of 4:4:3.

Compute the departmental gross income, expenses, and net income using the ratios given.

Departments

	Stereo	C.B.	TV Recorder-Player		Totals
Gross income (6:3:5)	_____	_____	_____	=	_____
Expenses (4:4:3)	_____	_____	_____	=	_____
Net income	_____	_____	_____	=	_____

PHASE III: CARSON'S HI FI DISCOUNT WAREHOUSE, INC.

Ed's store did very well and he decided to expand the volume and incorporate. The expansion capital he needed would come from selling a small portion of his corporate stock and from selling corporate bonds.

15. The corporation had some idle cash that was not needed in the immediate future so it was invested at 6%, earning interest compounded monthly. They deposited $50,000.

Determine the compound amount after 2 years.

$$\text{Compound amount after 2 years} = \underline{\hspace{1.5cm}}$$

16. The treasurer anticipates he will need $15,000 in 4 years. It is expected a certificate of deposit would provide interest of 9% compounded semi-annually.

What single present value would she invest so that the $15,000 would be available?

$$\text{Present value of } \$15,000 \text{ at } 9\% \text{ for 4 years} = \underline{\hspace{1.5cm}}$$

17. The addition of the more expensive merchandise caused Ed to sell some of it on an installment basis. The advertised terms of his best-selling item follow:

"Matochi component stereo system with 8-track tape recorder. Includes AM/FM stereo receiver, automatic full-size record changer, 2 speakers, 2 mikes, 1 blank tape, and simulated wood stand with slide-out shelf and storage space—$199.88 cash price. Convenient installment plan allows a low monthly payment of $22.50 for one year."

What is the approximate true or effective interest rate of this installment plan, rounded to one-tenth of 1%?

$$r = \frac{2mi}{P(N + 1)}$$

$$m = $$
$$i = $$
$$P = $$
$$N = $$
$$r = \underline{\quad}$$

18. Marge Harris purchased the system described in (17) and made 8 installment payments before she paid off the balance due shortly after the eighth installment.

What was her rebate of interest?

How much did she need to pay the balance due?

Rebate of interest = _____

Balance due = _____

19. On January 12, 19X0, Ed bought a $700 cash register with a life expectancy of 16 years. On January 14, he purchased some display shelves and cabinets that cost $450 and could be expected to last 20 years. Neither the cash register nor the store equipment would have any residual value. On March 20 he bought a small duplicating machine for $300. It had a life expectancy of 8 years and a trade-in value of $50.

Ed decided to use an accelerated depreciation method, but wanted the one that would give him the most *total* depreciation the first 5 years. Including the building, determine which method, double-declining balance or sum-of-the-years'-digits, he should use and how much the *book value* would be at the end of the five years—on December 31, 19X5. (The building would have a sales value after 25 years of $5,000.)

DOUBLE DECLINING BALANCE METHOD:

	Year	Cost	Deprec. Exp.	Accum. Deprec.	Book Value
Cash Register:	1	___	___	___	___
	2	___	___	___	___
	3	___	___	___	___
	4	___	___	___	___
	5	___	___	___	___

	Year	Cost	Deprec. Exp.	Accum. Deprec.	Book Value
Store Equipment:	1	___	___	___	___
	2	___	___	___	___
	3	___	___	___	___
	4	___	___	___	___
	5	___	___	___	___

	Year	Cost	Deprec. Exp.	Accum. Deprec.	Book Value
Duplicating Machine:	1	___	___	___	___
	2	___	___	___	___
	3	___	___	___	___
	4	___	___	___	___
	5	___	___	___	___

	Year	Cost	Deprec. Exp.	Accum. Deprec.	Book Value
Building:	1	___	___	___	___
	2	___	___	___	___
	3	___	___	___	___
	4	___	___	___	___
	5	___	___	___	___

	5th Year Accum. Deprec.	Book Value
Cash Register	___	___
Store Equipment	___	___
Duplicating Machine	___	___
Building	+ ___	+ ___
Totals	___	___

SUM-OF-THE-YEARS'-DIGITS METHOD:

	5th Year Accum. Deprec.	Book Value
Cash Register	___	___
Store Equipment	___	___
Duplicating Machine	___	___
Building	+ ___	+ ___
Totals	___	___

20. From the condensed statements for the past two years, compute the following for 19X5. (Express the answers accurate to one-tenth.)

A. Current Ratio
B. Accounts Receivable Turnover
C. Equity Ratio
D. Rate Earned on Stockholders' Equity

E. Horizontal Analysis on Merchandise Inventory
F. Vertical Analysis on Net Income

Balance Sheet
Assets

	19X4	19X5
Current assets		
Cash	80,000	104,000
Accounts receivable	65,000	80,000
Merchandise inventory	120,000	265,500
Total current assets	265,000	449,500
Plant assets		
Buildings (net)	196,000	196,000
Land	50,000	50,000
Total plant assets	246,000	246,000
Total assets	511,000	695,500

Liabilities

	19X4	19X5
Current liabilities		
Accounts payable	43,000	82,000
Long-term liabilities		
9% Bonds payable	250,000	250,000
Total liabilities	293,000	332,000
Stockholders' Equity		
Preferred, 8% stock, $50 par	15,000	15,000
Common stock, $100 par	170,000	275,000
Retained earnings	33,000	73,500
Total stockholders' equity	218,000	363,500
Total liabilities and stockholders' equity	511,000	695,500

Income Statement

	19X4	19X5
Net sales	350,000	465,000
Less: Cost of merchandise sold	120,000	190,000
Gross profit on sales	230,000	275,000
Operating expenses:		
Selling expenses	60,000	80,000
General expenses	30,000	40,000
Total operating expenses	90,000	120,000
Net income before taxes	140,000	155,000
Less: Income taxes	67,200	74,400
Net income	72,800	80,600

A. Current ratio $= \dfrac{\text{Current assets}}{\text{Current liabilities}} = $ ___ to 1

B. Accounts receivable turnover $= \dfrac{\text{Net sales}}{\text{Ave. accounts receivable}} = $ ___ times

C. Equity ratio $= \dfrac{\text{Stockholders' equity}}{\text{Total assets}} = $ ___ to 1

D. Rate earned on stockholders' equity $= \dfrac{\text{Net income after taxes}}{\text{Ave. stockholders' equity}}$

$= $ ___ %

E. Horizontal analysis on merchandise inventory $= $ \$ _____

_____ %

F. Vertical analysis on net income $= $ ___ %

Index